The Archaeology of Beekeeping

Alii faciunt ex viminibus rotundas, alii e ligno ac corticibus, alii ex arbore cava, alii fictiles, alii etiam ex ferulis quadratas longas pedes circiter ternos, latas pedem, sed ita, ubi parum sunt quae compleant, ut eas conangustent, in vasto loco inani ne despondeant animum.

Some make round hives out of withies, some make them of wood and bark, some from a hollow tree, some of earthenware, and others again from the fennel plant, making them rectangular, about three feet long and one foot across, except that, when the bees are too few to fill them, they reduce the size, so that the bees do not lose heart in a wide empty space.

Marcus Terentius Varro (116–27 B.C.), *De re rustica* 3.16.15

translation by A. J. Graham

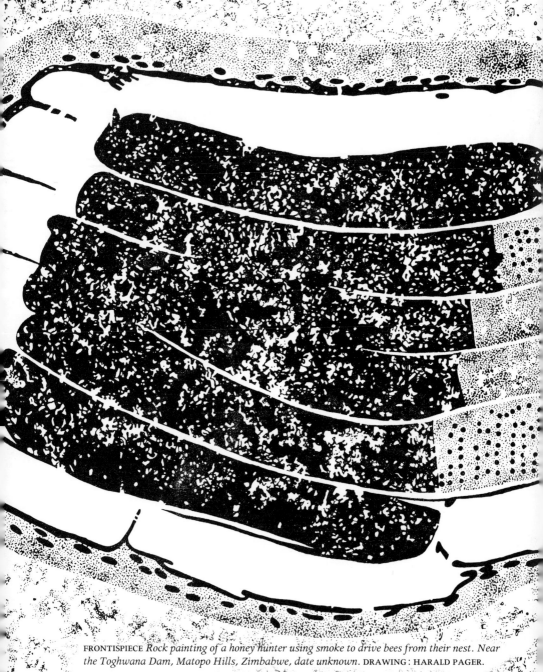

FRONTISPIECE *Rock painting of a honey hunter using smoke to drive bees from their nest. Near the Toghwana Dam, Matopo Hills, Zimbabwe, date unknown.* DRAWING: HARALD PAGER.

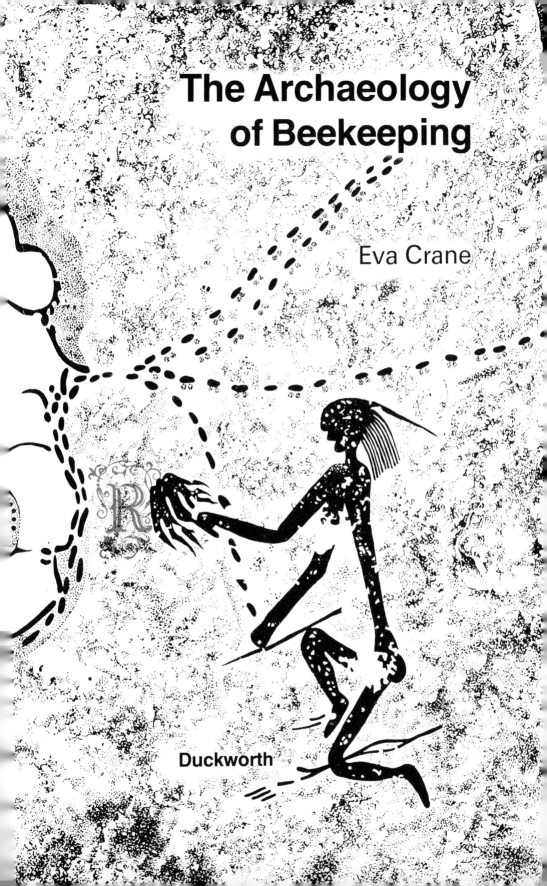
The Archaeology of Beekeeping

Eva Crane

Duckworth

First published in 1983 by
Gerald Duckworth & Co. Ltd.
The Old Piano Factory
43 Gloucester Crescent, London NW1

© 1983 by the International Bee Research Association

All rights reserved. No part of this publication may be reproduced, stored in a retrieval system, or transmitted, in any form or by any means, electronic, mechanical, photocopying, recording or otherwise, without the prior permission of the publisher.

ISBN 0 7156 1681 1

Book design by Alphabet & Image Ltd, Sherborne, Dorset

Filmset and printed in Great Britain by
BAS Printers Limited, Over Wallop, Hampshire

Preface

For more than thirty years my work has been concerned with scientific and technical developments involving bees, beekeeping and hive products, on a world-wide basis. Current advances are necessarily based on past experience, and I have always been interested in the development of beekeeping from its early beginnings. Through the years, as it became possible to travel more widely, I became fascinated by the variety of traditional beekeeping in the more remote areas of countries I visited. I took every opportunity to search out as much as I could, and this book presents some of the results of these searches. It also attempts to use the evidence accumulated, to formulate certain underlying patterns of beekeeping development in the four thousand years of its known existence.

Honey hunting, which preceded beekeeping, forms an important early part of these patterns. Wherever social bees exist, man exploits them for their honey by hunting or by beekeeping. As a result, almost every archaeological find from the past has its counterpart in practice and usage somewhere in the world today. The tropical and subtropical regions are of great importance here, and it is still possible to see at first hand how honey has been harvested at all stages of human development, from the Stone Age onwards. Europe is, however, the richest hunting ground for finds from the more recent past, and in this book the word archaeology is used in its current extended sense, incorporating objects of fairly recent date.

The hives and other equipment discussed here belong to traditional beekeeping, which was simple, involving minimal intervention by the beekeeper. I have indicated the important events that led to the metamorphosis of traditional into modern beekeeping, but the book can be followed without any knowledge of modern beekeeping techniques.

Through the ages, man's attitude to bees has often been close, caring, and even reverential. This is exemplified much more widely than by beekeeping artefacts themselves, and I therefore include in the book some examples of the use of bees and hives to embellish works of art and objects of everday life.

In writing this book, I have been able to make extensive use of a

Preface

number of Libraries and Museums, especially those of the International Bee Research Association. Buckinghamshire County Library (Chalfont St Peter Branch) has also obtained many publications for me. A great many individuals have contributed directly or indirectly, and I appreciate their help immensely. Some have enabled me to see material in out-of-the-way places, and others have generously shared their specialized knowledge with me, particularly Dr John Graham, Harald Pager, Penelope Papadopoulo, Marjorie Townley, Raoul Verhagen and Frank Vernon. Penelope Walker's help in organizing the text and illustrations was invaluable, as was her partnership in handling the results presented in Chapters 7 and 8. Page 162 lists others who made special contributions to the material on which these chapters are based.

Authors cited in the Bibliography include many more people to whose work I am indebted, as do the photographers and artists named in the titles of the illustrations. I am grateful for their permission to reproduce their work, and I hope that the few who could not be traced will not object to the use of their material. All photographs not acknowledged are my own. I am especially pleased with the close integration of the illustrations with the text, which owes much to the expertise of Anthony Birks-Hay.

I greatly value the support I have received from Colin Haycraft of Duckworth throughout the preparation of the book.

International Bee Research Association,	Eva Crane
Hill House, Gerrards Cross, Bucks, UK	1983

Contents

Preface		5
Introduction		9
1.	Bees and hives	13
2.	Rock paintings and honey hunting in prehistory	19
3.	Tombs and temples: pictures of beekeeping	35
4.	Horizontal hives from Ancient Greece and after	45
5.	Forest 'beekeeping' and the precursor of upright hives	77
6.	Upright hives in apiaries	91
7.	Bee boles: 600 years of skep beekeeping in Britain and Ireland	117
8.	Bee shelters and houses: Britain, Ireland and continental Europe	163
9.	Towards movable-frame beekeeping	196
10.	Bees in art and in everyday life	213
Appendixes		250
Bibliography		338
Index		353

Units and geographical names

Metric units are used where possible. However, for historical reasons, dimensions quoted in Chapters 7 and 8, Appendix 1, and certain other passages, are in British Imperial units. It would be too clumsy to provide metric conversions on each occasion, and the following list may be useful:

1 in. = 2.54 cm
1 ft = 30.5 cm
1 lb = 0.454 kg
1 gallon = 4 quarts = 4.55 litres
1 peck = 2 gallons = 9.10 litres
1 sq. in. = 6.45 sq. cm
1 mile = 1.61 km

Imperial USA units are different from the above.

Countries and their subdivisions (such as counties) are usually referred to by their current names, but when referring to a period before the creation of a country, an older name is used if more appropriate.

Table 1. Summary of the extent of beekeeping evidence

	From historical period concerned			Ancient methods still practised	Chapter
	Artefacts	Pictures	Writings		
Honey hunting	+	+ +	+	+ + +	2
Tree 'beekeeping'	+	+	+ +	+	5
Beekeeping:					
in Ancient Egypt	0	+	+	+ + +	3
in Ancient Greece and Crete	+	0	0	+	4,9
in Ancient Rome	0	0	+ +	+ + (elsewhere)	4
in medieval continental Europe	+	+	+ +	+	6,4
in Britain and Ireland 1200–1900	+ + +	+ +	+ + +	+	7,8,6,9
in Ancient Africa and Asia	0	0	+	+ + +	4
in prehistoric America	+	+	+	+	4

0 = none known; + = a few; + + = a fair number; + + + = many

Introduction

This book explores the history of man's association with bees, as evidenced by material objects that have so far been discovered and identified. Bees are 'kept', and their honey and wax harvested, all over the world. In many less developed regions, which modern travel has now made more accessible, traditional methods have survived unchanged for centuries. As a result, we can enrich our study of archaeological finds with studies of their equivalents, still used by living beekeepers and honey hunters. As we shall see, the past can thus literally come to life after hundreds or even thousands of years. Table 1 shows the amounts of evidence from different periods and regions.

From earliest times man has valued honey, the sweetest food known to him, and in this he followed his pre-human ancestors. Primitive man developed an affinity with the bees from which he took the honey—in spite of stings—and the bees were often revered as magical, or even divine. Over the centuries the symbolism has changed[74, 78*] but both man's affinity with bees and his practical exploitation of them have continued. The economic value of bees is often underestimated. A 1981 paper[242] expressed it thus: 'Historically the impact of sericulture and apiculture on world trade and economics cannot begin to be evaluated. Silk, honey, and other domestic insect products may rival steel in their all-time productive value, when measured over the entire span of their use by man.'

I hope that the rich heritage of archaeological material presented here—which dates from soon after the Ice Age—will interest many people besides beekeepers. The term beekeeping archaeology is interpreted rather broadly here; like industrial archaeology it includes quite recent remains. But I have not taken it beyond traditional, so-called 'fixed-comb' beekeeping. Modern beekeepers use movable-frame hives, of which the first effective type was devised in 1851. The further development of these hives and of associated technologies would interest beekeeping specialists rather than the general reader. Chapter 9 discusses the early steps that led up to the movable-frame hive of 1851.

This book is concerned in principle with material remains, but

*Superior figures in the text refer to the numbered bibliography.

Introduction

sometimes illustrations, written records and published literature have been used to provide links which are not otherwise available. The book throws new light on the history of beekeeping, and many of the finds described here have only been discovered within the past 25 years. I have tried to simplify an extremely complex story, and I have sometimes generalised where anomalies exist that could be examined in depth. I have also skirted round a good many questions that would need a detailed assessment in a systematic history. And I have dealt mostly with hives rather than with tools and other equipment, whose use needs detailed explanations. Beekeepers' protective clothing is dealt with elsewhere.[82] There is such a wealth of material that not all of it can be included. The archaeological remains in Britain and Ireland, described here for the first time, are very extensive, and therefore more space is devoted to this area than would be given in a world history of the development of beekeeping. Such a book still remains to be written, but in two other books I have given short accounts of the story of man's practical exploitation of bees, and also shown how bees have appeared to the minds of men through the ages.[74,78]

THE STORY IN BRIEF

Chapter 1 gives a brief summary of the life and activities of bees, so that all readers can understand what is going on inside the hive. The ancients knew little about this, but nevertheless kept bees successfully and harvested honey and wax from them.

Chapter 2 explores the earliest direct evidence of man's association with bees: paintings showing bees' nests being robbed of their honey. The first was found in Spain in the 1920s, and in recent years more have been discovered in Spain and in India, many others in Africa, and one in Australia.

Chapter 3 introduces bee *keeping,* and here we rely on four illustrations made between 2400 and 600 B.C. in Egypt—one is in a temple and three are in tombs. They show beekeepers at work, and honey being processed and packed. No more illustrations are known for the next 1600 years, after which illuminated manuscripts provide clues.

Actual hives used in Ancient Greece have been excavated, dating from 450 B.C. onwards. Chapter 4 describes them and their counterparts that are still in use. It also describes nine different types of hive known to Roman authors, and shows where such hives can still be found in use today. Many are horizontal cylinders, and this chapter surveys the world's horizontal hives, which are much more widespread than the upright hives familiar in northern Europe. Photographs indicate the great variety of materials and crafts that have been used to make them.

Chapter 5 focuses on northern Europe, where hollow trees in the extensive forests provided nesting places for colonies of bees. These nests were generally much more accessible to primitive man than those in rock crevices in drier lands farther south, and a form of tree beekeeping developed, in which the beekeeper owned and tended bee colonies in the living trees. The practice probably originated among Finno-Ugrian peoples around the middle Volga region, and spread through Russia as far as the Ural mountains in the east and Karelia and Germany in the west. There are also possible vestiges of it in England.

Chapter 6 describes the upright log hives derived from tree beekeeping; they are characteristic of the same region, and also of early beekeeping in North America and parts of northern Asia, for instance Korea. Many basket and other upright hives, with the open mouth down, were also used in northern Europe. They provide a spectacular variety of craftsmanship, and examples are described and illustrated.

Basket hives, often called skeps, reached the off-shore islands of Britain and Ireland, and structures were built there to shelter the skeps from adverse weather. Many of these structures were much less perishable than the skeps themselves: in 30 years of searching, interested people have located over 800 that still survive. Together they represent locations for around 3000 skeps, and are a unique archaeological monument to beekeeping over the past 700 years; details are recorded in full in a register maintained by the International Bee Research Association since 1952. Chapters 7 and 8 present the interesting results of a first analysis of these records, and details of individual sites are summarised in Appendix 1. I believe that no other region in the world has a comparable archaeological heritage from beekeeping of the past. Chapter 7 deals with *bee boles*, which are recesses built into walls to accommodate skeps— usually one in each, but occasionally two or more. Chapter 8 deals with other structures: elegant ones which I call *alcoves*, housing several skeps and often built into a wall; *bee shelters*, comprising one or more shelves, with a roof over; and *bee houses*, which had a space behind the hives where the beekeeper could work. German bee houses are also discussed.

Chapter 9 forms an epilogue to the evidence presented in earlier chapters, describing the scanty but possibly very important evidence of early hives that used a principle similar to the one on which the world's honey industry is now based: combs that could be lifted out individually and replaced.

Finally, Chapter 10 explores the archaeology of a much wider field: representations of bees in sculpture, glass, jewellery, coins, etc., and also objects that were made in the ancient world with honey and beeswax. This chapter shows the continuing affinity that man has felt with bees through the ages.

One great benefit to us of the survival of traditional beekeeping as a living craft in many parts of the less developed world today, is that hives and tools like those used centuries ago can still be seen. In writing this

Introduction

book I have been able to use the important collection held by the International Bee Research Association[305]—which, alas, is not on display for lack of funds. Appendix 2 gives details of more than seventy beekeeping museums where displays can be seen; information about access is included so that readers can seek out and see for themselves some of the material discussed in this book. Most of the museums are in European countries. There is an urgent need for material to be collected and preserved in continents where traditional methods are still used. In 1976 the First International Conference on Apiculture in Tropical Climates passed the following resolution, which was reconfirmed by the Second Conference in 1980:

> *The Conference, recognizing* that among many peoples in the tropics and subtropics honey hunting and beekeeping are still practised by methods which have remained virtually unchanged for centuries, *knowing* that these methods constitute a basis for further development, but *fearing* that being traditional they may be despised and therefore lost, *urges* national museums and appropriate cultural institutions to record and study in depth, and to preserve, both traditional methods and equipment, and the traditional beekeeper's knowledge about his bees and their environment.

The Third Conference will be held in 1984 in Africa, a continent which has material for the most splendid beekeeping museum. The variety of hive types still in use there, or used within living memory, is the richest in the world, and a collection of examples of materials and crafts used in different countries for beekeeping and honey hunting could preserve knowledge of them for future generations. A special exhibition will be held in the National Museum in Nairobi, the venue for the Conference.

FIG. 1 *An empty honeybee nest in a hollow tree, viewed from below. The nest itself would have looked much like this as far back as the Ice Age.*

1. Bees and hives

The bees that live in hives are, in most parts of the world, the European honeybee *Apis mellifera*. Except in modern frame hives, the bees construct their nest by building a group of parallel combs vertically downwards from the roof of the hive or nest cavity (Fig. 1). The distance between the combs is an inbuilt characteristic of the bees, and varies slightly for different species or subspecies, according to their body size. The spacing is most exactly followed for combs in which brood is reared— normally the central part of the group of combs, where the temperature can most easily be controlled. Bees show a greater tolerance in the spacing between the honey-storage combs above and around the brood nest. Most of the traditional hives with which this book is concerned were 'fixed-comb' hives, so called because all combs were attached to the hive itself, and the beekeeper could not remove and replace them.

Apis mellifera is very widely distributed (Fig. 2); there are temperate-zone races native to Europe, and subtropical and tropical-zone races

Bees and hives

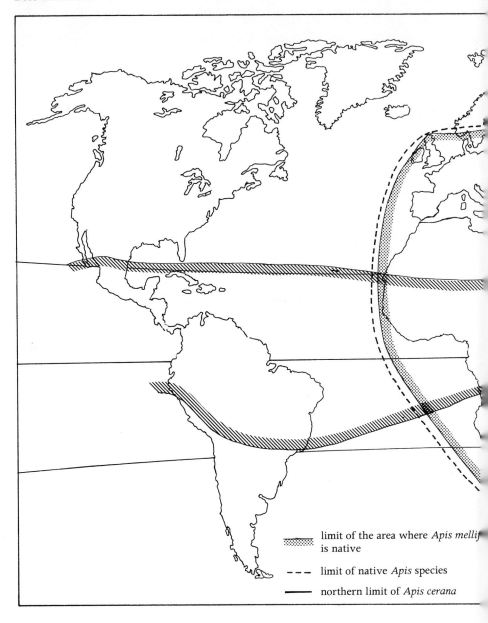

Bees and hives

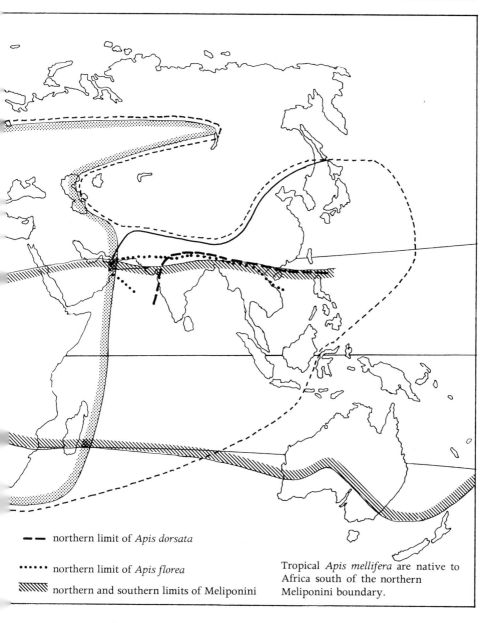

— — northern limit of *Apis dorsata*

•••••• northern limit of *Apis florea*

▨▨▨ northern and southern limits of Meliponini

Tropical *Apis mellifera* are native to Africa south of the northern Meliponini boundary.

FIG. 2 *World distribution of honey-producing bees. Honeybees (*Apis *species) are native only in the Old World, stingless bees (Meliponini) only in the tropics and warmer subtropics.*

native to western Asia and Africa.[278] No honeybees are native to the New World, but European races were taken by man: to the Americas from the 1620s onwards, and later to Australia, New Zealand and the Pacific area. In 1956 tropical-zone honeybees from Africa were taken to Brazil, where they hybridised with the 'European' bees already there and prospered so well that they spread rapidly—making news under such extravagant names as 'killer' bees.

In temperate zones, a colony of European honeybees in a large enough hive is constituted somewhat as follows:

	seasonal maximum	at onset of winter	at end of winter
queen (female reproductive)	1	1	1
drones (male reproductive)	300	0	0
workers (female non-reproductive)			
adults	50 000	25 000	13 000
eggs, larvae, pupae (brood)	35 000	500	2 000

The size of hive a colony requires varies throughout the season, both because of the varying number of bees and because of the need to store honey for use in the dearth season. Also, bees spread out more when the hive is hot, to prevent overheating, and contract into a small space when it is cold; in very cold winter weather they form quite a tight cluster, occupying less than a tenth of the space they use in the summer.

Traditional hives vary greatly in size (Table 2), for reasons that have not been fully explored. We shall see that some Roman beekeepers adjusted the size of their hives according to the needs of the colonies in them, as modern beekeepers do.

There is no need for a colony to fill all the space in a hive all the time; empty combs in a hive can stimulate nectar collection, although they are also vulnerable to damage by pests. The critical factor is the ability of the colony to maintain the brood temperature at a constant 34°–35°C (93°–95°F), otherwise the brood will die. Large, well ventilated hives help bees to maintain this temperature when it is hot outside. On the other hand, in cool or cold weather a small colony is better off in a small hive than in a large one. Bees store pollen immediately round the brood nest. In the active season, as honey is accumulated the bees store it in any available space outside the pollen. But when temperate-zone bees prepare for winter, they usually amass the honey in an arch above and around the central part of the nest where next season's brood will be reared.

Honeybees create new colonies by swarming: in spring, brood rearing increases rapidly, and when the population becomes high in early summer, the colony is likely to rear drones (males) and also several queens (females). When the new queens are nearly adult, a swarm issues from the nest or hive during the warm part of the day; it contains about half of

Table 2. Approximate size of various hives (for *Apis mellifera*)

	Volume (litres)	Chapter
Egyptian mud hives now used in Egypt (Egyptian honeybees build small colonies)	20*	3
Sloping cylindrical clay hives now used in Crete	20–40*	4
Recommended by authors in Ancient Rome:		
cylindrical hives 3 ft long, 1 ft diameter	44*	4
ferula hives $3 \times 1 \times 1$ ft	57*	4
Measurements of 19 log hives now used in Kenya[174]	40–130*	4
Recommended by authors in Britain:		
skeps 1593–1890	15–30	6,7
occasionally down to	9	
up to	36	
skeps (1870)[235] up to	50	6
Modern hives:		
10-frame Langstroth hive, one brood box (only)	40	
three such boxes (common usage in summer)	120	
six such boxes (of which 4 get filled with honey)	240	
Bees' choice and requirements:		
size of cavity containing natural nests[270]	12–443 (20–100 most common)	
maximum space occupied by bees in England[275]	120	

*Volumes calculated from external dimensions, allowing 1 in. (2.54 cm) for all walls.
The space available to adult bees is less than quoted, especially when the combs in it are full of brood, pollen, and/or honey.

the adult workers and the original queen that headed the colony. In the simplest case only one new queen survives in the colony left in the nest or hive. When she is a few days old she flies out and mates in the air with perhaps 10 to 15 drones, returns, and after a few days settles down to egg laying, and the colony continues. It will produce only perhaps half as much honey for the beekeeper as it would if he had prevented swarming; modern beekeeping management therefore aims at high populations without swarming, and beekeeping manuals explain how this can be achieved.

Meanwhile the swarm flies to a branch or some other support nearby, and clusters there. From the cluster, scout bees fly off and locate and inspect any cavities in the countryside around that might serve as a possible new nesting place, including empty hives. These scout bees can estimate the size (volume) of a cavity by walking over its inner surfaces; this takes them about 40 minutes.[270] In experiments with bees of European ancestry in New York State, USA, the commonest size chosen was about 35 litres, and almost all cavities selected were between 20 and 100 litres, although a few were larger. Some sizes for comparison with hives discussed in later chapters are given in Table 2.

It seems likely that, without hives provided by beekeepers, the number of colonies in any one area would more often be limited by a shortage of

Bees and hives

nesting sites than by a shortage of food.[271] By providing hives, a beekeeper remedies this deficiency, and can keep many colonies together in an apiary, although this is not how bees live naturally.

In many forms of traditional beekeeping, swarms were normally prized because they were relied on to populate new hives. The beekeeper watched for them, and as soon as a swarm clustered he 'took' it by shaking it into a container such as a basket or skep, and thence into a new hive.

Early beekeeping books quoted in Chapter 7 show that the English beekeeper deliberately used small hives in order to get early swarms to fill extra hives: 'A swarm of bees in May is worth a load of hay.' In the autumn the beekeeper killed the bees in half the skeps (the lightest because the bees would eat what little honey they had, and then die, and the heaviest because they would contain most honey), and he wintered the other half. In later centuries, as Chapter 6 shows, improved skeps were devised so that the bees need not be killed. Many other early hives were better than the small skeps, and did not necessitate killing the bees. The Romans, for instance, used much larger and better hives (Chapter 4). But, ironically, it is the traditional *skep* that is used throughout the world as the symbol of the colony of bees and, by implication, of harmony, industry and thrift; Chapter 10 provides an insight into this subject.

OTHER BEES

There are four species of honeybee altogether, three being native to tropical Asia. *Apis cerana* is like a smaller version of *Apis mellifera*, and both hives and honey harvest are proportionately smaller. This bee is native to the area south of the Himalayas, but also extends as far north as Japan and the mainland nearby[79] (Fig. 2). *Apis dorsata* is a large bee which builds a single vertical comb in the open, suspended from a rock overhang or a branch of a tree. It cannot be kept in a dark hive, and honey is still harvested from wild nests as in prehistoric times (Chapter 2). Finally, there is a very small species, *Apis florea*, which also builds a single-comb nest in the open.

As well as the *Apis* species, Fig. 2 shows the distribution of a separate group of bees from which honey is harvested (Chapter 4): the Meliponini, known as stingless bees.* They live in the tropics, but are native to the Americas and Australia as well as to Africa and Asia. It was their honey that Christopher Columbus referred to in his account of the 1492 voyage to America, and it was their wax that was used by early American Indians for casting their fabulous golden ornaments. We shall see that they have their own place in the archaeology of beekeeping.

*The family Apidae contains two subfamilies: the Bombinae which includes the bumble bees, and the Apinae which is divided into two tribes—the Meliponini (stingless bees), and the Apini which consists only of the four species of *Apis*, the honeybees.[205]

2. Rock paintings and honey hunting in prehistory

This chapter deals with the earliest archaeological evidence of man's association with bees—harvesting honey by hunting for wild nests of bees. The following rough time-scale shows, however, that bees were producing honey for many millions of years before man existed:

for 150–100 million years	flowering plants have existed and produced nectar and pollen
for 50–25 million years	solitary bees have existed, also early primates (monkeys)
for 20–10 million years	social bees have produced and stored honey
for a few million years	man has existed and has eaten honey
for ten thousand years	records have survived of man's exploitation of honey.

Evidence of the harvesting of honey and beeswax by prehistoric peoples comes from three main sources: rock paintings, examples of equipment used for similar operations in the recent past, and anthropological studies of recent and present-day operations by tribes that still hunt for honey. The evidence shows that honey was—and is—highly valued and considered worth harvesting in spite of the stings suffered in the process. All over the world the honey-hunting techniques that have been developed are fairly similar, because the problems encountered were similar: inaccessibility of nests, attacks by bees, and the problems of getting hold of the combs and carrying them away with honey dripping from them. Methods used by animals to solve these problems may well indicate how primitive man managed to get honey in the first few million years of his existence.[74]

Europe

The earliest direct evidence of honey hunting comes from rock paintings. There has been a suggestion of bee-connected paintings[230] in the palaeolithic art of Altamira in northern Spain, dating possibly from the last major

19

Rock paintings and honey hunting in prehistory

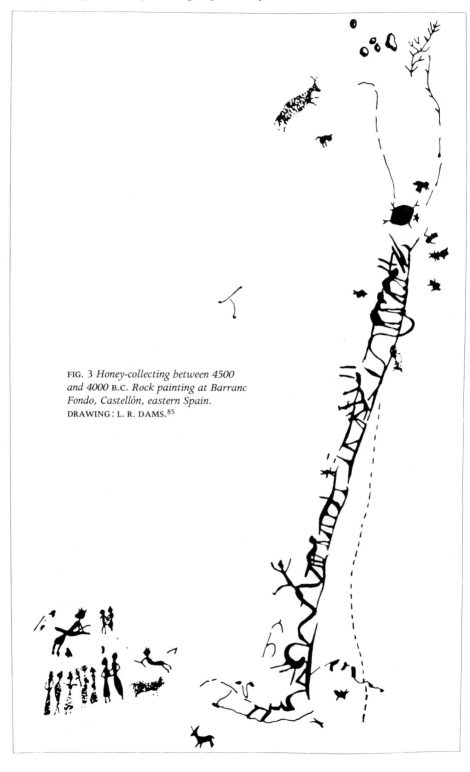

FIG. 3 *Honey-collecting between 4500 and 4000 B.C. Rock painting at Barranc Fondo, Castellón, eastern Spain.*
DRAWING: L. R. DAMS.[85]

glaciation of the Ice Age (30000–9000 B.C.). But mesolithic paintings in eastern Spain certainly show honey-hunting scenes. They have been dated provisionally to the period 8000–2000 B.C., on the grounds that associated paintings show no animals of the glacial period, that they provide some evidence of primitive agriculture, and that most of the human figures are unclothed, suggesting a climatic optimum such as occurred around 6000 B.C.[85]

Groups of flying bees can be seen in association with human or animal figures at a number of open rock shelters in Teruel Province, Spain;[85] sometimes what appears to be the bees' nest is depicted. Fig. 3 shows a complex honey-hunting scene at Barranc Fondo, Castellón, with five figures on a ladder leading to the nest, and a group of others below, who may have been waiting for their share of the honey. Honey hunters nowadays commonly distribute the combs to others with them, and there is a Biblical record of it too: when Samson found the lion's carcass with bees and honey in it: 'He scraped the honey into his hands and went on, eating as he went. When he came to his father and mother, he gave them some, and they ate it.' (Judges 14:9).

Another rock painting of the same period (Fig. 4) was discovered earlier at La Araña shelter, Bicorp, in eastern Spain.[144] It shows two figures on a ladder; the upper one, at the bees' nest, is carrying a container for the honey combs.

Among the rock paintings in the gorge El Mortero near Alacón, between Zaragoza and Valencia, are five rock paintings that show a person on a 'feathered' ladder or tree. In several there is something that could be a bees' nest. The largest human figure (Fig. 29, p. 36, of the report by Beltran[34]) could, as he says, be a honey hunter with a collecting bag; alternatively it could be someone picking fruit, and the 'nest' a bunch of dates.

Instructions for finding wild bees' nests were set out by Columella, who was a Spaniard, in Roman times, although by then beekeeping was well advanced (Chapter 4), and the bees were sought to populate hives. Columella (*De re rustica* 9.8.7-13) recommended that you should go into the woods and seek out a place where many bees collect water. You mark the backs of the bees, using plant stalks smeared with liquid red ochre. You wait, and according to the time between the bees flying off and returning, you know whether their nest is near or far away. If it is near you can find it by following the direction in which the bees fly off. If not, you bait a piece of hollow reed with honey and trap some bees in it; you allow one bee to escape and then follow its line of flight, repeating this until you come to the nest. The bee hunter is warned 'to choose the morning for the search, so that he may have the whole day to spy out the comings and goings of the bees'.

Honey hunting has continued until more recent times in Hungary,[25,209]

Rock paintings and honey hunting in prehistory

FIG. 4 *Another honey-collecting scene, around 6000 B.C., and the first to be recorded, in 1924. Rock painting in La Arana shelter, Bicorp, eastern Spain.* COPY BY E. HERNANDEZ-PACHECO. *Enlarged detail below.*

FIG. 5 *Honey hunter climbing a forked ladder to reach a bees' nest. Rock painting (date unknown) in Eland Cave, Drakensberg Mountains, Natal, South Africa.* DRAWING: H. PAGER.[225]

Romania,[28] and especially in the Carpathian Mountains.[133] In the forests of northern Europe honey hunting almost certainly started as soon as the climate allowed both bees and human beings to live in the region; it is well documented from the early Middle Ages onwards and is the subject of Chapter 5.

Africa

Africa has many more rock paintings connected with honey hunting than any other continent. We owe our knowledge of them to Harald Pager, who has described and analysed the paintings in various publications[225,226,227,228] and in a comprehensive unpublished report.[229] These paintings were made by Stone Age hunters whose cultural development was similar to that of the rock painters of eastern Spain, but few can be dated. Many depict scenes still familiar to the people who live in the same or nearby regions.

Very few of the paintings show the honey hunter taking honey; in one, shown in the frontispiece, near the Toghwana Dam in Zimbabwe[226] he is using smoke to stupefy the bees. Several paintings show ladders being used to reach the bees' nest; Figs 5 and 6 are from the Drakensberg Mountains, Natal, South Africa. Fig. 6 and the frontispiece show the

FIG. 6 *Complex of three bees' nests with ladders giving access to them. Rock painting (date unknown) in Anchor Shelter, Drakensberg Mountains, Natal, South Africa.* DRAWING: H. PAGER.[225]

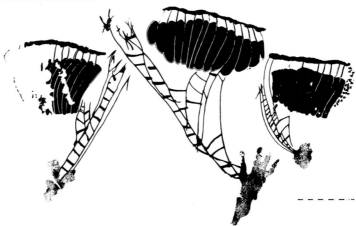

combs that constitute the nest, with their centre part dark and the outer/upper part light, which could represent cells of honey and empty cells or, more likely, cells of brood and of honey, respectively; honey is normally stored above the brood. Fig. 6 would then show, at the top of each nest, the roof from which the combs are suspended, then the honey, and below it the brood. These are life-like drawings.

These irregular oblong shapes, referred to above as combs, appear widely in the rock art of Zimbabwe, and two sets have been found in South Africa. Until Harald Pager presented the evidence for their interpretation as combs in bees' nests,[225,226] they had been referred to as 'formlings' or 'tectiforms', of unknown significance. Others besides the frontispiece picture show these shapes in an enclosure, with bees streaming in or out of it, for instance in Zombapata Cave, Zimbabwe.

There was a painting, unfortunately now destroyed, in the Ebusingata Cave, Royal Natal National Park, South Africa,[227,228] which showed a man near the top of a ladder leading to a bees' nest, with bees flying around. Two further paintings in Zimbabwe may possibly represent honey hunting—at Mimosa Farm, Salisbury,[130] and Makumbiri Farm, Msonedi area;[233] if so, the first shows one man on a ladder surrounded by bees which spread towards a tree top, a second man just below him and a third nearby. Close to this scene are huge formlings of the type described, which may have been painted later. The second shows five armed hunters, one carrying a collecting bag, and another up a ladder or tree, either netting birds (Petie's interpretation) or fighting off bees to get at their honey. It is similar in concept and vigour to Fig. 3. Ladders are shown in many other paintings, with or without human figures. Some of them may have provided access to bees' nests (as they do today), although certainly not all.

Many other paintings show a group of flying bees, like a swarm, and with some of these there is a group of roughly parallel 'catenary curves' (Fig. 7) which may be interpreted as the combs of the bees' nest.[225] These same curves or similar ones appear in a host of other paintings, but—like the formlings—until they were found associated with flying bees it was impossible even to guess what they represented. What are called 'comb patterns' (groups of juxtaposed circles looking like the openings of cells in a comb) occur widely in Namibia as petroglyphs.

Table 3 indicates countries in Africa where various bee motifs appear in rock paintings. More than four thousand rock art sites are known in South Africa and Zimbabwe, and these countries are correspondingly rich in nearly all the bee-associated subjects. I have seen paintings only at one site, in Ndedema Gorge in the Drakensberg Mountains; the bees have a dark body and white wings spread out as in flight, and the paintings give a vivid and realistic impression of activity round a hive entrance. The most recent discoveries I heard of, in

Rock paintings and honey hunting in prehistory

FIG. 7 *Catenary curves indicating a bees' nest. Botha's shelter, Ndedema Valley, Drakensberg Mountains, Natal, South Africa.*

November 1981, were paintings of a swarm of 376 bees in the Brandberg, Namibia,[231] and of a ladder in close association with two bees' nests in Tugela Shelter near Bergville, Natal.[201]

Rock art sites in general are much rarer in central Africa than in the southern part and in the Sahara, and rock paintings of bees are almost absent from these regions. However, this does not imply that there was no honey hunting, and it is in fact in the central parts of Africa that honey hunting still enriches people's life and diet today.

Table 3. Location of rock paintings in Africa that are, or may be, related to bees or honey hunting

	Comb patterns	Catenary curves	Formlings	Ladders	'Swarms' of bees	Honey hunting
Algeria		×				
Lesotho		?		×		
Libya				×		
Malawi		?	?	×		
Morocco		× ×				
Namibia	× × ×	× ×	×	×	×	
South Africa		× × ×	× ×	× × ×	×	× × and ?
Tanzania		×			?	
Tunisia			×	×		
Western Sahara	×					
Zimbabwe		×	× × ×	× × ×	× × ×	× and ?

× = one site, × × = several, × × × = many, ? = uncertain (evidence from Harald Page[229,231])

Rock paintings and honey hunting in prehistory

Honey hunting is a living craft among many African tribes, and their operations and equipment can still be seen, although generally in areas that not many travellers reach. Here are a few descriptions, and others are available.[74,78]

In *Zanzabuku*, Lewis Cotlow[65] described honey hunting by Mbuti pygmies in the Ituri Forest between the Congo River and Lake Albert. 'One thing will divert Pygmies from their hunting—the finding of a bees' nest with honey. Once I watched them scramble up a tall tree, where there was a hive [nest] hidden in a big hole. They used a heavy liana looped around the trunk to help them in their ascent, much as a telephone linesman uses a wide strap. Once up even with the hive, the first hunter enlarged the hole with his spear, as hundreds of bees swarmed about him. Then he plunged in his hand, brought it out full of honeycomb dripping with honey, and crammed the whole mess into his mouth. One after another, each hunter climbed the tree and ate his fill of honey.' Cotlow wrote in 1957; Ichikawa's ecological and sociological study published in 1981[152b] found that in the honey collecting season 80 per cent of the calorie intake of the Mbuti comes from honey; honey collection takes up to 6 to 7 hours a day. Ichikawa has also studied the Dorobo honey hunters in East Africa.[152a]

Colin M. Turnbull says that honey calls the pygmies more strongly than any other food; the honey season is one for abandoned merrymaking and rejoicing. His book *The forest people*[299] gives a full account of the hunting and dancing and singing; one of the songs has been quoted elsewhere, together with details of several honey-hunting film sequences from other parts of Africa.[74] Vessels for collecting honey were made of gourds, basket-work, wood and leather, or leather alone; some are shown in Fig. 8. Bushmen who occupied the mountains to the north of Knysna Forest in South Africa collected honey and carried it over the mountains in bags made from antelope skin, to barter for other goods from the Hottentots. Hence the region was called the Outeniqua, 'bags of the honey people'.

Honey hunting always needed the strongest climbing rope that could be made. In some areas a round stone was attached to one end, to be thrown over a branch of the tree, and such stones can still be found. Ladders were often made of lianas, or shorter ones of wood. If the nest was high up in a suitable tree, wooden climbing pegs might be knocked into the trunk.[135] I hope that before it is too late, a museum somewhere in Africa will create a collection of the honey vessels, and the ropes, ladders and other materials used for honey hunting. The most comprehensive description of the equipment was published by Seyffert in 1930.[273]

Some expert honey hunters in the Brandberg in Namibia (where the 'comb patterns' are found, Table 3) would observe the flight of bees to and from water, as Columella did. Once a hunter had established the direction as far as the horizon, he took up the trail, following the bee

FIG. 8 *Vessels used in Africa for collecting honey.*
The shallow basket (B71/80) is from Mali, and also the gourd vessel (B71/81) which is decorated with a representation of a Mali wicker hive (Fig. 59), used on its side. The wood and leather barrel (B68/17) is used by the Tharaka tribe in Kenya, as is the sisal climbing rope (B68/19). IBRA COLLECTION.

droppings which would be barely visible to the unpractised eye. If he was overtaken by night in a place too far from home, he set up camp and at first light again took up the spoor. It was said that he would know when he had got near to the nest from the colour of the droppings.

Asia

The honeybees hunted by prehistoric man in Europe and Africa are the same species as the bees now kept by beekeepers in most parts of the world: *Apis mellifera*. This species is not native to tropical Asia; instead, there are three species, all adapted to tropical conditions. The largest is *Apis dorsata*, known as the giant honeybee, which builds a very large single comb for its nest, hanging down from a tree or a rock ledge. It will not live in a dark hive, and its honey must still be harvested by hunting. The smallest is *Apis florea*, the little bee, which is rather like a tiny version of *Apis dorsata*. Its honey yield is small and—perhaps because of this—it is highly valued, especially for medicinal purposes. The third species, *Apis cerana*, is much more like the European honeybee, and is now kept in hives.

Five rock paintings are known, all in central India, and all showing *Apis dorsata* which gives the largest harvests. There are four in Pachmarhi region—two in Jambudwip Shelter (one of which is shown in Fig. 15), and one each at Sonbhadra and Imlikhoh—and one in Shelter IIIF-35/b at Bhimberkah. This last is mesolithic; the others are later. Four paintings show the shape of the comb, which is characteristic of *Apis dorsata*. In general the rock paintings known in India suggest a much less close observation of natural details than those of southern Africa.[1] The bees, shown in three of the honey-hunting scenes, are represented only by dots, not as realistic insects. Nevertheless, this is compensated by the realism with which the honey hunters are depicted, as they try to get at the honey while standing as far from the nest as possible. The right-hand person in Fig. 15 wields a very long, pronged wooden stick, such as is still used to prize comb out of a cleft in the rocks.[1] The painting in Imlikhoh shelter shows a fairly realistic burning brand being used as a smoker, but no ladder.

Three hundred years ago Robert Knox was held captive in Ceylon for nearly twenty years, and in 1681 he published an account of his life there.[176] He described the three species of honeybee. What we know today as *Apis cerana* bees 'build in hollow trees, or hollow holes in the ground . . . into which holes the men blow with their mouths, and Bees presently fly out. And then they put in their hands, and pull out the Combs, which they put in Pots or vessels, and carry away. They are not afraid of their stinging in the least, nor do they arm themselves with any cloths against them.' *Apis dorsata* bees

> make their Combs upon limbs of Trees, open and visible to the Eye, generally of a great height. At one time of year whole Towns, forty or fifty in company together will go out into the Woods, and gather this honey, and come home laden with it for their use. . . . When they meet with any swarms of Bees hanging on any Tree, they will hold Torches under to make them drop; and so catch them and carry them home. Which they boyl and eat, and esteem excellent food.
>
> [Veddas or Wild Men] have their bounds in the Woods among themselves, and one company of them is not to shoot nor gather honey or fruit beyond those bounds.
>
> They have a peculiar way by themselves of preserving Flesh. They cut a hollow Tree and put honey in it, and then fill it up with flesh, and stop it up with clay. Which lyes for a reserve to eat in time of want.

H. Williams, in *Ceylon—pearl of the east*,[319] describes the Veddas' delight in honey, and their hunting of honey from 'wild and inaccessible crags'. He adds:

> Small boys go through the motions of descending cliffs in search of honey in exact pantomime. They make small replicas of the bamboo ladders used by the men during this supremely dangerous pursuit and pretend to smoke out their prey exactly as their fathers do. On gaining the top of the supposed cliff, they brush off countless but happily imaginary bee stings before rushing off into the jungle, amid the yells of the other children, as if to dodge the attack of enraged bees.

The Veddas still live the same life as their Stone Age forebears. They do not grow corn, and their only food is flesh, and honey when they can get it. They have a small axe which they wear at their side to cut honey out of hollow trees. The women go hunting and honey collecting with the men.

C. N. Taylor, in *Odyssey of the islands*,[290] describes his experience in the Philippines, presumably with *Apis cerana*:

FIG. 9 *Honey hunter in Sumatra, Indonesia, climbing a tree to reach a high nest of* Apis dorsata, *using ropes as shown in Fig. 10.*
PHOTO: N. KOENIGER 1980.

FIG. 10 *Equipment used for collecting honey from nests of* Apis dorsata *in Sumatra, Indonesia (see Fig. 9). Funnel-shaped honey container (left), and climbing rope organized into a backpack (right).* DRAWN FROM PHOTO BY N. KOENIGER 1980.

For several days we lived on meat and wild honey . . . Bees were not hard to find, but they were usually located in the topmost branches of enormous gum trees. The trees ranged from seven to nine feet in diameter at their bases, their first branches being fifty feet or more above the ground. Climbing them, however, was a simple matter for the Ibilaos. First, they would cut bamboos and split the sections into pegs; then they would drive the pegs into the soft tree-trunk, one above another, to form a ladder to the first branch. Naked except for their gee-strings, they would go up, hand over hand, to the branch, where the bees could be seen swarming in and out of a knot-hole. A few puffs of cigar smoke blown into the hole; then a dozen bold strokes—and the honey would belong to us. If the robbers were stung, they paid no attention to it.

Returning to mainland Asia, and *Apis dorsata*, the honey-hunting equipment of the Gonds of Central India was shown in an exhibition at the British Museum in 1973, and also appears on the cover of the catalogue.[311] It comprised a ladder made from creeper, a bamboo stick with a bundle of grass tied to one end, a large wooden spatula with a bamboo handle, and a basket attached to a bamboo pole. The ladder was let down the face of the rock, and the collector—covered with a blanket and carrying a herb to protect himself from the bees—descended with his implements. He lit the grass to drive away the bees, then used the spatula to scrape the honey from the combs into the basket.

Fig. 9 shows a honey hunter in Sumatra on his climbing rope, attached rather elegantly to the tall tree trunk, to reach an *Apis dorsata* nest just visible at the top of the photograph. When fixing the rope, he uses a backpack (Fig. 10, right), and he puts the honey comb in a funnel-shaped vessel (Fig. 10, left).

An informative account of honey hunting in the swampy Sundarbans forest at the mouth of the Ganges[58] quotes honey yields as 3 to 14 kg per nest. Between 1963 and 1972, with 913 to 1495 honey collectors per season, there were 96 casualties due to carnivorous animals. I have heard that boys accompanying their fathers, to help and to learn the work, are the most likely to be caught by tigers.

Heinrich Harrer,[142] who gave a graphic description of Nepalese honey hunters working in the gorge of the River Kosi in Tibet,[74] told me that Tibetans were officially forbidden to take honey by their government, because their religion does not allow them to deprive animals of their food. Peter Aufschneiter also told me in a letter that Tibetans consider collecting honey a great sin because all the worms (brood) then die; they also profess not to have the skill. But the people who do collect the honey are also of Tibetan stock, although they come from Nepal. Beekeeping was entirely new to most people Aufschneiter encountered in Tibet.

In 1980 S. S. Strickland watched five honey-hunting expeditions by the Gurungs in Nepal, and Figs 11 to 14 are reproduced from his detailed report.[284a] Fig. 11 shows the group of *Apis dorsata* nests on the rock, with the smoke directed towards them. In Fig. 12 the ladder has been fixed in position (from above), and the honey collector is descending it with his long reaching-pole of bamboo, at the far end of which is fixed a knife for cutting the combs; the knife is clearer in Fig. 13. Strickland draws attention to the similarity between the action in Fig. 14, in which the long reaching-pole is being used to cut the comb, and the rock painting shown in Fig. 15, although there the reaching-pole appears to end in a three-pronged fork.

FAR LEFT: FIG. 11 *Honey hunting by the Gurungs, near Khilang, Nepal. Rising smoke drives* Apis dorsata *bees from the lower part of their nests on the rock face. At least 6 nests can be seen, light in colour near the curved lower edge, and darker above, where the comb is covered with bees.*

CENTRE LEFT: FIG. 12 *The honey collector climbs down a ladder of bamboo twine, his long reaching pole (with a knife at the far end) and his collecting basket being suspended by separate ropes.*

LEFT: FIG. 13 *With his reaching pole (vertical here, showing the knife at the lower end), and using a second pole to manoeuvre the lower rope of the collecting basket under the rocky overhang.*

ABOVE: FIG. 14 *Cutting a comb (on the left) with the reaching pole. Honey streams down from the comb in a fine thread.* PHOTOS: S. S. STRICKLAND 1980.[284a]

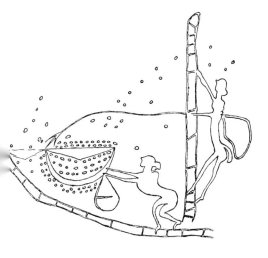

FIG. 15 *Collecting honey from a nest of* Apis dorsata *bees. Rock painting in white (about 500 B.C.) in Jambudwip Shelter at Pachmarhi, Central India.* DRAWING: D. H. GORDON.[130a]

Rock paintings and honey hunting in prehistory

Australia

Australia has no native honeybees, but it does have tropical social bees that also form permanent colonies and store honey: the stingless bees or Meliponini. These bees have only vestigial stings, and they defend their colonies in other—also very effective—ways. Different species bite or tickle, or burn the skin with a caustic fluid, or demoralise an intruder by crawling into his eyes, ears, nose and hair.

Nests of stingless bees have been raided for their honey wherever they occur in tropical Africa and Asia, and especially in tropical America and Australia where they were the only honey-producing bees (Fig. 2). In *Honey: a comprehensive survey* (1975)[74] I published a photograph of a rock painting by Australian aborigines at Secure Bay near Darwin, which shows a nest of stingless bees. I had always hoped that a rock painting showing a honey hunter at work would one day be discovered, and such a painting was found in 1972 by Percy Trevise,[298] in a very rugged remote area on headwaters of Coamey Creek, an affluent of the Laura River, about 200 miles north west of Cairns. Mr Trevise says:

> This is a totemic (sacred) site of Beehive Dreaming. The top 'hive' is about 21 ft long, the lower about 14 ft long. The hives are depicted as the imagined shape of the hive inside the hollow tree. Swarms of bees are leaving the narrow entrance tunnel of wax; some are just dots but some have wings, especially four much larger ones which are obviously queens.

In most parts of the world honey hunting, like the hunting of wild animals, was traditionally men's work. But among some Australian tribes the women were pre-eminently the food providers, and they hunted for honey as well as for roots, fruits and small game. In *Stone-age bushmen of today*[195] J. R. B. Love described the actions of a woman of the Worora, in the extreme north west of Australia, between Prince Regent River, the sea, and Walcott Inlet (Collier Bay):

> Keenly scanning the trees, one of the women will see bees entering a hole in a limb of a tree. These little native bees are smaller than a house fly. They do not sting, nor do they make the regular hexagonal cells of honey-comb that the European bees make. The honey cells of the native bees are globular, stuck together in no order. A good hive may contain a quart of honey, pollen, young bees, and wax, all of which are indiscriminately eaten . . . Honey is almost the only sweet known to the Worora, and they love it. Chopping out a wild bees' nest, with a crude tomahawk, while clinging to a high branch of a tree, is hard work, so, naturally, the Worora woman does not chop one out unless she is sure of finding honey in the nest. To test this she uses a long reed which is poked into the hole where the bees are entering. The reed is withdrawn, the end inspected and tasted. If the nest contains good honey, then the woman sets to work to chop it out . . . As a change from eating honey, or to make it go farther, when there has not been very much obtained, the people will mix honey in a bark bucket with water,

and then sop up this solution with a handful of pounded bark, which is dipped into the liquid, then sucked with relish.

One of the aborigines' uses of beeswax is mentioned in Chapter 10.

The Americas

Like Australia, the Americas have no native honeybees (*Apis*), but some of the tropical areas are rich in the social stingless bees or Meliponini (Fig. 2). These bees, especially species of *Trigona* and *Melipona*, have been widely hunted for their honey and also kept in hives. There is a comprehensive book on them.[266] North America has many rock paintings (pictographs) and petroglyphs, and Wellmann published a comprehensive study of them,[313] but there are no clear indications of any relationship with bees.

A detailed description of honey hunting by the Guayakis in Paraguay has also been published.[303] In 1968 the West Indies Mission in Surinam reported that the isolated Akoerio tribe of Indians 'know at least 32 different kinds of honey and rely heavily on it. The Trios and Wayanas [larger neighbouring tribes], by comparison, use very little'. A further report, on an attempt to relocate the Wayarekule Indians, said that after contacting these hunters and gatherers, it was very difficult to keep their attention for more than a few moments at a time because of their preoccupation with finding food: 'they were continually jumping up to crack nuts, scan the trees for the possible activity of honey bees etc. . . Before we arrived many of the people had gone off in search of honey and nuts.'

In tropical America there are also various species of social honey-storing wasps, notably of *Nectarina*, *Brachygastra* and *Polybia*.[206, 247] Characteristics of the honeys of these various insects are discussed elsewhere.[74]

Hives of European honeybees were shipped to North America from 1621 onwards.[78] The bees prospered, and gradually colonised North America except for the far north, a process that was speeded by the many nesting sites for wild colonies in hollow trees of the wooded areas, and in the later nineteenth century by direct shipments to the west coast. In British Columbia the bees spread rapidly through the forested regions.[300] Hunting these wild colonies became a popular—and profitable—pastime, because some had stored quite large amounts of honey. So, for the first time, honey hunting (of a non-native bee) developed in a culturally advanced society. The new honey hunter was better equipped than prehistoric man: he wore a veil to protect his face from bee stings, and he had a saw for cutting off a branch or trunk containing a bees' nest. The operators were called bee hunters, because they captured and hived the

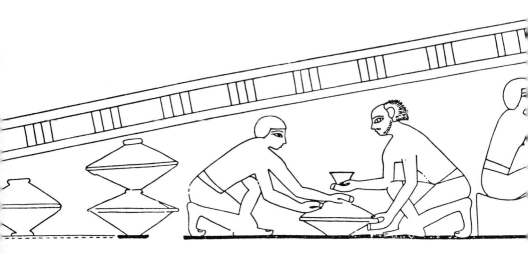

FIG. 16 *Wall-painting in the tomb of Rekhmire, West Bank, Luxor. The upper register shows honey being harvested from hives, and packed into containers. Egypt, c.1450 B.C.* REPRODUCED FROM N. DE DAVIES.[88]

bees as well as collecting the honey. They also included writers: Paul Dudley of Boston published a paper on bee hunting in 1721,[97] and there were books by Dr G. H. Edgell of Boston[104] and by L. G. Slater of Washington State.[276] J. Fenimore Cooper's *The bee-hunter* (1848)[63] is, however, disappointing as a source of information.

Columella marked bees to find their nest, and primitive hunters in Namibia used a hollow reed containing honey as a lure and trap. The new bee hunters devised various traps and lures for capturing test bees and releasing them at suitable points to 'line up' the site of the nest from several transects. Examples of the trap-boxes can still be found in North America—some probably in use—and there is one in the IBRA Collection (B53/135). The other items of equipment—saws, buckets and tubs—are hardly specialised enough to be part of beekeeping archaeology.

Honey hunting is thus contemporary with modern beekeeping. In the next chapter we shall see that beekeeping was already developed by the time that all but the very earliest of the honey-hunting scenes were painted in rock shelters.

3. Tombs and temples: pictures of beekeeping

THE EARLIEST RECORDED BEEKEEPING SCENES

We do not know when or where man first became a beekeeper, tending colonies of bees in hives of his own choosing, sited where he wanted them. The change from hunting for nests of bees with stored honey that he could take, to keeping bees in his own hives, probably took place independently in many parts of the world. It could well have been occasioned by a swarm settling in an empty basket or tub left upside down, or in a pot lying on its side. Or a hollow log with bees inside could have been carried home—to what was later called the apiary.

Some very early designs, such as those painted at Catal Hüyük in Anatolia and dated to about 7000 B.C., have been interpreted as depicting honey and brood combs, and bees foraging on flowers,[202a] but even if this is so, they provide no information about beekeeping—or even honey hunting.

Sherds from neolithic times found in Billa Surgam caves in Andra Pradesh, India, have been reconstructed into a pot, and it has been suggested[2] that this might have been used (like a gourd, see Fig. 66), as a hive for stingless bees. I do not think it very likely that this was the case.[83]

35

FIG. 17 *Earliest known representation of beekeeping and honey handling and packing. Egypt, c.2400* B.C. *From the sun temple of Neuserre, Abu Ghorab.* DRAWING FROM EGYPTIAN NATIONAL MUSEUM CATALOGUE.

The earliest certain evidence of beekeeping comes from Egypt.[83, 179] It consists of four scenes depicted around 2400 B.C., 1450 B.C. (two), and 600 B.C., respectively. These scenes are sufficiently similar to show that in 2400 B.C. beekeeping was already a well developed craft, and that it did not change very much in the next 1800 years. We know nothing of its earlier history or chronology. Peasant beekeepers in Egypt today use hives and methods that are fairly similar to those shown in the Ancient Egyptian scenes. There has, in fact, been relatively little change even in 4400 years.

The first scene is a stone bas-relief in the sun-temple of Neuserre (5th dynasty of the Old Kingdom) at Abu Ghorab, a few miles up the Nile from Cairo and within sight of the stepped pyramid of Zozer. The sun-temple still stands, but the bas-relief (Fig. 17) is in the Egyptian Museum, East Berlin. It is fragmentary, and does not add anything to the information about hives provided by the two scenes described below; it is most similar to that in Fig. 20. The section showing the handling of honey and the sealing of filled vessels is undamaged.[4]

The other three scenes are still in their original positions, in tombs on the West Bank opposite Luxor, 600 miles up the Nile. One of the pictures dated to about 1450 B.C. is shown in Fig. 16. It is a wall-painting near the entrance to the tomb (no. 100) of Rekhmire, a highly placed official of the 18th dynasty. The tomb, in the Valley of the Nobles, is open to the public, and when I saw it in 1978 it was complete, in good condition, and showing full, fresh-looking colours.

The other scene from about 1450 B.C. is a wall-painting in tomb 73, high on the hillside above the same valley. It is mostly destroyed, but what remains (Fig. 18) is similar to that in tomb 100.

The fourth scene is in tomb 279 where Pabesa of the Saite dynasty, 660–525 B.C., is buried. This tomb is kept sealed, and when I was there in

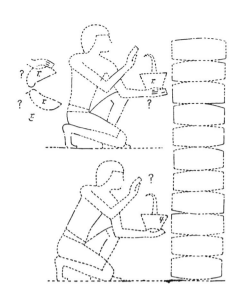

FIG. 18 *Less well preserved scene of harvesting honey. Egypt, c. 1450 B.C. Wall painting in tomb 73, West Bank, Luxor.* RECONSTRUCTION BY T. SÄVE-SÖDERBERGH.[262a] REPRODUCED WITH PERMISSION OF THE GRIFFITH INSTITUTE, ASHMOLEAN MUSEUM, OXFORD.

1978 the entrance was also blocked by boulders, but these were cleared and the tomb unsealed so that I could enter it. Inside I found the incised and painted relief (Fig. 20) on one of the stone pillars of a little courtyard. The right-hand edge of the pillar had unfortunately been broken off, probably—I was told—by soldiers of Alexander the Great.

The scenes show that horizontal hives were used, stacked one above the other. The hives were painted blue-grey, indicating that they were made of unbaked clay or mud. The large number of honey jars and pots in tomb 100 suggests that the honey was produced in some quantity. In both tombs the honey vessels are painted red, indicating that they were of baked pottery. The operations with the upright honey vessels are fairly easy to interpret (filling and possibly straining), but I did not understand the flattish vessels on the left of the scene in tomb 100 until 1980, when I visited Kashmir in northern India. There I found that the standard container for comb honey is a pair of common red pottery dishes 26 cm in diameter, one inverted over the other, and sealed together with mud (Fig. 19). Their shape (ratio of height to diameter) is the same as that

FIG. 19 *Pair of red pottery vessels from Kashmir, 1980, used as a container for comb honey. Compare the vessels on the left of the scene in Fig. 16.*

Tombs and temples: pictures of beekeeping

FIG. 20 *Beekeeper attending to his hives; above, honey being poured into a storage vessel. Egypt, 660–625 B.C. Incised and painted relief in the tomb of Pabesa, West Bank, Luxor.*

Tombs and temples: pictures of beekeeping

depicted in tomb 100 in 1450 B.C., and they stand on a similar low projecting base. I do not know any evidence of their use for honey elsewhere, at any time in the 3400 years since Rekhmire's tomb was built. In tomb 100, the beekeeper is putting combs into similar dishes, and they were in wide use in Ancient Egypt.

In tomb 101, of the same period, there is a painting, reproduced elsewhere,[74] of a man bringing food offerings to Pharaoh. He carries an open dish, rather similar to the red pottery ones, containing seven round honey combs, with several bees just above them. (An alternative interpretation is that the 'combs' are slices of melon.)

Reverting to the hives, there is not much evidence of progress in the 800 years or so between the scenes in Fig. 16 and Fig. 20. A smoker is shown in the earlier one only, and the hives there are more regular in shape, but there are only 3 of them; Fig. 20 shows 8 hives and vestiges of a further set on the right. Evidence from hives used in Egypt today is considered in the next section, and the four beekeeping scenes are discussed in further detail elsewhere.[83] Hives in current use that closely resemble those painted in 1450 B.C. in Rekhmire's tomb can be found in Malta, where they are made of baked clay (Fig. 73).

I have seen one curious find from Ancient Egypt that should be mentioned here, although its identity is puzzling. In the early 1970s, Harold Inglesent searched through objects in the Manchester Museum for any that might relate to bees. He found a clay cylinder (no. 296), inside which were a small lump of beeswax (identified by gas-liquid chromatography), pollen grains, and part of the hind leg of a honeybee. It originated in the Middle Kingdom workmen's town of Kahun, i.e. about 1900 B.C., and will be described in the catalogue of all the Kahun finds.[87a] The cylinder is 38 cm long, 9 cm in diameter at one end and 7 cm at the other; its walls are thickened towards the narrower end, so that the internal diameters are 6 cm and 1 cm, respectively. Mr Inglesent regards the cylinder as a small hive.[153a] Conical hives of woven material are referred to in Chapter 4 (see Fig. 80), but this pipe is extraordinarily narrow for such a hive, and its volume is only 0.36 litres, whereas that of the smallest hive in Table 2 is 9 litres.

PEASANT BEEKEEPING IN EGYPT TODAY

The native honeybee in Egypt is *Apis mellifera lamarckii*, which tends to build smaller colonies than European bees, and to defend them more vigorously—it is what beekeepers call an 'aggressive' bee. European bees have been introduced into Egypt and are kept in modern frame hives, where they build larger colonies and produce much more honey per hive. But the old-style beekeeping is still carried out in many parts of the country on a very considerable scale, with the native bees in long cylindrical mud hives stacked together. The shape of the hives is somewhat different from those

Tombs and temples: pictures of beekeeping

FIG. 21 *Bank of 400 cylindrical hives of unbaked mud near Assyut, Middle Egypt, 1980.*

depicted in the tombs. Those I measured in Middle Egypt were, externally, about 120 cm long, i.e. 8 times the diameter (15 cm); measurements published in the 1920s[203] were 137 × 17 cm, which gives the same ratio. For the ancient hives shown in Fig. 16 and Fig. 20 the ratio is only 1.5, and 2.9 to 3.7, respectively.

The hives are nowadays constructed by a sort of Swiss-roll technique. A set of laths strung together (like miniature close chestnut paling) is laid on the ground; a thick layer of Nile valley mud is spread over the laths and the whole rolled up, sometimes round a convenient pipe, and left to dry. The laths are then removed and another layer of mud added, especially to strengthen the join. These hives are stacked 8 to 10 high, forming banks that contain 300 to 500 hives each (Fig. 21). The hives round the edges of a stack are often left empty, probably because they are less well insulated from the heat than the inner rows. Summer temperatures are extremely high in Egypt (40°C, 100°F, or more), but thermal insulation is not the only reason for stacking the hives close together. Land is very precious, and hives can be placed only where it is not used for cultivation, on road verges for instance.

Tombs and temples: pictures of beekeeping

One traditional beekeeping area in the Nile valley is just north of Assyut. In 1979, I counted 34 stacks of hives by the roadside in a stretch of 27 km, representing some 10 000 hives—370 per km (600 per mile). All faced south, not to catch the sun as in Britain (Chapter 7), but because the prevailing wind is from the north. Another important centre, near Tanta in the Nile delta, has similar numbers of hives. It must be remembered that the Nile valley and delta, together with a few oases, are the only parts of Egypt where vegetation grows and bees can live.

The system of management used nowadays with these mud hives is advanced for fixed-comb hives.* It would be tempting to assume that beekeepers in Ancient Egypt used the same system, but no hives from Ancient Egypt are known to have survived, and we do not know whether this was so or not. Some written references to beekeeping survive from Ancient Egypt,[83] and a great many to honey, mentioned later in this chapter, but none of these describe how the hives were managed. Fig. 20 gives the impression that the bees are flying out from the opposite end to that being opened by the beekeeper (the back). In Fig. 16 it is impossible to say. With the pipe hives today honey is still taken out from the back, and dishes of food for the bees are inserted there if necessary. Opening a hive at the back and blowing smoke into it tends to make the bees move away to the front and fly out, but the stack of hives protects the beekeeper from these flying bees. If necessary, brood combs are removed for inspection from the front, their attachments to the hive having been cut with a special knife. They are replaced, spaced at the correct distance apart, by fixing them in position with a little forked twig.

In Fig. 16 the combs just removed from the hive are round, as they would be if the bees had built their combs *across* the hive. The pipe-hive beekeepers today ensure that this is so: they prime an empty hive for a new swarm by propping up a few combs from another hive, fixing them cross-ways near the flight entrance (and at the natural distance apart) with a forked twig. The brood nest is started in these combs, and honey is stored towards the back. Beekeeping with long horizontal hives worked from the back, or from either back or front, can develop much further than that with hives such as skeps (Chapters 4 and 6) and vertical logs that are open only at the bottom. The horizontal hives formed the basis of beekeeping in Ancient Greece and Rome (Chapter 4), but beekeeping north of the Alps developed separately (Chapter 6), and lacked many of the facilities available to beekeepers of the classical world.

Beekeepers in many other countries prefer combs built across the hive. In Kashmir, where the red pottery honey-comb containers have survived (Fig. 19), there are words to describe comb built along the hive and across

---

*The introduction, by L. L. Langstroth in 1851, of suspended wooden frames to hold the combs represented a major breakthrough (Chapter 9), because each framed comb could be removed at will, like a file in a suspension filing system.

it, the latter being *soechegan* (*soeche* = round loaf of bread, *gan* = colony of bees). In Jordan the round combs are said to be built by *kameri* bees, *kameri* being the word for the (round) moon.

In 1740 the French traveller de Maillet (quoted by Wildman in 1768)[318] described an unusual type of migratory beekeeping in Egypt. At the end of October, when the plants in Upper Egypt had finished flowering and the honey had been harvested, the hives were placed on boats built for the purpose and floated down the Nile. The boats were halted at places with many flowering plants, and the bees released to forage. The bees were moved on downstream, storing honey continually as the flowering season progressed northwards, until in February they reached Cairo, and the honey was sold. This same practice may have been followed in Ancient Egypt, but we have no direct evidence of it.

OTHER EVIDENCE FROM EGYPT

Hives were certainly moved about in Ancient Egypt, although their weight must have made this difficult—I could not lift an empty pipe hive, which must weigh 50 kg or more. A papyrus from around 250 B.C. contains a petition from some beekeepers in the Fayum oasis to Zenon, an official whose records still exist, begging for donkeys to move their hives away from an area due for irrigation flooding. There is also a record from A.D. 16 about beekeepers owning several hundred hives, some of which had been sealed up by an ill-disposed person. Another fragment in the Zenon archives, dated 8 October 256 B.C., says 'The 5000 hives are . . .'[83] so beekeeping was on a large scale. Other evidence is provided by the amounts of honey used for sacrifices. For instance one list of offerings made to the Nile god by Rameses III (1198–1167 B.C.) amounted to about 15 tons of honey.[78] Honey was the most common ingredient of medicaments in Ancient Egypt—possibly to make other ingredients more palatable—and is mentioned some 500 times in the 900 remedies that are known. Honey is included in an even earlier prescription on a clay tablet found at Nippur in Iraq, dated to around 2000 B.C.[74]

It was reported that two jars labelled 'honey of good quality' were found in Tutankhamun's tomb, but it was not possible to confirm their identity by analysis.[185] In the Agricultural Museum at Dokki, Cairo, there are two 'honey pots' from New Kingdom tombs (c. 1400 B.C., contemporary with Rekhmire) with their contents still in them. The identity of a piece of comb honey from another tomb nearby, of the same period, has been confirmed by pollen analysis; the shrivelled pollen grains show that it came from trees (rare in Egypt now), which probably included limes and *Mimusops Schimperi*,[322] related to the trees that produce gutta percha.

In the Museum at Dokki there are also 'honey feast cakes' from the same period (p. 239), together with objects involving beeswax.

Tombs and temples: pictures of beekeeping

Throughout Egypt, bees are portrayed in countless temples, tombs and monuments, and on objects in many museums. These bees had little or nothing to do with beekeeping, but were part of the hieroglyph in the Pharaoh's titulary throughout the dynastic period, from about 3000 to 350 B.C. With the bee, which symbolised Lower Egypt, was a sedge plant symbolising Upper Egypt, which is thought to be the *Scirpus* reed (Fig. 22). The changing symbolism of bees through the centuries is explored in Chapter 10.

FIG. 22 *Incised relief at Karnak temple, Luxor, showing the bee and sedge, part of the hieroglyph in the titulary of the Pharaoh of Ancient Egypt.*

FIG. 23 *Illustration in a manuscript (Exultet Roll) made between A.D. 1070 and 1100 at Montecassino, Italy; see page 53. On the left, a hive of wooden boards has been opened from the side to cut out honeycombs—a development since Roman times. On the right two other beekeepers are trying to get a swarm into an upturned hive.* PHOTO: BIBLIOTECA VATICANA.

4. Horizontal hives from Ancient Greece and after

From Ancient Greece we can see actual examples of pottery hives that were used, so we know what they were made of and their size and shape, and we can deduce the systems of beekeeping and honey harvesting that could have been used with them. Hives of other materials may also have been used, but we have no direct evidence. No hives of any description are known from Ancient Crete, nor any illustrations of them. The same is true for Ancient Rome, but several Roman authors wrote about beekeeping,[112] and I shall show that certain hives still in use closely resemble the hives they described. We can therefore learn quite a lot about Roman beekeeping.

TERRACOTTA HIVES EXCAVATED IN GREECE

Red pottery hives, somewhat similar in shape to the three from 1450 B.C. in Egypt, have been found in archaeological excavations at five sites in Greece. We know that the vessels are hives because the same type is used by beekeepers today on some of the Greek islands. Also, fragments from one site (Vari) were tested for the presence of beeswax with positive results.[131,164] In 1980 I found similar hives in use in Kashmir, which received its cultural influences from the Middle East; Fig. 19 shows Kashmir honey containers that are similar to Ancient Egyptian ones.

The most prolific site has been the agora (market place) in Athens. Seven pots, now identified as hives, were found during excavations there and are now in the Agora Museum; one is shown in Fig. 26. Their dates range from about 400 B.C. to Roman times. Two complete hives were found at Marathon, placed mouth to mouth to make a coffin for a little boy who died between 200 and 100 B.C.[163] Fragments of others were found at Trachones (Figs 24, 25), just south of Athens and west of Mount Hymettus, and still more, farther south at Vari,[131] where the Mount Hymettus range slopes down to the sea. Three others, dated to A.D. 500–600, were found in the Justinian fortress at Corinth. These places are all in mainland Greece, in or near Attica, and hives of this type are unknown there now. But they are still used on certain smaller islands of

ABOVE: FIG. 24 *The nearer hive is similar to Fig. 25; the hive at the back was currently in use on Antiparos, Cyclades, in 1973.*
PHOTO: M. I. GEROULANOS.

FIG. 26 *Red pottery hive (BELOW) from about 400 B.C. found in the Agora in Athens (44 cm long; diameter 39 cm at mouth). Agora Museum cat. no. A/P11017.* PHOTO: AMERICAN SCHOOL OF CLASSICAL STUDIES AT ATHENS: AGORA EXCAVATIONS.

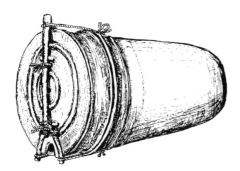

FIG. 25 *The drawing (ABOVE) of a restored hive from Trachones, with lid and extension ring, shows the conjectured method of fastening, using cord. The flight entrance is the notch at the bottom of the lid.* REPRODUCED FROM J. E. JONES ET AL.[164]

the Cyclades, to the south east of Attica, and one is shown in Fig. 24.

These thimble-shaped hives show internal 'combing', scratches made as by drawing a comb across the surface of the wet clay before firing. The combing goes about half-way round the hive, the lines being parallel to the circumference. When the hive lies flat on the ground, as in Fig. 24, with the combed surface at the top, it gives the bees something rough to attach their combs to. The hives now used in Greece do not have this

FIGS 27 AND 28 *Two views of a hive in Kashmir, similar to Fig. 26, but with flight holes in the rounded end, 1980.*

combing, but I found it in the one in Kashmir (Figs 27, 28) that I was able to inspect closely. In this, and also in shape, these hives have changed more in Greece than in Kashmir, although it does not seem that the changes in Greece represented advances.

The Greek hives open at only one end and look like the Egyptian ones illustrated in Fig. 16, except that they widen out towards this end. Flight holes for the bees are made in a lid fitted on to the open end. The hives from Ancient Greece have a mouth diameter of 30 to 39 cm (12 to 15 in.) and are 36 to 59 cm long (14 to 23 in.). The later hives, from Corinth, are in general longer than the earlier ones, the ratio between length and mouth diameter increasing from 1 to 2.[83] Kashmir hives follow the earlier Greek style, but—a feature unknown from ancient or modern Greece—flight holes for the bees are provided in the closed end, so that smoke can be applied when the lid is removed, to drive the bees through the hive and out of the flight holes. The hives, like most traditional types in Kashmir, are kept on shelves in the thickness of the house wall, and they are operated from inside the house, the bees flying from the other (outside) end of the hive (Fig. 34). The bees native to Kashmir are *Apis cerana*, but a form almost as large as *A. mellifera*.

Horizontal hives from Ancient Greece and after

FIG. 29 *'Cannon' pottery hive open at both ends, in current use in eastern Crete (1979); the diameter at the front is larger than at the back.*

In Crete, which might perhaps be expected to represent an intermediate stage of development between Egypt and Greece, a traditional hive still used (at the eastern end of the island, and in some islands of the Cyclades) is a long tapered horizontal cylinder, open at both ends, known as a cannon (Fig. 29); the ends are closed with discs of pine bark (compare also with Fig. 43). Specimens I measured were 64 to 73 cm long (25 to 29 in.) and had an external diameter at the larger end of between 29 and 37 cm (11 and 15 in.)—measurements in the same range as the later Ancient Greek thimble-shaped hives. I found no evidence of internal combing. Information on the use of these hives is available.[260]

A hive that can be opened at the back and operated on while the bees are smoked to and out of the front (as in Figs 21, 28, etc.), is a considerable advance on a hive with only one opening. If we accept the evidence of present-day traditional hives, hives opening at both ends were used in Crete[260] as well as in Egypt. It was explained in Chapter 3 how Egyptian beekeepers use a movable-comb technique to ensure the building of combs *across* the hive, and this is also done in Crete.[260] (In the Radfan mountains of Arabia, a horizontal wooden hive is used with comb-guides cut on the top surface *along* the hive; beekeepers in some other countries, too, prefer combs built along the hive (Fig. 48).)

The facility for using a hive that can be opened at the back, and the facility for spacing and orienting the combs, thus seem to have been used and lost again in ancient times. Another development that was subsequently lost was the use of an extension ring at the mouth of Ancient Greek hives, to increase their capacity and serve as a honey chamber (Fig. 25). Such hive extensions are currently used on pottery hives in Malta (Fig. 73) and in Morocco (Fig. 33) for instance, and on sloping cylindrical (cannon) hives in Crete, but not now on the thimble-shaped hives. The whole situation is so complex, and so much is still unknown—or unco-ordinated—that one hesitates to draw conclusions.

Horizontal hives from Ancient Greece and after

LEFT: FIG. 30 *A woven horizontal hive used in Kashmir; the flight entrance can be seen at the centre of the front end.*
PHOTO: F. A. SHAH 1980.

BELOW: FIG. 31 *Agricultural basket (dark) used as an extension at the back; it may be added to any hive in Kashmir.*

LEFT: FIG. 32 *Mud hive used in Kashmir in 1980, much less elongated than those in Egypt.*

The Greek 'thimble' hives used in Kashmir (Figs 27, 28) have several flight holes at the rounded end, so converting them into the more convenient type. In common with the various types of traditional hive I saw there (e.g. Figs 30, 32)—all of which were placed on shelves in the house walls—a shallow agricultural basket was added at the back (inner) end as an extension in summer (Fig. 31). A deeper extension ring of

FIG. 33 *Pottery hive and extension, incurved to give telescopic fitting, Gharb, Morocco, 1963. With the extension, the high ratio of length to diameter shows more similarity to currently used Egyptian mud hives than do most other Mediterranean hives.*

49

FIG. 34 *Hive entrances in a house wall in Kashmir (one above each of two upper windows, one to the right of the small boys, and one midway up, at the side of the house), 1980. Apis cerana is used in all the hives in Kashmir that are illustrated.*

unbaked mud was used similarly, upon occasion. Fig. 34 shows the appearance of the flight entrances from outside a Kashmir house, and the same house had thirteen other hives.

Traditional hives in Iran, shown in photographs by Pourasghar,[239] look almost identical to the many different hives I saw in Kashmir; I believe that this is but one facet of the widespread cultural influence from Persia

Horizontal hives from Ancient Greece and after

that still exists in Kashmir. Iran (Persia), with Turkey, probably has a greater variety of traditional hives than any other area, and this may perhaps provide a clue to the question posed at the very end of Chapter 6, about the origin of the hives that were depicted in Ancient Egypt (Chapter 3).

It is tempting to think that the hives shown in tomb 100 (Fig. 16) may have been the predecessors of the Greek 'thimble' hives, with only one opening, and that those shown in tomb 279 (Fig. 20) had the flight entrance at the front, as do present Egyptian mud pipe hives and the Crete pottery ones, both of which open at both ends. Hives rather like those in tomb 279 are still used in the High Simien Mountains in Ethiopia (Fig. 35); to get the honey, the back end (opposite the flight hole) is broken off, and cemented back on with mud. These hives are the most primitive type that I know. They are made of mud and cow dung, whereas the Nile mud seems to have adhesive properties that make additives unnecessary.

One advance that was not lost, although it seems never to have been made in Egypt, was the change from unbaked mud to baked clay or terracotta, as a hive material. In Greece the large sloping cylinder hives are carried on pack animals for migration, and they must be more suitable for the purpose than the hives transported on donkeys in Ancient Egypt, which were presumably made of sun-dried mud (Chapter 3). Mud hives are also quite easily broken, and the replacement of unbaked mud by baked clay as a hive material represented a real advance in lightness and durability.

FIG. 35 *Primitive hive of mud, dung and chopped straw from the High Simien Mountains in Ethiopia.* IBRA COLLECTION B72/15.

ROMAN HIVES AND THEIR SURVIVAL

Just as heavy hives of mud were superseded in some areas by lighter ones of baked clay, so these in turn were superseded in some places by even lighter and thus more convenient hives of woven wicker. We do not know whether this took place in Ancient Greece or not; beekeeping books that might have told us were written by Philiscus of Thasos and by Aristomachus of Soli, but they have not survived, and we know of them only from a comment by Pliny.[83] Beekeepers in Ancient Greece may well

FIG. 36 *Slovenian hive fronts painted in the traditional style by Milan Plešec, and currently on sale from IBRA.*

have used still other types of hive; Roman writings refer to at least nine types, and it is hard to believe that the Romans invented most of them.

1. Log hives were referred to by both Columella (a hollow tree) and Palladius (a hollow trunk).

2. Hives made of cork bark were very highly regarded by the Elder Pliny, Palladius and Columella, and were mentioned also by Varro and Virgil.

3. Columella and Palladius referred to hives of wooden boards.

4. Hives of woven wicker were also favoured, and there is a passage written at some date between A.D. 150 and 400, by an author known as the pseudo-Quintilian: 'It gave me pleasure to weave the pliant withies with spring twigs, and to fill the gaping fissures with clinging mud, to prevent either the heat of summer or the cold of winter from penetrating. . . .' The application of a protective coating of mud, sometimes mixed with cow dung, is known as clooming.

5. Varro and the Elder Pliny wrote of hives constructed of fennel (*Foeniculum vulgare*), also referred to as ferula, which were generally approved.

6. Columella quoted Celsus as condemning hives of dung on the grounds that they were liable to catch fire; perhaps these were like the High Simien hives in Ethiopia (Fig. 35).

7. Earthenware hives were mentioned but not favoured, being 'burnt by the summer heat and frozen by the winter's cold' (Columella). There is no reference to extension rings.

8. Columella deprecated brick hives, because they 'cannot be moved when circumstances demand it'.

9. Finally, the Elder Pliny referred twice to a transparent hive, saying in one passage that it was made from transparent stone (mica?), and in the other from 'transparent horn used for lanterns'.

Although no hives seem to have survived from Ancient Rome, hives of most of the nine types are still in use today, in more distant regions that

FIG. 37 *Painted Slovenian hive with curved sides, made in 1831.* IBRA COLLECTION B54/4.

were within or without the Roman Empire. We cannot be certain whether knowledge of them was transmitted during (or since) Roman times, or whether they came into use independently in different places.

Horizontal log hives (1) are used in many parts of tropical Africa and elsewhere, and are discussed in the next section. Again, the Nile Valley would have provided a likely route for transmission of ideas, but since bees adopt hollow tree cavities as nesting sites, their use as hives in tropical Africa may well have arisen independently.

Hives made of cork bark (2), roughly cylindrical in the shape of the bark as it is cut off the tree, and joined up again with wooden pegs, are still common anywhere in North Africa where cork oak trees (*Quercus suber*) grow; the hives are laid in rows on the ground. In Spain and Portugal such cork hives, and more elegant ones, are used upright (Figs 100–102).

The nearest equivalent to the Roman hives of wooden boards (3) that still exists is probably the type used in Slovenia in northern Yugoslavia. There, a tradition has grown up that the front board of each hive is painted with a motif based on a biblical story or folktale.[22,215,252] The earliest of these hive fronts that is dated was made in 1758; it shows a Madonna. Fig. 36 demonstrates that this tradition is still alive, and Fig. 37 depicts an early—and, unusually, rounded—hive with St George and the dragon on its front board. The hives are kept in a special type of bee house (Fig. 209).

A splendid record of wooden board hives as they were around A.D. 1000 in central Italy, and probably much as they were in Ancient Rome, is provided in a number of illuminated manuscripts known as the Exultet Rolls. The word *Exultet* (Praise) starts the hymn in praise of the paschal candle that was (and is) blessed during the Roman Catholic service on Holy Saturday. As the priest read the manuscript aloud during the service, he unrolled it in front of the congregation, so that they saw the paintings. Twenty of the surviving manuscripts, made in monasteries in Capua, Gaeta, Montecassino, Pisa and elsewhere in Italy, are ornamented with beekeeping scenes, painted to show the source of the wax for the candle— the bees 'who produce posterity, rejoice in offspring, yet retain their virginity'.

FIG. 38 *Illustration in a manuscript (Exultet Roll) made between* A.D. *1100 and 1200 at Benevento, Italy. Nine hives of wooden boards are stacked in 3 tiers on a stand lifted off the ground by wooden legs.* REPRINTED BY PERMISSION OF PRINCETON UNIVERSITY PRESS (COPYRIGHT HOLDER).[21]

Fig. 38 shows nine hives stacked in 3 tiers as recommended in the quotation below from Columella. In Fig. 23 one hive is being opened and another is held to catch a swarm. I was interested to find a few somewhat similar hives in Corsica in 1963 (Fig. 58).

Woven wicker has been a common way of constructing traditional hives in many parts of the world, and these can assume many different shapes. Horizontal cylindrical woven hives of cane or wicker (4) are found in many parts of Africa (Fig. 39), the Middle East (Fig. 40) and in China (Fig. 62). Bulbous horizontal basket hives with a sort of Ali Baba shape (like water pots, as we shall see) are in use in Kashmir and elsewhere, set horizontally (Fig. 32).

Hives made of fennel (5) were also laid horizontally, and were probably square in cross-section. Varro gave measurements for fennel hives, about 3 ft long and 1 ft across (90 and 30cm). In 1979 I measured three present-day ones made in Sicily (Fig. 41), and they were slightly smaller, about 33 in. long and 9 in. across (82 and 23 cm).

Mud and dung hives (6) have already been mentioned (Fig. 35). Earthenware hives (7), which are widely spread, are discussed again in some detail in the next section. Brick hives (8) are more puzzling—I think they may have been walls that contained cylindrical pottery hives (see p. 194).

The transparent hives (9) I have, alas, been unable to find in current use. I would guess that they were rather like the board hives, with transparent panels set into them; the mica or horn pieces available must surely have been quite small.

RIGHT: FIG. 39 *Merchant with a camel load of horizontal hives woven from cane. Chiadma, Morocco.* PHOTO: R. VERHAGEN.

All the hives were probably laid horizontally, and all except those of boards and of fennel stems (3 and 5) were more or less cylindrical, with a removable closure at each end, or sometimes only at the back. With some cylindrical hives the back closure could also be moved forward into the hive to reduce its length so that, as Varro wrote: 'when the bees are too few to fill [the hives] the [beekeepers] reduce the size, so that the [bees] do not lose heart in a wide empty space.' Pliny was more explicit: 'it is very advantageous . . . for the lid at the back to be movable, so that it can be pushed forward into the hive, in case the hive is large or the bees' labour unproductive.'

FIG. 40 *Horizontal woven hives, Beirut, Lebanon, showing also the clay lid with an entrance hole.* PHOTO: BROTHER ADAM.

FIG. 41 *Ferula hive in Sicily.* PHOTO: A. ZAPPI-RECORDATI.

FIG. 42 *Woven hive of split cane, cylindrical in shape but 'most narrow in the middle', as described in Roman times by Varro; used in Ethiopia, 1972.* IBRA COLLECTION B72/14. PHOTO: J. MASON.

Another way of providing a small space for bees to start working in, perhaps for a new swarm, was mentioned by Varro immediately after the passage just quoted: 'it seems that they [some beekeepers] make [hives] most narrow in the middle, in order to imitate the shape of the bees.' The description has puzzled readers who have pictured the hives standing vertically. For horizontal hives, it presents no problem, and such hives are still used; Fig. 42 shows a woven example from Ethiopia. Pottery hives of a similar shape have been illustrated, but I do not know where they were found.

This completes the tally of hives described by the Romans. Columella gives explicit instructions on the siting of hives, and we shall find that the method is still *de rigueur* in some places, 2000 years later; see p. 75.

> A platform made of stone is constructed across the whole apiary, three feet high and three feet wide.... On it are placed the hives, whether, as Celsus prefers, they are built of bricks, or, as I prefer, they are walled round except at the front and back; or again—as is the virtually universal practice among those who take trouble in the matter—they are arranged in a row and held firm by bricks or concrete, in such a way that the individual hives are contained by two narrow walls, and both ends are left free. For the hives must sometimes be opened at the front, from which the bees issue forth, and, much more frequently, at the back, through which the colony is repeatedly tended.... Three hives placed one above the other is plenty, since even then the beekeeper does not inspect the top row with sufficient ease. The fronts of the hives, which provide the bees with their entrance, should be lower than the backs, so that rainwater does not flow in, and, if it has chanced to penetrate, does not stand in the hives but flows out through the entrance. For these reasons it is useful to protect the hives with a roof above.

HORIZONTAL HIVES IN THE WORLD AT LARGE

Hives similar to those known in the classical Ancient World are still used today in various parts of Africa and Asia—and even in America. Outside northern and western Europe the most widespread type of traditional hive is without doubt the horizontal 'cylinder' of round or other cross-section. It is made of wood, bark or other plant materials or, in dry areas

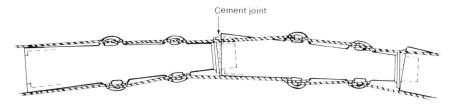

FIG. 43 *Drawing of pottery water pipes from the palace at Knossos in Crete. Each is tapered, and has a collar which is cemented into the mouth of the next pipe; compare Fig. 29. The handles are used for tying the pipes together.* REPRODUCED FROM A. EVANS,[110] FIG. 1. *(A photograph of two of the pipes is shown in his fig. 2.)*

like Mediterranean lands and the Near East, of unbaked dried mud or baked clay.[75]

Whereas it is easy to identify the prototype of a vertical log hive with a hollow tree (Chapter 5), we do not know the prototype of the more universal horizontal hive. Suggestions have included a hollow horizontal branch of a tree, a drum, and a water pipe. Both drums and water pipes were used in early civilisations, and some of the latter found in Crete (Fig. 43) suggest a possible origin for the sloping cannon hives (Fig. 29). (Horizontal cylindrical wooden hives are traditional in Réunion, Mauritius, Rodrigues and Agalega in the Indian Ocean; there the hive is not a *cannon* but a *bombard*; one with a square cross-section is a square bombard! Fig. 44 shows a bombard protected with banana leaves. Bees were introduced to these islands after 1600.)

Some of the horizontal hives are derived from those used in the Ancient World that have already been discussed. But it is hard to believe that all such hives have this origin, because of their very wide distribution—in the past as well as today. Horizontal cylinders are traditional not only in many parts of Africa, but also as far away as China, Bali (Indonesia) and Central America where they are used for stingless bees. Along with this wide distribution, we find a rich variety of materials and craftsmanship.

We are fortunate in that traditional hives survive in many parts of the tropical world, and we can thus learn at first hand how they are made and how they are used. As an example of the variety within only a small

FIG. 44 *Horizontal log hive with cover of banana leaves, Réunion, Indian Ocean, 1982. The large stone holds the cover down.*

country in Africa, Börje Svensson told me that in 1980 he found the following hives among various ethnic groups in Guinea-Bissau on the west coast, the first three being used all the year round and the others seasonally:

> hollowed trunks of the cibe palm (*Borassus flabellifer*) or oil palm (*Elaeis guineensis*)
> upright large clay pots, with a smaller clay pot inverted above as a honey chamber
> (in use, but no longer made) wine barrels imported from Portugal, used on their side, sometimes with an extension as a honey chamber at one end
> hives woven from bamboo leaves and reinforced with cow dung
> hives made from leaves of the tara palm (*Raphia* sp.), pegged together with short lengths of the midrib of the leaf, the whole coated with cow dung or clay
> hives made by coiled work, using grass.

All these hives had a volume between 20 and 50 litres. Bees are also encouraged to build their nests inside houses, and holes are made in the walls for them to enter by.

I shall now discuss horizontal hives made from different materials and by different techniques in various parts of the world.[75]

FIG. 45 *Horizontal log hives on the southern slopes of the Caucasus (Nukha, Georgia, USSR), 1930.* PHOTO: E. F. PHILLIPS.

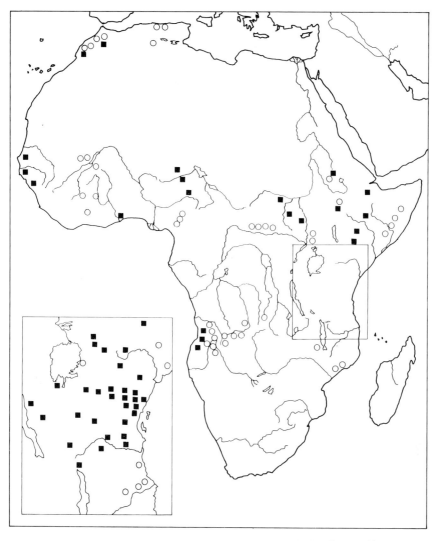

FIG. 46 *Distribution of horizontal log* ■ *and bark (including cork)* ○ *hives in Africa, as recorded by C. Seyffert in 1930.*[273]

Wooden hives

Hollowed from a tree trunk

A hive can be made from a section of tree trunk, provided suitable trees are available which do not have other more important economic uses (such as fruit production), and also tools to remove the log from the rest of the tree and to hollow it out.

Horizontal log hives are used in parts of the Near and Middle East, the Caucasus (Fig. 45), parts of Asia, for instance China,[172] many

ABOVE: FIG. 47 *Hollowing a log to make a hive. Tharaka tribe, Kenya, 1967.*

ABOVE RIGHT: FIG. 48 *Hive hewn from wood, wider at the back than the front. Both end closures are decorated; internally, grooves have been cut along the upper surface, to serve as comb guides. Hive, $116 \times 45 \times 19$ cm internally, used in the 1960s in the Radfan mountains of Arabia.*
IBRA COLLECTION B68/16.
PHOTO: J. MASON 1982.

RIGHT: FIG. 49 *Log hive used by the Pokot tribe, Kenya. It is cut horizontally into two halves, which are separated to take the honey.*

parts of Africa (Fig. 46) and Central America. Fig. 47 shows the process of hollowing the log. This produces the simplest type of log hive, in which the ends are plugged with wooden stoppers. In Oman, date palm trunks are used; these are also sometimes available in Egypt, but I could not ascertain that they were ever used for hives there. In some areas the log is cut into a rectangular shape; the example from Arabia (Fig. 48) is only 11 cm high ($4\frac{1}{3}$ in.) internally. Sometimes the log is cut in half longitudinally and fixed together again; it is then split open along the cut when the honey is removed. The Pokot tribe in the Rift Valley in Kenya use such hives (Fig. 49).

In most of tropical Africa wooden hives cannot be set on the ground

Horizontal hives from Ancient Greece and after

FIG. 50 *Log hives hung in a tree, Kenya.* PHOTO: K. I. KIGATIIRA.

because of ants, termites and mammals such as the honey badger (*Mellivora capensis*). The hives are usually hung from the branch of a tree, by a rope or by a forked stick, or wedged in the angle between branches (Fig. 50).

Horizontal log hives are also used in the Himalayan region; Figs 218 and 219 show a type that is built into house walls in Afghanistan.

We must now make a digression to Central America, where honey hunting and beekeeping with hives were already developed in prehistoric times, although not many records of it survive. There are no native honeybees in America, and the bees used were various species of stingless bee (Meliponini); they build rather amorphous nests from which honey

ABOVE: FIG. 51 *End view of log hives of stingless bees (*Melipona beecheii*) stacked on a A-frame rack under a thatched roof, Yucatán, Mexico. An empty hive lies in the foreground.* LEFT: FIG. 52 *A moment in the minor bee ceremony (1978) at the same hives; a prayer of thanks is offered for the honey harvest.* PHOTOS: E. C. WEAVER.[310]

and wax were harvested. The earliest surviving written account is by Bishop Diego de Landa who arrived in Yucatán in 1549, 7 years after the Spanish conquest.[181,296] He said:

> There are two kinds of bees and both are very much smaller than ours. The larger kind of these breeds in hives, which are very small. They do not make honeycomb as ours do, but a kind of small blisters like walnuts of wax all joined one to the other and full of honey. To cut them away they do nothing more than to open the hives and to break away these blisters with a small stick, and thus the honey runs out and they take the wax when they please. The rest breed in the woods in hollows of trees and of stones and there they search for the wax, in which and in honey this land abounds and the honey is very good. . . . These bees do not sting nor do they do harm when the honeycombs are cut out.

In many areas the people had not advanced from honey hunting to beekeeping when the Spaniards came (this was certainly true of Honduras), and today some are still at the honey-hunting stage. The traditional hives in Yucatán for the 'larger kind' of bee, *Melipona beecheii*, are shown in Figs 51 and 52. They are hollowed-out logs, from *Vitex gaumeri* and other trees, with wooden stoppers at the two ends and the flight hole midway along the side of the hive.

Fig. 51 shows how the hives are housed in a single layer on each side of an A-shaped rack under a thatched roof. I do not know of the use of

Horizontal hives from Ancient Greece and after

such racks for supporting log hives anywhere in the Old World, nor of the stacking of hives in a bank (as has been customary in Egypt, e.g. Fig. 21) in the New World. Fig. 52 shows the same hives as Fig. 51, but on its own the picture could almost be interpreted as the end view of a bank of hives. Probably the Yucatecan hives, and their positioning, were developed on the peninsula by stages that we shall never know. But accumulated evidence shows that there must have been contact between the Mediterranean and America in very early times, and it is worth raising the possibility that the Yucatecan system could have originated in some early traveller's imperfect description (or a description imperfectly remembered) of stacked horizontal cylinder hives in Egypt or elsewhere.

Apart from the evidence of both beekeeping and honey hunting in Yucatán,[256] there are indications from many areas that honey hunting was widespread, probably wherever stingless bees were found (Fig. 2), and that it extended to nests of some social wasps.[74] The Yucatecan Maya apparently did not use the beeswax, and even had a prejudice against burning it,[256] but after the Spaniards arrived it was much sought after for candles. The extensive use of beeswax for lost-wax gold casting in Colombia is referred to in Chapter 10.

Maya beekeeping operations were under the protection of ancient bee gods (Fig. 53), whose blessing was invoked at special ceremonies. In 1978 the ceremony was recorded and described in detail,[310] on possibly the last occasion when it will ever be performed in its entirety. When honeybees were first taken to Yucatán, less than a hundred years ago, they seemed to the Maya beekeepers so large that it was believed they must be the bee gods themselves. I found stylised bees and bee gods in two of the only three Maya manuscripts not destroyed by the Spaniards, Codex Tro-Cortesianus and Codex Troana, both now in the Archaeological Museum in Madrid. Fig. 54 shows some examples.

RIGHT: FIG. 53 *Statue of the bee god Ah Mucan Cab, in the Museum in Mérida. Yucatán, Mexico. The god holds a cluster of honeycombs in his hands; two hives are on the right.* REPRODUCED FROM B. AND R. DARCHEN.[86]

BELOW: FIG. 54 *Bees in the Codex Tro-Cortesianus, on display in the Archaeological Museum, Madrid; the god-like figure on the right, holding the honeycombs, may be compared with Fig. 53.* REPRODUCED FROM A. M. TOZZER AND G. M. ALLEN,[297] PLATE 2, WITH PERMISSION FROM THE PEABODY MUSEUM, HARVARD UNIVERSITY.

Horizontal hives from Ancient Greece and after

In 1676 William Dampier reported both wild and hive bees in Tabasco, a Mexican province south west of Yucatán. He also said that some of the log hives there were set upright;[256] I do not know any other record of this.

Bark and cork hives

In the New World, traditional wooden hives for stingless bees seem to have been limited to hollowed logs. Small horizontal log hives are used for stingless bees also in Africa, for instance by the Chaga tribe in Tanzania.[134,273] But for honeybees in Africa, bark hives—which are lighter in weight—are also common. A complete section of bark of the required length is removed from a tree (Fig. 55); the debarked tree dies, so the operation is a wasteful one. In Tanzania the bark of *Julbernardia* or *Brachystegia* is used—sometimes only the inner bark, turned inside out and sewn together. When bound with straw as a protection against the sun, these hives are superb examples of craftsmanship (Fig. 56). Bark hives may be 3 to 5 ft (90 to 150 cm) long and 12 to 18 in. (30 to 45 cm) in diameter. *Bee-keeping in Zambia*[274] gives a detailed description of the construction and use of bark hives in that country; the bark of *Cryptosepalum pseudotaxus* is preferred, being smooth, thin, and very easy to work. The ends of the hive may be closed with a piece of bark or a circular pad made of coiled rope.

ABOVE: FIG. 56 *Finished bark hive protected with straw, and with a hooked stick for hanging in a tree.* PHOTOS: F. G. SMITH.

LEFT: FIG. 55 *Tree in Tanzania stripped of its bark to make a hive, 1950; this treatment kills the tree (*Julbernardia globiflora*).*

Horizontal hives from Ancient Greece and after

In western parts of North Africa, and in the Iberian peninsula, the cork oak (*Quercus suber*) grows. It has the unusual characteristic that when the outer bark is removed, the living layers beneath are left undamaged. In due course a new layer of cork is formed, which in turn can be harvested. Cork has been a favoured material for hives, horizontal in Africa and vertical in Spain and Portugal (Fig. 102). The join(s) are sometimes secured with wooden pegs. Such a hive needs no special treatment or protection, and provides dry housing for the bees, well insulated against heat and cold.

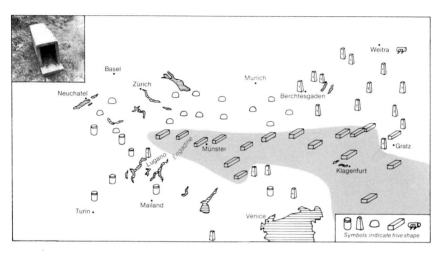

FIG. 57 *Map showing the area in and around the Alps where traditional hives were horizontal, not upright as in surrounding districts.* AFTER L. ARMBRUSTER,[6] INSET: FIG. 58 *Horizontal wooden hive at Sorbollano, Corsica, 1963, similar to the 'tunnel' hives used in parts of the Alps.*

Hives of wooden boards

North of the Sahara, in Morocco and Algeria, hives are made of strips of wood put together to form a long box, and also of ferula stems such as the Romans also used (Fig. 41); joints are sealed up with clay. Where beekeepers had ready access to wooden boards—and more recently to manufactured boxes in which goods were delivered—they have often used them for housing bees.

The region of horizontal wooden hives in Europe includes Slovenia where the hive fronts are painted (Fig. 36), and has a spur stretching westwards through the Alps as far as the Rhine-Rhône watershed (Fig. 57). In the Alps some of these hives are called 'tunnel' hives; I found a few rather like them in Corsica (Fig. 58). Such hives continue the Roman line we have already traced up to A.D. 1000 when the Exultet Rolls were made (Figs 38, 23).

65

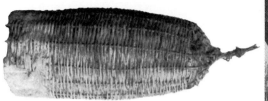

ABOVE: FIG 59 *Woven wicker hive from Mali, where it is used horizontally. A honey container from Mali (Fig. 8) depicts such a hive. In use it would be cloomed with cow dung or other material.* IBRA COLLECTION B72/35.

ABOVE: FIG. 60 *Similar woven wicker hive in Belgium, 1952, used* upright. PHOTO: D. S. VANDEPUTTE. LEFT: FIG. 61 *Woven hive with straw protection, used horizontally in Cameroon, now in the Norwegian Beekeeping Museum at Billingstad, Oslo.*

Hives made from stems and twigs

All over the world, baskets and other containers have been made from plant stems and pliable twigs and roots ('pliant withies' of Roman times): woven, coiled, plaited or bound together in other ways. The same techniques have been used for basket hives, according to materials and skills locally available.

Woven hives

Horizontal woven hives are used across almost the whole breadth of north equatorial Africa, from Guinea and Mali in the west, through Nigeria, Cameroon and Chad, to Sudan and Ethiopia; in general the log and bark hives replace them near the east coast. Fig. 59 shows a woven Mali hive without its coating of mud and dung. Its end closure is of dried mud. Fig. 8 shows a gourd used as a honey pot, on which a drawing of one of these hives is burned into the surface. If this hive were stood

FIG. 62 *Chinese horizontal woven hive, Hong Kong, 1968. There are five entrance holes (small, to prevent attacks by hornets) in the middle.* IBRA COLLECTION B68/14.

FIG. 63 *Woven cane hives on a platform, protected from the heat only by roof tiles, Menorca, 1981.*
PHOTO: J. E. H. STRETTON.

upright on its open end, it would look very much like a Belgian wicker skep, Fig. 60. The hive in Fig. 61, with its conical straw 'hat', is used horizontally in Cameroon; set vertically it would look like a simplified form of the Dutch skeps, Figs 111, 112. In Rwanda, a roof of banana leaves is used to protect the cylindrical woven hives.

Woven cylindrical hives are found in the west and east of Asia—in Turkey and China (Fig. 62), for instance—but not, I think, in many areas in between, which are less rich in traditional beekeeping than Africa.

In the discussion of Roman hives, above, the reference (4) 'to fill the gaping fissures with clinging mud' is puzzling to people who have seen only close-woven upright wicker hives. But many horizontal wicker hives seem to be more loosely constructed. This is true of some still used by a farmer in Menorca, one of the Balearic Islands in the Mediterranean (Fig. 63). Substantial stakes of cane are used along the length of the hive, and leaves of canes or reeds are woven loosely round the circumference; the gaps are impregnated with cow dung and made smooth and bee-tight inside. The end closures are of stone, 16 to 24 cm (6 to 10 in.) in diameter. The hives are about 150 cm (5 ft) long (said to be a full arm's reach from each end, when smoothing the inner surface).* Like the Egyptian hives, their length is around 8 times their diameter.

A woven hive that is 'most narrow in the middle', as described by Varro, is shown in Fig. 42.

*But many pottery and log hives (which do not need clooming) have a similar length. I suspect that bees prospered in such hives *because the beekeeper could not reach from one end to the other* and so, when he took the honey, he was forced to leave intact a few combs that contained brood and the queen.

FIG. 64 *Five hives of coiled straw in a tree in The Gambia, West Africa; their owner is inspecting them. The flight entrance to each hive is in the centre of one end; the other end is removed for harvesting the honey.* PHOTO: O. ÄNGEBY 1980.

FIG. 65 *Detail of African hive in which the binding cord is plaited (1976); Beekeeping Museum, Tilff, Belgium.*

Hives of bound stems

Hives with a square cross-section, made of thick fennel stems (Fig. 41), have already been discussed. Cylindrical hives are readily made by twined work, using stiff and substantial plant stems lengthwise, and binding them tightly together at intervals with more pliant stems. Fig. 130 shows vertical hives, made (not very well) by this technique.

Hives of coiled work

Coiled work is commonly associated with upright straw skeps, but it is also used in some parts of Africa for horizontal hives. In Algeria such hives are laid on the ground, but in The Gambia they are suspended in trees, tilted at all sorts of angles (Fig. 64). Coiled-straw horizontal hives are, however, not very widely used. It may be that there is too great a wealth of other materials that yield more durable hives, or that involve less work in making up.

Hives of plaited work

Plaited work is also not commonly used for hives, but Fig. 65 shows an example which incorporates this construction.

Horizontal hives from Ancient Greece and after

FIG. 66 *Gourd hive for stingless bees under the porch roof of an old colonial house in São Paulo Province, Brazil, 1973.*

FIG. 67 *Hive in Sudan made from leaves of the doum palm (*Hyphaene thebaica*).* PHOTO: S. E. RASHAD.

Hives of leaves, fruits, nuts and flowers

This last group of plant materials may sound unpromising, but they are used for hives in various places. Some leaves, such as banana, provide excellent protection; Fig. 44 shows a hive covered with them. In Sudan very long cylindrical hives are made of leaves of the doum palm, the ends being closed with a plug of leaf fibres; they are hung from trees in a slanting position (Fig. 67).

Wherever gourds grow to a suitable size, they constitute a ready-made expendable container for a swarm of bees. The pulpy contents must be removed, for instance by drying and then inserting small stones and shaking them around. A small hole is left for the bees to use, and the gourd is often hung up in the shade of a house (Fig. 66). It is broken to harvest the honey. In Africa gourds are used as hives for honeybees in Malawi, Nigeria, Togo and probably elsewhere.

The only hives I know that incorporate nuts are used in Bali, Indonesia, for the native honeybee *Apis cerana* which does not need as much space as the European *Apis mellifera*. The hive is a horizontal cylinder constructed of the bark and flower spathe of the coconut palm, each end being closed with half a coconut shell. Fig. 68 shows several hanging on a house

ABOVE: FIG. 68 *Hives occupied by the Asiatic honeybee* Apis cerana *in Bali, Indonesia, made from parts of the coconut palm, and hanging under the eave of a house, 1972. The flight entrance is bored in the half-coconut shell that is used concave side outwards as an end closure.* PHOTO: R. VERHAGEN.

RIGHT: FIG. 69 *Hive in a tree north of Inhambane, Mozambique, 1977, which may have been constructed from the hide of an animal (or from bark).* PHOTO: R. GUY.

wall. With this attractive hive in a beautiful tropical island, we leave our survey of horizontal hives around the world, and return to clay hives in the Mediterranean and the Middle East, which was the centre of development in ancient times.

There is just one postscript. In my 1977 article on traditional hives[75] I said that whereas many plant materials, and earth in various forms, were used for hives, animal products (except dung) were not, although many other containers were made from hides. Fig. 69 shows a photograph sent to me in response to this statement; the photographer could not be certain whether the hive was made from animal hides or from bark. Perhaps some reader may be able to settle the issue.

BELOW: FIG. 70 *Hive of sun-dried clay in Sichem, Jordan, 1952.* PHOTO: BROTHER ADAM. *They are often stacked like the mud hives in Egypt (Fig. 21).*
RIGHT: FIG. 71 *Ornamented example of a Maltese hive, 1969.* IBRA COLLECTION B69/71. *A cylindrical extension is sometimes added at the wide end. The holes at the neck end are for ventilation.*

RIGHT: FIG. 72 *Hives at Abougosh, Israel, rather like water pots laid on their side, but with the far end removable, 1952.* PHOTO: BROTHER ADAM.

Hives of earth treated in various ways

We have already encountered hives made of sun-dried mud or clay, alone or mixed with dung and possibly straw. Hives of baked clay (referred to as pottery or earthenware, or terracotta if red) have been discussed, and we shall return to them. Stone and brick constructions have been used for hives in places where less heavy and cumbersome materials were not available (Fig. 131), and they have been widely used for structures to shelter hives, which are discussed in Chapters 7 and 8.

Hives of unbaked clay or mud
This material is heavy, breakable, and to us it appears inconvenient, but it costs nothing, and the making requires much less time and skill than hollowing out a log or constructing a weathertight basket. It seems to me the most primitive of all hive materials, and its continuous use in Egypt— one presumes throughout a period of 4500 years or more—is impressive. It is also still used in some other Mediterranean countries, Jordan for example (Fig. 70).

Hives of baked clay
In Roman times pottery hives were not favoured by Columella, but they have survived in use throughout most of the Mediterranean lands (Fig. 74), and they also occur sporadically much farther afield. One reason for their popularity in dry areas where plant materials are scarce may be the almost universal availability of the materials—clay and water—coupled

BELOW: FIG. 73 *Three Maltese clay hives; see text for details.* PHOTO: K. STEVENS.

ABOVE: FIG. 74 *Cylindrical clay hives in Cyprus, 1977. The Roman recommendation of three tiers on a platform is still adhered to.* PHOTO: SUSAN NEIL.

with the greater durability of a baked clay vessel than an unbaked one.

Many pottery hives in Mediterranean lands are simple pottery cylinders. (I have seen a bank of mud hives near Alexandria in Egypt, interspersed with pottery hives made by placing two commercially made short drainage pipes end to end.) The large 'cannon' hive still used in Crete is shown in Fig. 29, and narrow cylinders incurved at one end so that several can be used together are shown in Fig. 33. In Malta, hives may be shaped like a squat bottle without a base (Fig. 71), and Fig. 72 shows hives in Israel that are intermediate between this hive and an ordinary water pot. Some other Maltese clay hives are shown in Fig. 73. The hive at the back consists of three units joined together (compare Fig. 33). The front hive marked R is a simplified form of Fig. 71, and that on the left is almost indistinguishable from the hives in Fig. 16, drawn in Rekhmire's tomb in 1450 B.C.; these were, however, made of unbaked mud.

Water pots laid on their side are used in Sri Lanka and other parts of Asia, where beekeeping did not develop in the rich and intense way it did in Africa and Europe; in many areas it was unknown until after the Second World War. Honey was often used for medicinal purposes rather than as a food, and was harvested more by hunting wild nests than by keeping bees in hives. In the central belt of Africa—often in areas with woven hives—pottery hives of various other shapes are also used. In Central America, pairs of small pots have been used for stingless bees (Fig. 75).

Spanish bee walls

Spain has a group of interesting archaeological remains known as bee walls (*muros de abejas*) in which horizontal pottery hives are embedded.

Horizontal hives from Ancient Greece and after

The hive is commonly known as *horno*, which means oven or kiln, and is also used generally for a cavity where bees nest. An alternative word is *jacente* or *yacente*, which derives from the fact that the hives are recumbent (lying down). There is clear evidence that both horizontal and vertical cylinders were used as hives in the same apiaries a few hundred years ago: an engraving of such an apiary in 1720 is reproduced in Fig. 79. The upright hive, stood in the open, is known as a *peon*; this was the word for a foot soldier, with the same derivation as 'pawn'.

Upright hives are common throughout Spain (Chapter 6); of the bee walls containing the horizontal hives that concern us here, not many now survive, and they are difficult to find. Otto Erup from Denmark,[107,108] and my husband and I, located twelve examples on our respective searches in and around the upper Ebro valley, in an area bounded roughly by Logroño, Pamplona, Huesca, Zaragoza and Lerin. A register of the information collected is kept at IBRA.

FIG. 75 *Hives used for stingless bees near Huejutla, Hidalgo, Mexico, each made from a small pot with a lid, 1972. Some are decorated.* PHOTO: R. DARCHEN.

ABOVE: FIG. 76 *Bee wall near Zaragoza, Aragón, Spain, with 56 hives, at the top of an enclosed apiary (such as that in Fig. 79); the enclosure was about 25 m wide and stretched 28 m (80, 90 ft) down the hill (1961). An apiary near Huesca had a similar layout, the wall having 3 groups of 3 rows of 7 hives, 63 in all.*
LEFT: FIG. 77 *Ruins of bee wall above Bargota, Aragón, Spain, 1961, showing the three rows of hives, typically of 7 each. There may well have been two other such groups, 63 hives in all.*
BELOW: FIG. 78 *Bee wall near Bargota, Aragón, Spain, 1961. There are 42 occupied hives on the left of the building and 42 empty hives on the right, without their front closures.*

Horizontal hives from Ancient Greece and after

We recorded four bee walls near the village of Bargota, and Fig. 77 shows the remaining end of one of them, which was clearly part of a building; the holes visible are the open ends of cylindrical pottery hives. In Fig. 76 a similar structure is sited at the top of a walled apiary, facing south east; the wall was in ruins. Fig. 78 shows a complete occupied building on another site.

Records from earlier years make it clear that there were once many of these bee walls in the area. Most showed no sign of being part of a wall round an apiary, but each had an enclosed space at the back of the hives where the beekeeper would remove the circular stones that closed them, and harvest the honey or do whatever else was necessary. Many of the empty hives that I inspected showed traces of combs, most having been built more or less (but not exactly) across the hive; one at least had combs built along the length of the hive. The hives shown in Fig. 76, which I measured, were about 5 ft (150 cm) long and 10 or 11 in. (25 to 28 cm) in internal diameter—the same length as those in Menorca referred to earlier, and the same as many hives in tropical Africa. They fit well with the recommendation of Roman authors to build three tiers of hives, and I believe that also here, in Europe, we are probably seeing apiaries fairly similar to those the Roman authors described and recommended.

It may well be that the enclosure at the back of the wall—which was sometimes a narrow passage, and sometimes quite spacious—is a post-Roman addition. In fact it creates a bee house, and the inclusion of the Spanish bee walls here instead of in Chapter 8 is largely a matter of convenience.

In 1974 Mr F. Beaugey reported a bee wall at Lerin, between Logroño and Pamplona, which contained 20 hives in three rows facing south east,

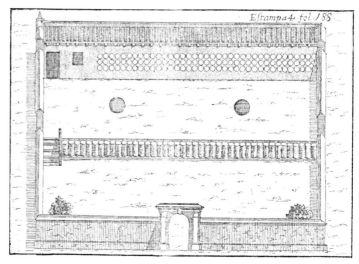

FIG. 79 *General view of apiary of horizontal hives (*jacentes*) in a wall at the upper end, with upright hives (*peones*) against the wall of a terrace, halfway up. The hives face south east.*
REPRODUCED FROM Economia general de la case de campo . . . BY L. LIGER AND D. F. DE LA TORRE (1720).[191]

Horizontal hives from Ancient Greece and after

but of another type. They were made of woven cane, plastered with clay and cow dung, and were tapered cylinders, their diameter being about 50 cm (20 in.) at the back and only about 2.5 cm (1 in.) at the front (flight hole) end. The wide back was covered with a carefully shaped stone lid fitted with a handle for easy removal (Fig. 80). I did not know of any other horizontal hives of this almost conical shape, but while this book was in proof I heard that they are used in the north of Iraq.[175a] They are woven with twigs and cloomed with organic matter and mud, about 50 cm long and 25 cm in diameter at one end and $\frac{1}{2}$ to 1 in. (1.25 to 2.5 cm) at the other, which contains the flight entrance. Three tiers of the hives are used, either in a specially built wall or in house walls, in which case the honey is harvested inside the house, from the wide end of the hives.

L. Armbruster, a prolific writer in the 1930s to 1950s on early beekeeping,[18] contributed his knowledge and views to the question of these bee walls,[17,19,109] which he saw as a link with bee boles (Chapter 7). Other articles include the following comments. Muro de Cameros, a village of not more than 200 inhabitants near Logroño, had 3 bee walls in 1958 (and 3 others in ruins), and all the villages round about had 2 or 3, but their use was dying out. Two of the buildings found by Erup[108] south of Zaragoza were like a rectangular bee house with 3 rows of six flight entrances (in one, all are above the level of the top of the entrance door). Some of the walls Erup[108] found contained hives of square, not round, cross-section, about 70 cm long (28 in.) and 24 cm (9$\frac{1}{2}$ in.) wide and high. An article[57] entitled *Los hornales* gives other information.

Two possible explanations for the survival of Roman-type beekeeping in this upright-hive area of Europe may be suggested. It could result from the fact that Columella, one of the Romans who wrote about beekeeping and who was almost certainly himself a beekeeper, was a Spaniard. The area concerned is also one of the farthest points reached by the Moors in their conquest of Spain in the eighth century. We have seen that the practice of building horizontal hives into a bank or wall still survives in various parts of the Middle East and North Africa, so the system could have been brought to Spain by the Moors, and have survived in this remote northern area while it disappeared farther south.

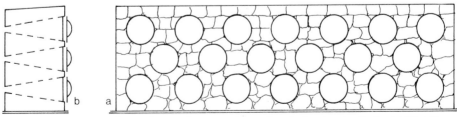

FIG. 80 *Disposition of conical woven hives in a bee wall at Lerin, Aragón, Spain. (a) The 20 hives from inside the building. (b) The shape of the hives, and handles of the stone closures; bees fly out to the left, and honey is harvested from the right.* DRAWING AFTER F. BEAUGEY 1975.

FIG. 81 *Massive door to a 'tree hive' in Białowieska National Park, Poland.* PHOTO: S. BLANK-WEISSBERG (1937).[43]

5. Forest 'beekeeping' and the precursor of upright hives

NORTHERN CONTINENTAL EUROPE

Most of the archaeological remains of early beekeeping discussed so far are from relatively dry, warm areas round the Mediterranean. In the extensive mixed forests of the cooler, wetter climates of northern Europe there was an intermediate stage between honey hunting and beekeeping. The trees provided good forage for bees, and also nesting sites for them in hollow trunks and branches. In the course of time certain men specialised in dealing with the bees nesting in trees, and in harvesting their honey. In

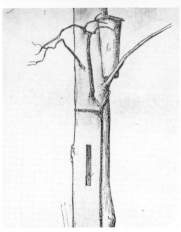

ABOVE: FIG. 82 *Tree in Cheremessia, Russia, showing a slit-like door to a bees' nest, whose flight entrance is on the left. Above is a climbing rope, and at the top a separate log for another colony of bees, fixed up as a bird nesting box would be.*
DRAWING: AGATHON REINHOLM 1884. COPYRIGHT NATIONAL MUSEUM OF FINLAND, HELSINKI.

RIGHT: FIG. 83 *One bear, trying to raid a bees' nest (door in tree marked* a*), has released a counterweight and been swung aloft. If the archer hits this bear, it is likely to fall on to the array of spikes below. A second bear is approaching the nest* b; *the trap is set but not yet sprung.*
REPRODUCED FROM J. G. KRÜNITZ (1774).[178]

Russia such a man was a *bortnik* (*bort'* is a hollow tree trunk), in Polish *bartnik*; in German he was a *Zeidler* (the verb *zeideln* has the meaning 'to cut the combs out', possibly connected with *schneiden*, to cut). Certain trees containing bees' nests would be the property of a forest 'beekeeper'—there is no special term for him in English—and he would cut his ownership mark on them.[3]

Finding a small cavity in a hollow tree, the forest beekeeper enlarged it, and in many regions he cut a tall narrow door to provide better access to the combs inside (Figs 81, 82). In the swarming season he checked his bee trees to see which had been occupied by swarms; he also observed the flight of bees and followed them from a bait he provided (as honey hunters did), hoping they would lead him to a nest that no one else knew of. At the end of the summer he cut out the honey combs to get his harvest. After the winter he cleaned out the cavity, and if he could he would collect wax combs in the spring, when they were white and fetched the highest price. He did what he could to prevent bears gaining access to the nest—bears seem to have been a great trouble, and many horrific devices were invented to kill them (Fig. 83).

It was generally easier for a honey hunter to take the harvest from a bees' nest in a tree than from a nest in the rocks like those shown in

Forest 'beekeeping' and the precursor of upright hives

Chapter 2. A tree cavity, even if high up, could be reached by climbing the tree, or by throwing a climbing rope over a branch, and access to the nest could be improved. Nevertheless, climbing the trees and working up them always brought the risk of an accident, and there must have been many injuries and deaths.

The next stage in the development towards beekeeping proper took place when and where tree cavities became scarce. Separate hollowed logs were then fixed high up in the trees (Fig. 82), and looked after in the same way. This was followed in turn by the use of similar logs, stood upright on the ground in an 'apiary'.

The very success of forest beekeeping in living trees may have delayed the development of apiary beekeeping. It certainly led to the use of upright hives, in contrast to the horizontal hives in the drier Mediterranean cultures.

Forest beekeeping probably first developed in the middle course of the Volga and the lower course of its tributary the Kama. The area, which includes Kazan, lies between Moscow and Bashkir (in the Urals), and is well north of the Caspian Sea. A branch of the Finno-Ugrian peoples lived there at the end of their neolithic period, around 2000–1000 B.C., when the climate was mild and supported rich deciduous forests. The forests probably included limes, willows, oaks, hazel, poplars and pines, but they were not all dense woodland, and raspberries, brambles and herbaceous plants would also have provided bees with nectar and pollen.

Forest beekeeping spread with the Finno-Ugrian peoples, for instance eastward as far as the Urals, and westward to Karelia and Estonia on the Baltic. From these peoples the beekeeping practices were passed on to the Slavs, and from the western Slavs to the Germans. Between 1200 and 1900—although not necessarily for the whole period—large quantities of honey and wax were produced in this way: by Volga-Finnic tribes, by Slavs (Russians, Bashkirs, Churashes, Poles), by Baltic peoples (Estonians, Livs, Letts, Lithuanians), and by Germans. Comprehensive accounts have been published about Cheremessia, written in Finnish[138,139,245] and German,[159] and about Estonia, in Estonian.[192] To the south and east, for instance in the Carpathians[26] and the forests of what are now Hungary, Slovakia, Romania and the Ukraine, honey hunting in general persisted without any development into forest beekeeping, although there are occasional tree ownership marks and other signs of 'keeping' bees in trees.[133]

The tradition of forest beekeeping has survived in some areas, and trees being worked in the old way can still be seen. For instance, there are about forty in the Białowiezka National Park in Poland,[308] and many more in the Bashkir Beekeeping Reserve in the Ural mountains[234] and elsewhere in the USSR. An extensive collection of photographs of tree beekeeping in Poland as a still living craft was published in 1937;[43] Figs 81, 87 and 88 are from this

FIG. 84 *Part of a tree trunk in which bees nested, showing the doorway made by a forest beekeeper 2000 years ago.*
PHOTO: MUZEUM IM. JANA DZIERŻONA.

source. This Polish book shows that some log hives were used horizontally as well as vertically, with a door as in the vertical hives (see below), not with end closures like the hives in Chapters 3 and 4. Forest beekeepers had their own charter in Poland,[202] as elsewhere.

Forest beekeeping, although well documented, has left little in the way of artefacts from the distant past. The earliest I know was found in an area that was in eastern Germany but is now in Poland. Part of a tree in which a colony of bees had been housed and managed was taken from the river Oder near Opeln, south east of Breslau (Wrocław), in 1901. It was investigated by F. Sprotte,[200] and carbon dating showed that the tree was growing in the first century B.C. Five metres above the tree's roots was a cavity, with a more or less vertical door (47 × 12 cm) cut in the trunk and, about 50° to the right of the door, a flight entrance 4 cm^2 in area. There was a ridge of what appeared to be propolis on the edge of the hole; this fact, and other evidence cited, suggest that the door was about 3 cm thick, and in two parts, which is typical for such tree hives in Poland today. The section of the tree containing the cavity (Fig. 84) is now in the Dzierzon Museum in Kluczbork nearby (see Appendix 2, under PO3).

The equipment used for working at the cavities occupied by bees varied from region to region, but was widely similar in principle. In 1981 I was shown a complete set of forest beekeeper's equipment in the Ethnology Department of the National Museum of Finland in Helsinki (Fig. 85). This was brought from Cheremessia around 1900 by Albert Hämäläinen, then Professor of Ethnology at Helsinki; the Cheremessians were a Finno-Ugrian people living to the west of the middle Volga-Kama region—the very heartland of forest beekeeping. All the ropes shown were made from hemp except the strong climbing rope which was made from the inner bark of the lime *Tilia cordata*; as in Africa (Chapter 2), a man's life and safety depended on this rope, and here it was in fact two ropes, sewn lightly together.

Forest beekeeping in Russia

The most comprehensive account of contemporary records of beekeeping in medieval Russian forests is Dorothy Galton's book *Survey of a thousand years of beekeeping in Russia*[119] and much of the information that follows is taken from it.

Very large amounts of honey and beeswax were produced, and much of the honey was converted into alcohol, as mead. The inhabitants of Belgorod north of the Black Sea are said to have sunk tubs of mead into the ground when the town was besieged in 997, to make their enemies believe that they drew unlimited supplies of food and drink out of the earth and could never be starved into submission. A Persian manuscript of the same period reported that the Slav land had very much honey, from which wine and other drinks were made. In 946, by command of Olga, so much mead was brought for the funeral feast of her son Igor that five thousand of the enemy were incapacitated, and killed by Olga's followers. Mead was used to celebrate births and to cement trade agreements and treaties. In 996, three hundred vessels of mead, each supposed to weigh a ton and a half when full, were provided for a seven-day feast to celebrate a victory. Warriors took great quantities of mead with them on service; one story tells how in 1489 Tatars came upon a place which had been a Russian camp, and found so much mead there that ten thousand of them got drunk—and fell prey to the Russians.

Records show that forest beekeeping was carried on all over ancient Russia, in Novgorod, Moscow, Nizhny Novgorod (now Gor'ky), Smolensk, and many lesser known places. Princes, monasteries and boyars owned great areas of bee woods; peasants worked for them as *bortniks*, and also themselves owned bee trees, which they marked, and for which they paid their landlords rent. In the eleventh and twelfth centuries the trees and the bees in them were safeguarded by laws which imposed different penalties for destroying: a tree with bees in it, a bees' nest complete with bees, the bees if they had no queen, the bees only, a tree only, and an empty hole in a tree.

Tree beekeeping was of very great value to the economy. In 1136, when

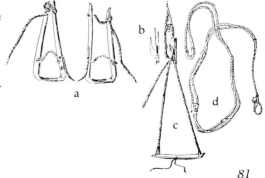

FIG. 85 *Equipment of a Cheremessian forest beekeeper, c. 1900*. DRAWING: A. HÄMÄLÄINEN.[138]
(a) *Contraptions to fit over the shoes, the curved spikes (below) being pushed into the tree when climbing it.*
(b) *Hook from which beekeeper was slung.*
(c) *Seat slung from the hook* b.
(d) *Rope made from the inner bark of* Tilia cordata.

FIG. 86 *Modern forest beekeeper in Turkey, with climbing rope—and binoculars.* PHOTO: R. VERHAGEN 1970.

a charter was given to St John's Church in Novgorod, the church's upkeep was paid for by customs dues on beeswax, and the amount concerned must have been nearly 400 tons a year. This probably represented the production of bees nesting in about a million trees, and it would have been only part of the beeswax going into Novgorod. Several thousand tons of honey must have been harvested with this wax, most of it probably being used for making mead.

There was extensive tree beekeeping in the Moscow lands in the fifteenth and sixteenth centuries. Access routes to the bee trees were *bortnye ukhozh'ya*, for which a possible translation is 'bee walk'. Bee walks were mentioned in many wills and deeds, and lands with them were often bequeathed to monasteries or sold. On change of ownership of a bee walk, the beekeepers were sometimes transferred too; for instance in 1488 a *bortnik* Ondreyko Tilitsin was given to a monastery with the village in which he lived. There are records in 1390 and 1472 of a Grand Prince parting with land but retaining the rights of his beekeepers and beaver hunters to go to the walks. Lands with bee walks were mortgaged or exchanged. In a charter of 1565 the people of one village were forbidden to cut down bee trees with or without bees. There were frequent disputes about ownership and rights in bee forests.

In some places artificially hollowed log hives were placed in trees, parallel to the trunk, either attached to the tree or stood on a specially constructed platform. This was necessitated by a decline in the number of suitable cavities in the trees themselves. (Similarly, we put up bird nesting boxes.) Log hives were also set up in apiaries on the ground from the end of the fourteenth century, particularly in monasteries, but this was not common even by the seventeenth century. The Turkish people have some affinities with the Finno-Ugrian groups, and Fig. 86 shows a forest beekeeper in Turkey in 1970, with his climbing rope, axe—and binoculars. He was visiting one of his hives fixed high up in a tree.

Forest 'beekeeping' and the precursor of upright hives

Bortniks preferred oak trees for the bees to live in, but limes were also used. They used cavities between 5 and 25 m above ground, sometimes two or three in one tree. Artificially hollowed logs for hanging in trees might be of willow or poplar, or—in the sixteenth and seventeenth centuries—of pine. They were placed high up as a protection against bears and also because swarms were known to settle high up. Trees with straight trunks were chosen, and the lower boughs were lopped.

Bee trees might be 120 m apart or more, even over 1 km, and in large forests the *bortnik* might have to walk 20 km to visit 5 to 20 cavities. As few as 10 per cent of the cavities were likely to be occupied by bees at any one time; the rest awaited flying swarms in search of nesting sites. Taxes were paid according to occupancy by bees. One record refers to 266 tax-paying bee walks, each with (on average) 6 trees with bees in, 36 trees previously occupied by bees but not now, and 'countless barren' trees where cavities had not so far been occupied.

The *bortniks* were peasants living off the land, by growing crops, fishing, snaring birds and hunting animals—particularly beavers. The beekeeping was seasonal work. The *bortnik* was said to 'go to the woods'; when he found a tree with bees in, he would put a mark on it with an axe, in the shape of a hare's ear, a pitchfork, a bow, a goat's horn, a circle, an oval, or a pattern of grooves (Fig. 89).

In the early summer the *bortnik* would keep a look-out for swarms, and identify the cavities where they settled. He would take the honey in late summer, after the end of flowering. Figs 87 and 88 show a Polish forest beekeeper on his way to work. Various means were used to climb to the bees' nest, and a bucket or other vessel was carried to put the honey

FIG. 89 *Ownership marks used on bee trees in Cheremessia near the Volga in what is now USSR.* REPRODUCED FROM A. HÄMÄLÄINEN (1934).[139]

LEFT: FIG 87 *Polish forest beekeeper carrying his climbing rope and* RIGHT: FIG. 88, *also his seat.* PHOTO: S. BLANK-WEISSBERG (1937).[43]

Forest 'beekeeping' and the precursor of upright hives

combs in. The door was removed from the cavity, the honey combs removed (with or without smoke) and the door replaced. In Poland the door was commonly in two parts which could be removed separately, to give access to the upper or lower part of the nest. In many areas, including Poland, White Russia and the Ukraine, the flight hole and the beekeeper's door were often at 90° to one another. But in the eastern Baltic and eastern Russian regions the flight-hole was in the plank that constituted the door.[133] In Slovakia the door was on the same side as the flight hole but above it.

Forest beekeeping in Germany

Forest beekeeping achieved a high level of organisation in medieval Germany, especially in parts of Brandenburg (round Berlin), Saxony and Bavaria (Nürnberg), and parts of Pomerania and Silesia (now in Poland). Fig. 90 shows honey being harvested. Two trees are being worked; on the right a beekeeper is operating from a ladder, using a tool to cut out honey combs; the door has already been removed. The beekeeper on the left is sitting in a sort of bosun's chair like that in Fig. 85; he smokes the bees with a pipe, and uses a tool to cut out combs which are visible through the open doorway. On his left is a bag for catching swarms; the supporting framework has a loop to fit on to a long pole, and the draw-string at the bottom allows the swarm to be dropped out into a hive. Three containers at the bottom, a tub, basket and bag, are ready to receive the honey combs, which the left-hand operator will presumably drop down individually to the assistant with upstretched arms.

There are records of *Zeidler* before A.D. 1000, and their privileges and responsibilities were set out in great detail.[13,36,147,189,312] A few incidents in the Nürnberg area are quoted here as examples. One of the centres for *Zeidler*, where they held their own court, was the market town of Feucht, 15 km south east of Nürnberg. In 1427 Feucht and surrounding land were sold to the city of Nürnberg, and in 1431 the Emperor Siegismund ordered the building of fortifications for the town, presumably because of its importance. The *Zeidler* courts dealt with a wide variety of cases—not only beekeeping—and many records of the sessions still survive. Most of the courts probably did not retain their original functions much beyond the sixteenth century, but the last court at Feucht was held on 1 September 1779. It was clearly an enjoyable event for the dignitaries from Nürnberg, who rode and drove out to the town, drank coffee, attended a special service in the church, and were given a banquet.[268] This occasion can be compared with formal gatherings of the City of London guilds, bodies that were similarly powerful in the Middle Ages in controlling their respective crafts.

A ruler of forest lands stood to benefit (in dues and in other ways) if

FIG. 90 *German forest beekeepers at work in the eighteenth century; see text for details.* REPRODUCED FROM J. G. KRÜNITZ (1774).[178]

peasants on his land got a harvest of honey and wax, but there was a conflict of interest between the foresters and the *Zeidler* on account of the damage done to the trees, for instance in making cavities for bees to nest in. In 1712 a *Zeidler* in Franconia complained to his court that the forester in his area would not allocate any trees for his use, and he would therefore have to buy log hives. If he did this, he would then be allowed to hang them in the trees; it was the making of holes that was objected to. This is the sort of way in which tree beekeeping was gradually superseded by

Forest 'beekeeping' and the precursor of upright hives

beekeeping with separate hives, and in due course the hives were kept on the ground in 'apiaries' rather than up in the trees. Change was slow, and even in 1772 there were reported to be 20 000 'hives' (probably in trees) in the Prussian forests. A castle at Feucht, said to have been built in the reign of Charlemagne around 800, and renovated in 1370, still has a medieval *Zeidler* carved in stone above the gateway; he carries a cross-bow and arrows, like the *Zeidler* shown in Fig. 92. One piece of equipment associated with *Zeidler* was a short-handled axe (*Imkerbeil*) used for hollowing out trees. The cutting edge was extended upwards into a halberd-like point, used for levering off the wooden door fitted to the hole. Long-handled axes, possibly derived from these, were used by skep beekeepers in Germany.[44,255,265,269] Quite large numbers of these axes have been preserved in German museums; see Fig. 91.

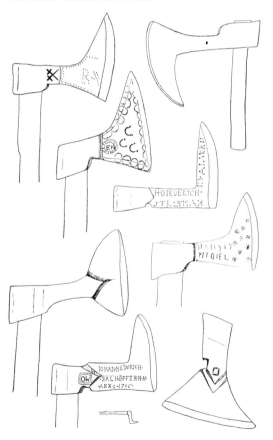

FIG. 91 *Some axe-heads that belonged to skep beekeepers in the Lüneburg area, north Germany. Forest beekeepers used axes that were similar, but had shorter handles.* M. BÖCKER.[44]

FIG. 92 Zeidler *similar to the one portrayed in the castle at Feucht.* REPRODUCED FROM De Buchigulariis, ALTDORF (1723).

Forest 'beekeeping' and the precursor of upright hives

ENGLAND
Forest beekeeping

Forest beekeeping is much less well documented in England* than in continental Europe, and it cannot have taken place on such a large scale. Before A.D. 1000 there were laws relating to the forests,[70,115] and these had plenty to say about swarms, honey and wax, and their ownership. But they seemed to be concerned with honey hunting rather than with the beekeeping in living trees discussed above. The same is true of entries in the Domesday Book, as the following examples show.

> [Worcestershire:] The same Bishop holds Breodum . . . woods 2 miles long and $1\frac{1}{2}$ miles wide from which the Bishop has 10/- and all the produce of the chase, honey and other things . . . The same Bishop holds Fledberie . . . there are . . . woods 2 miles long and $\frac{1}{2}$ mile wide of which the Bishop has all the issues arising from hunting and honey and the supply of wood to the saltworks of Wich [Droitwich] and 4/-.
>
> [Sussex:] Wood for 70 hogs, and 20 hogs for rent, and 2 sextaries of honey. [A sextary of honey is likely to have weighed about 1.7 lb (0.75 kg).]

In 1978 a gamekeeper at Newlands Farm, Sowley, Hampshire, drew the attention of Frank Vernon to an oak tree in which a door had been cut to give access to a cavity which had a separate flight hole. After enquiries and exploration a few more were found at Ladycross (where they can still be seen) near the middle of the Forest; Sowley is close to its southern edge. The New Forest is one of the Royal Forests of England, where all land and trees are the property of the crown. What about bees and honey? The doors found in the trees would have been well camouflaged, and unlikely to attract the attention of a Head Keeper or Surveyor making his rounds of the various forest paths and looking for any signs of misdemeanour. The method used for opening up the nest was to make two saw cuts cross-grain, one above and one below the nest cavity, and then to split off the wood with a chisel at each end of the saw cuts. Removal of the large piece of wood—complete with bark—would allow the beekeeper to take all the honey combs. The entire piece of wood (or door) was nailed back into position, and on future occasions the nails could be withdrawn and the door prised open again. Figs 93 and 94 show examples.

Apart from the camouflaging—made necessary by the clandestine nature of the operation—the whole set-up is very similar to that on the continent, and several questions arise. How long has the practice existed in England? How widespread has it been? How did it arise? The direct evidence is of recent use—in the 1970s and during the past 60 years or so. We have not yet been able to trace the practice back farther. It may have started during the 1914–18 war-time shortage of food, by someone who

---

*It is even less well documented for Ireland, but I believe that many of the mentions of bees and honey in the ancient laws of Ireland[168] could as well refer to forest beekeeping as to the use of colonies in hives.

had read about (or seen) the practice in the larger forests of Europe, or devised a similar method independently. But it could well be very old, dating back to the creation of the Forest in the late eleventh century. Domesday entries for Hampshire include the following. For Edlingen (now Eling): 'Into the Forest were taken sixteen dwellings for villeins and three for bordars from the pannage and a yearly produce of three sextars of honey—all of which is now taken from the manor . . .'. For the Manor of Wallop: 'Formerly the reeve had the honey and pasture towards paying his farm, but now the foresters enjoy this and the reeves nothing. The honey and pastures in the King's Forest are worth 10s each'. Although not explicit, this seems to imply that gathering honey from bees in trees was a continuing operation from year to year.

'Bee gardens'

There are other survivals from beekeeping in the past in the New Forest: bee gardens or bee beds. Large areas of the Forest have flowering heather in August and September, and beekeepers— in 1635 those living 'within the perambulation of the Forest'—were allowed to place hives where the bees could benefit from it.

Skeps on the ground could be knocked over by deer (maintained for the royal hunt), livestock (ponies and pigs which were allowed to forage for acorns), and wild animals such as badgers which like to eat honey. Since the whole forest was maintained as a royal hunting preserve, erection of fences or other enclosures was forbidden, and the laws on this were strict and stringently enforced. Apiaries were therefore set up where protection already existed, usually in an enclosure bounded by banks of earth—of varying dates and origins. Whatever the origin of the enclosure, the beekeeper would probably have camped out with his bees and made temporary brushwood barriers across gaps in an enclosing bank.

Older Ordnance Survey maps mark certain enclosures with the name 'bee garden'. I owe my knowledge of these and other bee gardens to Frank Vernon, and to W. A. Humby who was a Forest Keeper at Deerleap, Kingshat, from 1916 to 1968 (his father and grandfather were also Foresters, and the family memory goes back to before 1850). Together we have visited over a dozen sites of interest. We do not know their age or the extent of their use for bees, but it seems proper to put them on record. One (referred to as a bee garden by Howard Sumner[287]) is an Iron Age fort in the Denny Lodge area, marked Church Place on the 1979 1:25 000 Ordnance Survey map of the New Forest (grid letters SU, reference 334 069). On the B3056 road towards Lyndhurst are much more recent enclosures, at Matley Passage (332 073) and Matley Ridge (327 073), which Mr Humby recalls by the name bee garden. Some 10 and 13 miles east of this group are Applemore and Dibden Inclosure. The bank enclosing the bee garden at

FIG. 93 *Section from the tree at Sowley in the New Forest, containing the cavity in which bees had been kept. The two saw cuts can be seen, and the gap between them where the wood had been refitted to make a (large) door.*

FIG. 94 *Living oak tree at Ladycross in the New Forest, with the round flight hole leading to a bees' nest and, to the right of it, the beekeeper's door. The tree was near a path, so the door was a small one to escape detection. This nest was worked regularly for a period within living memory.* PHOTO: F. G. VERNON 1978.

Applemore (about a mile west of the houses, 398 077) is now all but gone, as a result of war-time ploughing. It can just be traced at the very boundary of the Forest at the south edge of a Roman road. Dibden Inclosure (408 057) is to the right of a car park in the base of a gravel pit, about half a mile along the B3054 inside the Forest boundary. This was known as the bee bed(s).

In the extreme south east of the Forest, at Rowdown (430 015) is an area with no signs of an enclosure now, but Mr Humby remembers skeps being brought there by horse and cart from Southampton. On the western side of the Forest, to the north of the road linking the A31 to the M27 and east of Appleslade Bottom, are Castle Piece and King's Garden. Castle Piece (199 089) is a hill fort in Roe Inclosure, which is now wooded but has a defensive bank that is fairly clear. King's Garden (213 092), 2 miles farther east, is in open land with little to be seen. To the west again is Ibsley Common, Gorley (176 106), with a bee garden referred to by Howard Sumner,[286] but which we could not locate.

Forest 'beekeeping' and the precursor of upright hives

Out of the Forest, in Dorset, was a rectangular medieval enclosure 86 ft by 92 ft (069 049), which is the subject of a note by Howard Sumner[288] entitled *The 'bee-garden', Holt Heath*. Sumner said in 1931:

> Comparison with 'bee-gardens' on Ibsley Common which, according to local memory, were in use eighty years ago, shows that this Holt Heath earthwork is of an altogether different type. Enquiries made of two old inhabitants elicited the following information: They and their fathers before them had lived on Holt Heath, and had always known the earthwork called 'the bee-garden' but had never heard tell that bee skeps had been put here. . . . I suggest that this name was given to an earthwork of unknown purpose by folk who had distant acquaintance with the New Forest bee-gardens, which, according to local tradition, were used by bee-keepers of Cranborne Chase.

A housing estate has now been built over the area, so this site is lost.

Several other places in the Forest (as elsewhere in Britain) have bee-associated names. Bisterne Close Farm, Burley (SU 227 029) had a field called 'bee garden' on an old Ordnance Survey map, and a mile to the south is Anthony's Bee Bottom (226 015). North of Hatchett Gate are Honey Hill (365 045) and Little Honeyhill Wood (365 037).

Bee garden seems to be a New Forest name, but at least one other is marked on an Ordnance Survey map—a triple-banked rectangular earthwork on Chobham Common in Surrey (SU 974 643), which may possibly have been a small Iron Age fort. In appearance it is very much like some of the New Forest enclosures, and may well have been used in a similar way for bees. A detailed description of it was published in 1930,[121] with the comment:

> The enclosure is quite unlike the remains of those in the New Forest which were in use eighty years ago. These are quite small and only cover an area of 16 ft each way. [Some of those I was shown were much larger than this; the Chobham enclosure is around 100 ft across.] . . . That the bee garden theory is wrong is shown by the name being associated, both locally and on the Ordnance Survey map, with a prehistoric defensive earthwork of a well-known type which lies a mile away to the west in Albury Bottom.

We shall probably never be able to confirm that all the enclosures referred to were used for hives. Enclosures of any sort are likely to have been used in different ways, according to what was needed at different periods. In Chapter 7 we present much firmer and more extensive evidence on the siting of hives—in recesses in walls. Meanwhile, in Chapter 6, we follow the course of the hives that superseded the tree trunks and the logs cut from them and set upright.

6. Upright hives in apiaries

The area we are concerned with in this chapter is, by and large, Europe north of the Pyrenees, the Alps and the Caucasus, but including also the Balkans. Almost all the types of hives used in this area are similar to types described in Chapter 4 as being used horizontally, but here they have been used standing upright. Some of them were ultimately derived from the forest beekeeping of Chapter 5, and others were baskets used upside down. Fig. 95 shows the regions involved; there was little indigenous beekeeping in northern Asia (beyond the Himalayas and the Gobi Desert) but upright log hives reappear in Korea and beyond (Fig. 98).

Chapter 5 described how the forest beekeeper cut a door in the side of the living tree (Fig. 82); this led to upright log hives similarly treated, and in Germany it finally gave rise to movable-frame hives (Chapter 9) *opened at the back*. Another stream of development was simpler in concept but had much wider consequences. If a hollow log open at one end (or a common basket or other container of a suitable size) was stood on its open mouth, it could be used as a hive for bees, provided there were cracks or holes where the bees could enter. Use of this type of hive probably arose independently in different places, as the result of a swarm of bees taking possession of such an inverted receptacle. (In a similar way, farther south, common water pots laid on their side came to be used as hives.)

In Germany an upright hive used open end down is known as a *Stülper*. There is no equivalent English word; the word skep means an inverted *basket* used as a hive. To inspect the interior of these hives the beekeeper must tilt the hive over, and then—for instance using smoke—get the bees to move away from the bottoms of the combs so that these are exposed. To harvest honey from a tall log, he may then cut or break away the lower parts of the combs containing honey, leaving the old combs containing the brood nest intact above. The honey combs he removes will have been freshly built in the current season, in the space left after last year's harvesting. With skeps, some more elegant operations were devised, but at its simplest, skep beekeeping used small hives (Table 2), and honey was harvested by killing all the bees in some of them each year.

A movement in England in the early nineteenth century, 'humanity to

Upright hives in apiaries

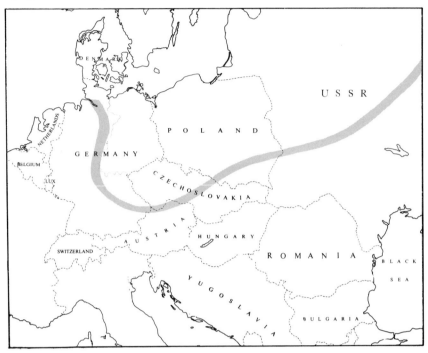

FIG. 95 *Map showing the boundary in Europe between the region of forest beekeeping (to the north) and skep beekeeping (to the south).* AFTER L. ARMBRUSTER.[8]

honeybees', played a significant part in the final development of movable-frame beekeeping. In Mediterranean areas, where beekeepers were working horizontal hives from the back, in some places even removing and replacing combs (Chapter 3), there was much less need to devise better methods.

The coiled-straw skep has become accepted as the archetype of the beehive (Chapter 10), even in parts of the world where it was rarely if ever used, and I think the reason for this was that many beekeepers in the skep area of Europe wrote books and spread knowledge about bees and skeps.

Here, I shall discuss types of upright hives according to their materials and construction, as in Chapter 4.[75]

UPRIGHT WOODEN HIVES

Simple log hives

Two simple upright log hives preserved in peat bogs in northern Germany have been excavated. The earlier was found in 1970[323] at Gristede, 25 km from Oldenburg in north-west Lower Saxony. It had latterly been used to line a well, and inside it was ceramic dated to A.D. 100–200, so the log

Upright hives in apiaries

would be as old or older. It was 1 m high (39 in.) and 31 to 44 cm in diameter (12 to 18 in.); a horizontal slit near the base is regarded as the flight entrance hole. The hive is now in the Forschungsstelle für Siedlungsarchäologie in Rastede nearby. The other hive (Fig. 96) was found in the 1920s by H. Diekmann[95] in a bog at Vehne-Moor, about 25 to 30 km from both Oldenburg and Gristede. It was dated to A.D. 400–500 and was in three parts, which were easily reassembled. The hive was made of beechwood, 1 m high (39 in.) and 30 cm diameter (12 in.), and a cover at the top was fixed on with wooden pegs. Beneath the hive(s) were remains of beeswax comb and of willow rods, which may have had some significance. There were two flight holes at different levels, a feature that can still be found in some upright log and basket hives in parts of northern Europe.

Two very much earlier hollow oak logs found in a Bronze Age settlement at Berlin-Lichtefelde have been discussed as possible hives, but their large size seems to make this unlikely. The archaeologists who found them in 1957–1960 believed them to be sacrificial wells; one contained about a hundred pottery vessels similar to those of the New Bronze Age (1100 B.C.), and charcoal in a nearby refuse pit gave a date of 1080 ± 55 B.C. It was later suggested that this log was a hive,[165,187,188,212] and it had a hole that might have been a flight entrance. But it is 145 cm long and 80 cm in diameter, and weighs about 300 kg (650 lb),[123] so it would have been very difficult to move. There was a grid across it, of fairly substantial pieces of wood and brushwood, on which the vessels were piled.

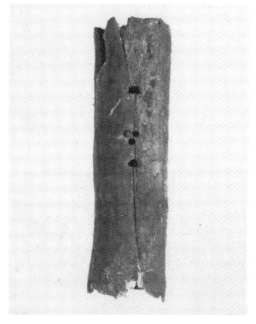

FIG. 96 *German beechwood hive from* A.D. *400 to 500 found at Vehne-Moor, near Oldenburg, and now in the Staatliches Museum für Naturkunde und Vorgeschichte in Oldenburg, which supplied the photograph.*

Upright hives in apiaries

Upright log hives have been used continuously since early times in forested areas of north temperate Europe, especially in earlier strongholds of forest beekeeping (Fig. 95), and occasional examples can still be found at similar latitudes right round the world: Fig. 98 shows one in Korea, used for native *Apis cerana*; these bees are also housed in them on Tsushima Island between Korea and Japan.[154] In North America settlers found plentiful woodlands in many areas; log hives were used there more commonly than skeps—and also hives made of wooden boards (Fig. 97).

*Upright log hives are used as widely apart as North America (*FIG. 97, *apiary in North Carolina,* ABOVE*) and Korea (*FIG. 98, LEFT*).*
PHOTOS: W. A. STEPHEN 1960; SANGYŌ-YŌHŌ 1(2): 73(1934).

Upright hives in apiaries

FIG. 99 *Upright wooden hives, each with a large flat stone for the roof, Severette, Mende, France, 1962.*

When American beekeeping history before 1851 becomes an accepted subject for research, a study of such hives and their use might yield interesting information.

The shape of a log hive is determined by the type of tree used, the typical straight tall hives being from conifers. In the south of France much more squat log hives were made from sweet chestnut (*Castanea sativa*). A large flat stone was used as a roof; such a cover is common for various types of hive (Fig. 99), where suitable stones are available.

Bark hives

Cork is by far the most common type of bark used for upright hives. Cork hives are still used today in Spain (Fig. 101), Portugal (Fig. 100), and where the cork oak grows on the north side of the Pyrenees. These are warm, weather-tight hives, lighter in weight than logs. They are used in a great variety of sizes, and sometimes also of shapes, to suit the pieces of cork available. Since the hives stand separately, there is no need for the

95

ABOVE: FIG. 100 *Neat row of cork hives near Port Alegre, Portugal, 1961.*

LEFT: FIG. 101 *Cork hive constructed with pegs, from central Spain, 1962.* IBRA COLLECTION B70/6.

BELOW: FIG. 102 *Square-section cork hives, Coto Doñana, Spain, 1963.* PHOTO: R. VERHAGEN.

FIG. 103 *Log hives in Poland showing door in side (below flight hole)*. PHOTO: S. KIRKOR 1960.

uniformity that is necessary for hives stacked together horizontally. In Portugal, however, they are often neatly and precisely made, as in some parts of Spain (Fig. 102).

Log hives with doors

The tree trunk in Fig. 84 shows that the practice of using a door to reach a nest of bees in a hollow tree was a very early one. In Russia, Poland and Germany, the provision of a door was continued with upright log hives cut from trees (Fig. 103). Since these hives were worked from the side (or back), there was no need to upend them to take the honey, as with simple log hives; heavier logs could therefore be used. Some were quite massive, and the greater thickness of wood was used for carved ornamentation, especially in Poland. Initially, perhaps, a protective figure was carved into the wood (comparable with the *Bannkorb* discussed below under skeps). Since the logs themselves were not moved or handled, quite elaborate treatment was in order, and interesting—even grotesque— figures were made (Fig. 104), an anatomical feature sometimes being made into the flight entrance. There is a magnificent display of these near Poznan in Poland;[177] see Appendix 2, under PO1.

In the nineteenth century attempts were made in various countries to devise a hive that would give the beekeeper more control over his bees, and over the arrangement of their combs (Chapter 9). The back-opening log hives led J. Dzierzon in Germany (in an area now in Poland), among others, to devise back-opening frame hives.[259] These hives could conveniently be operated in a bee house, but were not suitable for use out of doors. This is, I think, an important factor in the continued use of the

Upright hives in apiaries

bee house in Germany and other German-speaking areas (Chapter 8). Only within the last decade or so have German beekeepers shifted substantially to using top-opening hives standing individually out of doors, such as are used in other countries where beekeeping is well advanced. In Russia, Prokopovich also invented a back-opening frame hive (Fig. 237), but beekeeping there was revolutionised after 1917 along with other operations, and top-opening frame hives became the standard.

FIG. 104 *Ornate hive carved from a log. The cavity is in the body of the man, and the flight entrance just above his right hand. There is an access door at the back.* PHOTO: J. MILKA.

FIG. 105 *Wicker skeps depicted in Sebastian Münster's* Cosmographia, *Bern, 1545.*

BASKET HIVES USED OPEN END DOWN (SKEPS)

The most common methods of making baskets were developed very early: weaving by 5000 B.C., and coiled work even earlier. Both methods have been used for skeps, and in principle hives of either construction could have been made earlier than log hives, which needed wood-cutting tools, but I know of no evidence that this was so.

Basketry may have preceded pottery in some civilisations, but the earliest hives were probably of sun-dried mud, and basket hives may well have come into use later than either pottery or log hives. As far as I know, coiled work was not used for skeps until straw was available, and its adaptation to other fibres (such as sedges by New Forest gipsies, Fig. 127) arose afterwards. New finds may throw fresh light on this sequence.

Wicker skeps

Part of what is believed to be a wicker skep survives from A.D. 0–200 (Fig. 106). It was dug up in the 1970s by W. Haarnagel in the Feddersen

LEFT: FIG. 106 *Woven wicker-work from the period* A.D. *0 to 200; probably the top part of a skep—the earliest known.* PHOTO: W. HAARNAGEL. ABOVE: FIG. 107 *Wicker skep, showing the crownpiece at the top, from a miniature in the Luttrell Psalter (fourteenth century).* DRAWING: D. HODGES.

Wierde, a peat bog near Wilhelmshaven on the North Sea coast of Lower Saxony.[258,272] This wicker object is very similar in construction to more recent examples of wicker hives. The hive is now in the Institut für Marschen- und Wurtenforschung, Wilhelmshaven. It has a special characteristic common to wicker skeps, which is shown in English and other medieval manuscripts (Fig. 107), and early printed books (Fig. 105): a short stick at the top (the crownpiece), to which the incurved stakes were fastened. The origin of the crownpiece (German *Holzfuss*) was established by Armbruster.[5] He found a wicker skep from Serbia that had four main stakes which comprised a whorl of thin branches of a spruce tree. These were *still attached* to a portion of the main branch, which protruded at the top of the skep and constituted the crownpiece. During weaving, the stakes were bent over, and subsidiary stakes were inserted between the main ones as the diameter of the skep increased. In subsequent developments the crownpiece was retained even when the stakes were not attached to it, sometimes becoming little more than a vestige; for instance the crownpiece of the skep in Fig. 59 fell out after the photograph was taken.

FIG. 108 *Four woven wicker skeps near the Antwerp–Brussels autoroute, 1965. They are cloomed with cow dung, and the flight entrances are at different heights.*

FIG. 109 *Cloomed conical skep woven in the traditional pattern, Vianden, Luxembourg, 1953.* IBRA COLLECTION B53/173.

FIG. 110 *The last known wicker skeps in Britain, in Herefordshire in the 1880s. Each is protected by a straw hackle; that on the right has nearly disintegrated and also the clooming, so the weaving can be seen.* PHOTO: A. WATKINS (HEREFORD CITY LIBRARY, NO. 1654).

Not many wicker skeps survive in use in Europe: Fig. 108 shows some I spotted behind an Esso station in Belgium in 1965. Their last recorded use in Britain was in the 1880s (Fig. 110).

Wicker skeps were not all the same shape. Some were almost completely conical (Fig. 109)—the shape shown in French carvings made about 1120 (Figs 244, 245). Most wicker skeps were fairly tall in relation to their width, and like many other tall upright hives they were often fitted with cross-sticks or spiles, to give support to the combs. Wicker hives were cloomed with clay and cow dung or a similar mixture, to keep out the weather. In the Netherlands skep making became a sophisticated art, and protective layers of straw or other materials were fastened over the skeps in decorative ways (Figs 111, 112).

Two styles of Dutch woven skeps, made by shaping an outer protective covering: LEFT: FIG. 111 *Bishop's mitre;* RIGHT: FIG. 112 *Bridal hive.* PHOTOS: RIJKSMUSEUM, ARNHEM.

FIG. 113 *Harry Wilson, a skep maker in Farndale, North Yorkshire, 1953; he is sewing the coils of straw together with brier (bramble), a common binding material. The girth (compare Fig. 115) is quite a narrow ring.* PHOTO: MRS TEASDALE.

Basic skep of coiled straw

This kind of skep became the cult symbol of the bee colony, and also of industry, thrift and the good attributes of a human community (Chapter 10). Coiled straw can produce a much closer, weather-tight construction than wicker, and some skeps preserved from the early twentieth century are beautiful examples of craftsmanship. It could also be used for making hives of various shapes and increasing complexity.

The earliest attested remains of part of a coiled-straw skep were found during an excavation at Coppergate, York, in 1980; with it were many bees, which have been confirmed to be honeybees. This find is dated to the twelfth century.[170]

In making a coiled-straw skep (Fig. 113), certain tools were used, and these are shown in Fig. 115. The coil of straw was maintained at a constant thickness by feeding it through a 'girth', often a piece of cow horn or a piece of old leather sewn into a tube. As each straw in the bundle came to an end, another had to be fed into the girth. Long straw was therefore preferred, and wheat was often considered the best. The binding material might be split cane, or briers—vines of bramble denuded of their

FIG. 114 *Apiary of straw skeps in a shelter on the Lüneburger Heide, 1952; some skeps had been in use for 150 years.*

thorns, for instance by pulling the vine through a piece of old boot reinforced with wires. The binding thread was passed through a hole made in the existing straw work with an awl—which might be a piece of chicken bone (Fig. 115), such as was used in neolithic times, or a metal spike set in a wooden handle.

There has recently been a revival of skep making as a craft, and instructions have been published in English[221,283] and Dutch.[47]

The straw skep is believed to have originated among Germanic tribes, somewhere in the region west of the Elbe.[264,272] One of the most famous areas of skep beekeeping is the Lüneburger Heide, heathland (*Calluna vulgaris* and junipers) near Hannover in north Germany.[125] Hives are beautifully made of coiled straw, and I have seen apiaries there in which some of the skeps had been in use for 150 years (Fig. 114). Farther west, a

FIG. 115 *Tools for making coiled-straw skeps.* IBRA COLLECTION. *(a) Girth made from a cow's horn. B53/38a. (b) The start of a skep using a leather girth. B53/38b. (c) Awl made from a chicken bone, for piercing the straw to insert the binding material. B53/38c. (d) Cleaver for splitting cane or brier into 3 or 4 strands to use as binding thread. B68/68. (a) and (c) were made, and used up to 1935, by J. Featherstone in North Yorkshire; (d) is Dutch.*

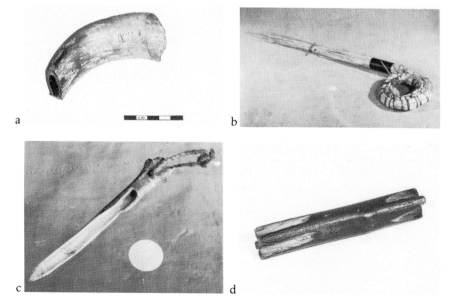

a b

c d

FIG. 116 *Dutch beekeepers at a bee market, waiting for purchasers to buy their skeps of bees, which are closed at the bottom with sacking.* PHOTO: HET VRIJE VOLK (ROTTERDAM).

feature of Dutch skep beekeeping has been the sale of skeps of bees at bee markets (Fig. 116) held at regular dates in different places.[33] Straw skeps in northern Europe were often taller—perhaps in imitation of log hives—than French or English skeps; sizes and shapes of English skeps are listed in Table 7.

The siting and sheltering of skeps in Britain and Ireland are discussed in some detail in Chapters 7 and 8; in common with other countries, each skep was commonly stood on a separate stand, or several were placed in a row on a bench (Fig. 252). Separate skep stands often had a projection to serve as an alighting board for the bees, and a shallow groove provided an entrance into the skep. The majority of stands were probably of wood (Fig. 126), but most that survive are of stone and slate. That shown in Fig. 117 was one of several I found in use in 1953 at Bee Stones Farm on the

ABOVE: FIG. 117 *Skep stand of dressed stone in use in 1953, in North Yorkshire. It has an 'alighting' projection for the bees, the centre part of which is hollowed to give them an entrance channel into the skep.* IBRA COLLECTION B53/187.

RIGHT: FIG. 118 *Skep with a new hackle of rye straw, made to reach below the bottom of the skep, and kept in place with an iron hackle ring. Landbrugsmuseet, Denmark, 1954.* COPYRIGHT UNIVERSITY OF READING, INSTITUTE OF AGRICULTURAL HISTORY AND MUSEUM OF ENGLISH RURAL LIFE.

FIG. 119 *Straw skeps at Sutton-in-Craven, North Yorkshire, in the late nineteenth century. Each is on a stand raised on a box and surmounted by a damaged crock. The beekeeper, known as Grandfather Barret, died about 1895, aged 92. Photographer unknown.*

North Yorkshire moors, on calling to enquire about the origin of the name of the house.

Often the bees would enter the skep where an unevenness in its construction left a gap at the bottom, or the straw might be cut to make a gap. Fig. 126 shows a wooden base ring with the entrance cut in it. Near Goathland in North Yorkshire I found 'snecks' of metal used to fit over a gap, which provided several small entrances such as were recommended by Roman authors[112] as being easily defended by the bees.

Protection from the weather might consist of a layer of sacking or similar material surmounted by a discarded earthenware cream pan (Fig. 119), a piece of turf, or a hackle: a bundle of straw passed through a hackle ring, and splayed out to make a rain-shedding conical straw thatch (Fig. 118).

Taking the honey from skeps

If the bees were to be killed to take the honey, it was common practice to take the lightest skeps (where there was least chance of their surviving the winter) and the heaviest ones (with most honey), and to overwinter those of medium weight—say half the total number. The skeps emptied at harvest time would be filled with early swarms next summer.

Bees could be drowned by immersing the skep in water, and William Lawson in 1618 recommended that, if clean water was used, this should afterwards be used for making bochet (mead).[184] Alternatively the bees

FIG. 120 *Driving bees in the late nineteenth century.* PHOTO: A WATKINS (HEREFORD CITY LIBRARY, NO. 2613).

were killed by sulphur fumes—for instance by digging a pit and throwing several lighted sulphur darts into it; the skep was stood on top, and the suffocated bees dropped off into the pit. I found sulphur darts in use in High Farndale, North Yorkshire, in 1953. Thomas Tusser, in 1557,[301] said the bees should be knocked out of a skep (by thumping it) into a fire below. If heat were applied in the skep itself, the wax combs would be endangered.

Combs of honey cut out of the skep could be eaten as they were, or strained through a cloth bag, known in Yorkshire as a honey poke (cf. 'to buy a pig in a poke'). The 'run honey' that came through first, without pressing, was the best quality, but more could be got by squeezing or wringing the bag. What was left was put into water (enough to float an egg with a piece the size of a shilling above the liquid—a primitive hydrometer) and fermented to make mead. Finally the wax was melted down, strained and made into blocks for future use.

Honey could be harvested from a skep without killing the bees, by 'driving' (Fig. 120). This could only be done in the autumn, when there was little brood in the hive, because it is very difficult to induce bees to leave any brood they are covering. The skep to be harvested was inverted (supported in a bucket), and an empty skep was placed above it, in contact at one point, and elsewhere held away from it at an angle, as in Fig. 120. A skewer and a pair of driving irons (Fig. 121) were used to fix the skeps in

FIG. 121 *Driving irons and skewer, Hampshire. The skewer fastened the two skeps together at the back, and the two irons kept them apart at the front, as in Fig. 120.* IBRA COLLECTION B53/57.

Upright hives in apiaries

position. The bees were induced to leave the lower skep by a rhythmic thumping with the hands on its sides, which made the bees walk en masse up into the upper skep. If the lower skep contained only sealed honey, the bees went easily enough. (In spring, driving could be used to move some of the bees into the upper skep to start a new colony; the queen was watched for, and the skep without her could be given a queen in a cage (Fig. 122). After the bees had got accustomed to her scent, the beekeeper would release her among them.)

The operation of driving was described by a number of early English authors (for instance William Lawson,[184] Henry Best[37]), and it has survived until quite recently as a demonstration (or even a competition) at honey shows.

I remember visiting a beekeepers' class in the mountains outside Madrid in spring 1960 and watching a lesson on driving bees to make extra colonies, using the local upright cork hives (Fig. 101); the thumping of the lower hive was accompanied by cries such as are used to urge donkeys on.

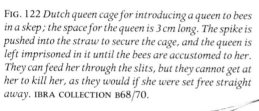

FIG. 122 *Dutch queen cage for introducing a queen to bees in a skep; the space for the queen is 3 cm long. The spike is pushed into the straw to secure the cage, and the queen is left imprisoned in it until the bees are accustomed to her. They can feed her through the slits, but they cannot get at her to kill her, as they would if she were set free straight away.* IBRA COLLECTION B68/70.

More complex coiled-straw skeps in Britain

Coiled-straw work could be built up into various shapes that were strong and rigid. Here we explore especially the elaborations used in Britain in beekeepers' attempts over the centuries to make hives that were more convenient to use.

One simple improvement was to insert a straw ring (an eke) as an extension below the skep (Fig. 123) when the honey flow started; the bees would then build comb down into the eke and store honey there, which could be harvested by cutting the combs between the skep and eke (as with a cheese wire). The name eke is probably connected with the use of the word to mean a supplement. Ekes were used in the Middle Ages, and Edmund Southerne's book (1593)[281] mentions them. Henry Best, a farmer in Yorkshire, kept records and accounts in 1641 (published 1857),[37] and he said that a skep normally has 17 or 18 'wreathes' of straw and an eke 5 wreathes.

Upright hives in apiaries

ABOVE: FIG. 123 *A 'four-ring' eke for putting below a skep to enlarge its capacity, used up to 1954 by its maker, Harry Wilson, in Farndale, North Yorkshire, where the name used was not eke but 'imp'.* IBRA COLLECTION B54/57.

ABOVE: FIG. 125 *I have taken the cap off a skep, to show the bees and their partly constructed honeycombs inside.* PHOTO: V. N. R. BASSETT.

LEFT: FIG. 124 *Skep and cap made by a gipsy at Pulham, Dorset, in the traditional pattern, probably around 1950. The workmanship does not match up to Harry Wilson's (Fig. 113).* IBRA COLLECTION B53/168, 169.

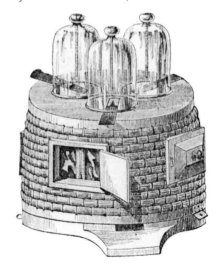

RIGHT: FIG. 126 *This skep has 3 slits in its wooden top for bell jars which are in place. In use they are protected by a large straw cap. The wooden skep base has a channel cut in it to provide a bee entrance to the skep, which has observation windows.* REPRODUCED FROM A. NEIGHBOUR (1878 ED.)[218]

Upright hives in apiaries

As an alternative to using an eke below the skep—which must have involved rather troublesome and messy honey harvesting—a 'cap' might be added at the top (Fig. 124). The cap was similar to, but smaller than, the skep itself, and a hole provided at the centre of the top of the skep gave the bees access to it. With luck, the queen would not go through the hole into the cap, so there would be no brood, but the bees would build combs there (Fig. 125) and these could be harvested when they were full of honey and the bees had sealed them. The combs were cut out, or the whole cap might be sold—high quality pre-packaged comb honey.

By the late eighteenth century comb honey was being sold in more elegant containers than the straw caps. Skeps were made with a flat top, often of wood, and one or more holes through it gave the bees extra space in the form of bell jars or glasses (Fig. 126),[218,318] which were covered over to keep them warm. Honey in a well filled bell glass fetched a premium price. Here is a description written in 1788 by Richard Hoy,[150a] of the filling and removal of the glasses:

> At the Time of Year when the Lime-trees begin to blow, which is about the last ten Days in June, until the 18th or 20th of July, your Glasses will be full, as that is the Season when the Bees lay in their Store of Honey; at this Time you will see two or three Bees standing before the Mouth of their Hive, drumming their Wings, and sounding their Trumps for Joy of their Harvest coming Home; and if it is a very fine Day, and no Sign of Rain, you need not be afraid of them, for they are all in good Humour; other Bees will be standing at the Sides of the Hive, in Order to keep the Way clear of all Robbers, such as Wasps, Hornets, strange Bees, &c.
>
> When your Glasses are all filled with Honey, and sealed up, take them off, if they should be fastened, run a thin Knife under them, and take them away some distance from the Hive; mind and stop up all the Holes with the Stoppers, and shut your Slide, or have other Glasses to put on, in the Room of those you take away, in Order that the Bees may lick up all the loose Honey that may run, by cutting of the Glasses: roll or twist a Cloth round the Glass which you take off, to keep the Glass dark, that the Bees may run up to the Top; for they are all in Confusion for the Loss of their Queen; then sweep them off with a Feather as they come up the Glass, and they will fly Home to their Habitation.
>
> Tie your Glasses over with Paper, to prevent the Bees from robbing them of the Honey: they will keep good in Glasses two Years, if neither the Frost nor Damps get at them.

The skep in Fig. 126 is an illustration from George Neighbour's book (1878),[218] and is believed to have been made by him. If so it must have been before 1898, when his firm in High Holborn closed down.

Some other skeps

Although straw was the usual material for coiled-work skeps, other

FIG. 127 *Sedge (*Cynosurus cristatus *or* Agrostis *sp.) was used by a New Forest gipsy to make this neat and weathertight skep.* IBRA COLLECTION B53/56.

materials such as sedges and reeds could be used. In England gipsies in the New Forest, who were allowed to sell objects they made from materials growing wild (Chapter 5), used sedges, and Fig. 127 shows a beautifully made skep, known locally as a 'bee pot'.

There are excellent sources of information on skeps in continental Europe, published in the relevant languages. The rich variety of skeps in France, and their ancestry and names, have been well documented by Elisée Legros,[186] and Philippe Marchenay has published interesting illustrations.[198] Bruno Schier has discussed and stressed (probably too much) the German contribution; his book,[264] published in 1936, was reprinted with an appendix in 1972. In some areas, notably the Lüneburger Heide,[125] holy and other figures were used on skeps to 'ban' (protect them from) evil—hence the name *Bannkorb* (Fig. 128). They probably have similar cultural origins to the carved log hives mentioned earlier.[264] Swiss skeps are covered by Melchior Sooder,[280] and L. Armbruster[6,8,12,18] has dealt with many other areas in the skep belt of Europe (Fig. 95).

The skep beekeepers in Europe had less control over the bees than some of the beekeepers in Mediterranean lands who used horizontal cylindrical

FIG. 128 *A German* Bannkorb *in which the coils of straw are used in an unusual way to frame a holy statue. Known as the 'Bee Madonna', it was found in Röddensen bei Lehrte, Hannover, but its origin is unknown.* PHOTO: BILDVERLAG FREIBURG IM BREISGAU NO. 1981.

Upright hives in apiaries

hives (Chapter 4). From the seventeenth century onwards, enquiring minds in several European countries, especially England, considered how greater control could be achieved. The development of a practicable movable-frame hive in the USA in 1851, which gave the beekeeper complete control (Chapter 9), had its genesis in the experiments of the late seventeenth century, and it seems to me possible that these experiments may have been made simply because the skep was so unadaptable and stultifying. If these same experimenters had been accustomed to more versatile horizontal hives, the impetus to devise new types of hive might have been less.

On a lighter note, I finish this section on straw skeps with a picture of Swedish beekeepers working with them in 1910 (Fig. 129). It may be compared with Brueghel's drawing of Dutch beekeepers in 1565 (Fig. 242).

FIG. 129 *Anders Paulander, a Swedish folklore scholar, driving bees (compare Fig. 120) at Knislinge in 1910. A new Scandinavian hackle is shown in Fig. 118. The middle beekeeper (only) wears a veil and gloves.* PHOTO BY PERMISSION OF PAULANDER'S GRANDSON.

OTHER TRADITIONAL UPRIGHT HIVES

Most upright hives were of wood or basket work, but a few other types are mentioned below, following the sequence used in Chapter 4 for the much wider variety of horizontal hives.

Stems of plants that are rigid enough, such as some reeds, are used for upright hives, fastened together by twined work to form a cylinder. The resulting construction may not be a very firm one (Fig. 130). I do not know of any upright hives made from fruits, leaves and flowers; horizontal hives that incorporate them (Chapter 4) use tropical plants, which have no counterparts in the north-temperate region of upright hives.

Lastly, we must consider earth in various forms. I have not heard of any traditional vertical hives of unbaked clay, mud or brick. In a stony treeless

ABOVE: FIG. 130 *Upright cylindrical hives of reeds (*Arundo donax*), Almuñicar, Andalusia, Spain, 1973. Such hives are bound together only at the two ends and in the middle; they usually look shabby, and provide many entrances for the bees.* PHOTO: R. VERHAGEN.

LEFT: FIG. 131 *Stone hives on the Dalmatian island of Brać.* PHOTO: V. RITTERMAN.[251]

BELOW: FIG. 132 *Upright clay hive at Urselina, São Jonge, Azores.* PHOTO: A. CASSOLA DE SOUSA 1954.

Upright hives in apiaries

Dalmatian island—Brač—there are many hives of stone, some of which have existed for at least 200 years (Fig. 131); the island has very good bee forage. The traveller Abbate Alberto Fortis wrote in 1776:[251]

> The hives on the island are built of slate slabs, which are carefully cemented together at the joints; the topmost slab, however, which serves as a cover may be removed at will; to prevent strong winds from blowing the hives over, stones are placed on top of them. The entrance to the hives is extremely small. Such hives are extremely numerous here; Count Evelio has some hundreds of them.

Use of the hives had ceased by 1942–43.[251] The stone hives were arranged in rows running east to west, one above the other in terraces, with the entrances facing south, and apiaries contained 100 or more hives. Some of the roofs were flat, and others were gabled with a variable pitch. The roof was removed when the honey was collected; the operation of the hives is described in Chapter 9.

Finally, we come to pottery hives, which provide some oddities and puzzles. Tall upright clay cylindrical hives are unusual, but Fig. 132 shows one in the Azores, which are Portuguese territory. In Portugal I have seen hives of a similar shape, made of concrete; the traditional hive shape and orientation often tend to be followed when a new material is available, or when the old materials are no longer available. These hives, and those of reeds (Fig. 130), suggest scarcity or absence of the more satisfactory cork bark (Fig. 100).

From Coorg in western India I brought back a black pottery hive in 1980 (IBRA B80/4) made like—or from—a round cooking pot, with six entrance holes bored in the side after firing. It was used mouth down, like a skep, for *Apis cerana* bees. A somewhat similar hive, with a separate cover, is used in Burma. Beekeeping is a recent development in most of tropical Asia, so these hives may be recent adaptations.

In the museum in Devizes, Wiltshire, England, there are some pots with a group of small holes in the side which have been regarded as hives, and it is difficult to see what other purpose such vessels would have served, but

FIG. 133 *Romano-British vessel that may or may not have been a hive, in the Museum of the Wiltshire Archaeological and Natural History Society at Devizes.*

Upright hives in apiaries

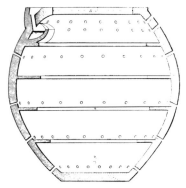 

LEFT: FIG. 134 *So-called bronze hive, as depicted by T. L. Donaldson, 1827.*[96] RIGHT: FIG. 135 *A similar vessel, 48 cm (19 in.) high, with a maximum diameter of 55 cm (22 in.). It has been identified by Dr A. J. Graham as a jar for dormice. There are well spaced-out ventilation holes, and internal ridges on the sides for food: 'When a lid has been placed on the jar, the dormice grow fat in the darkness' (Varro,* De re rustica *3.15.2).* PHOTO: MUSEO ARCHEOLOGICO NAZIONALE, NAPLES 24245.

they are very small for hives. They are most puzzling, and close inspection does not help much. One (E32, Fig. 133) has six holes in the side, of which three are intact and have diameters of 7, 8 and 10 mm, three others being similar or smaller; a bee would need 8 mm at least. It was found[84] during excavations at Casterley Camp, Wiltshire, in 1909–12, and is shown in plate 4, fig. 1 of the report. The description is:

> Jar of brown ware, with polished leathery surface. It has six holes drilled through the side after baking. Height $10\frac{3}{4}$ in. (27.5 cm), rim diameter $4\frac{3}{4}$ in. (12 cm), base $4\frac{1}{2}$ in. (11.5 cm).

The pot is dated to Iron Age C, between 100 B.C. and A.D. 100. Pot E35, of the same source and date, is even smaller, and has four holes in the base, 10, 11, 11 and 12 mm in diameter. One might think in terms of Roman influence, but no Roman hives we know were as small as this or of this shape. The same Museum has other fragments of both sides and bases, with similar groups of small holes, some from sites that pre-date the Roman occupation of Britain.[254]

I have also inspected an Iron Age pot (*c.* 50 B.C.) that has been suggested as a hive, in the Museum at the Roman villa at Rockbourne, Hants. It is about 12 in. (30 cm) high and 8 in. (20 cm) wide at its greatest diameter, with three holes in the side, 6 to 7 mm in diameter. A fragment of a pot from a Roman kiln at Sloden Enclosure, not very far away, has holes that were made before firing (two through the side and one through the rim). This has also been suggested as part of a hive[199] but Frank Vernon measured the holes and found they were not more than 3 mm across, too small for bees to pass through.

It is of course possible that pots such as these were used for bees, not standing mouth down with bees flying through the holes in the side, but lying on their side with bees flying in and out through the mouth. Fig. 71 shows a traditional Maltese hive with holes in the narrow neck that look like bee entrances, but in fact the bees use gaps between the closure across the wide end and the circular rim. At present all these English pots and fragments must remain in question.

One supposed hive has, however, been dealt with satisfactorily. It is a vessel that has been referred to and illustrated in a great many books as a bronze hive from Roman times (Fig. 134), preserved in the museum at Naples, or alternatively at Pompeii. I failed to find it at either museum, but Dr A. J. Graham succeeded in tracking it down in Naples, and established that it is neither made of bronze nor a hive. It is a pottery vessel that was used for housing dormice;[132] it comes from Pompeii, and is therefore dated to A.D. 79 when the town was buried (Fig. 135). H. M. Ransome's book[243] shows in fig. 14 a 'Cretan beehive from Phaistos', but this is not a hive either.

HORIZONTAL AND UPRIGHT HIVES

In this and preceding chapters a distinction has been made several times between horizontal and upright hives in traditional beekeeping. *Horizontal hives* are very widespread (Chapters 3 and 4), though their prototype can only be guessed at. *Upright hives* belong to a relatively small northern region (in Europe and Asia), and their prototype is clearly a standing tree trunk (Chapter 5). Traditional hives show a rich variation; moreover almost every type is found in northern Europe/Asia being used upright, and elsewhere being used horizontally. Although the type and construction of a hive is determined by local materials, whether it was used horizontally or upright depends on the geographical locality.*

This dichotomy into horizontal and vertical traditional hives seems to have persisted throughout recorded times, and the following is a tentative explanation of it on the basis of direct and indirect effects of latitude on the honeybees.

The relatively small region of traditional vertical hives is the *only* high-latitude temperate area where beekeeping was practised early; there were no native hive bees at latitudes beyond 50°N in America or beyond 50°S in any continent. High latitudes have a relatively long winter period when bees cannot collect food. Colonies of bees can survive only if they can rear brood in cool spring weather, store a large amount of food in summer, and in the winter form a cluster in the hive so that their energy requirements

*While this division holds in principle, there are various exceptions. For instance log hives in northern forests are sometimes used horizontally, and a drawing in a Dresden manuscript *Sachsenspiegel*, probably made in Meissen before 1375, shows long plank hives in a stack about six high, superficially resembling those in the Exultet Rolls.[21] In the tropics, pot hives are sometimes used mouth downwards, like a skep.

Upright hives in apiaries

are minimal and their metabolic heat is conserved. These processes may all be best achieved if honey is stored above the brood nest, as in a vertical hive; the honey provides a thermal buffer which reduces loss of heat from the brood, and the rising heat helps the bees to evaporate water in the storage cells when they are making honey. It may well be that there has been genetic selection of bees, both for choice of nesting cavities that were tall rather than wide, and for comb building that conforms to such cavities: the bees thus did better in them.

At low latitudes, on the other hand, temperatures are usually high enough for bees to forage during most of the year. A dearth period is due not to cold but to drought or to heavy rains, and is relatively short. Tropical bees do not need such large honey stores to survive their dearth period—which they may even evade by migrating to a new environment where food is available. Bees in a horizontal hive can avoid overheating by spreading out to the two ends. At higher temperatures bees readily secrete wax and can thus build combs with less expenditure of energy, and in any part of the hive, and they can store honey there. Heat may well soften and break tall combs in a vertical hive under the weight they bear.

In very early times beekeepers were already able to remove the honey from a horizontal hive, a few combs at a time, from one or other end (Chapter 3). At the cooler high latitudes it was more complicated for beekeepers to take honey from traditional upright hives, because honey was stored above the brood nest, between it and the attachment of the top of the combs to the hive itself. A door cut in the side of the tree (Chapter 5) was one expedient used. The use of a cap above a skep (Chapter 6) was another way. Removable top-bars made operations much easier, and Chapter 9 discusses the efforts to use them to advantage, which led to modern movable-frame hives. It has been suggested earlier that these were the outcome of the *difficulties* of using vertical hives.

This analysis, if valid, may provide food for thought about types of hives suitable for modern beekeeping in developing countries of the tropics and subtropics: improvements and innovations in horizontal hives may be more important than the introduction of modern vertical hives. A corollary is the surprisingly advanced nature of the beekeeping that is presented to us in the earliest known representations of it, in Ancient Egypt (Chapter 3). There must, somewhere, have been an earlier, developmental stage, presumably in a Middle Eastern civilisation, but we have no real clue as to when or where this stage occurred, or as to its nature.

FIG. 136 *A small bee bole in a pre-1625 brick wall at Willow Farm, Oakley Green, Windsor, Berkshire (449).* PHOTO: J. HARDING 1965.

7. Bee boles: 600 years of skep beekeeping in Britain and Ireland

BEE BOLES IN RELATION TO BEEKEEPING IN GENERAL, 1300–1900

Skeps of coiled straw (and before them wicker skeps) were the usual hives for bees in much of north-western Europe from the Middle Ages onwards (e.g. Figs 108, 119). Without protection from rain, skeps would quickly rot away, and protection was provided in several ways:

(a) The skeps were set on a wooden bench, or singly on wooden or stone stands. Wicker skeps were smeared all over (cloomed) with a thick layer of cow dung. Straw skeps were more weather-tight, and were provided only with a cover to shed the rain—for instance a straw hackle (Fig. 118) or a discarded earthenware cream pan (Fig. 119), often with some insulation below it.

(b) One or more rows of skeps were housed in an open-fronted *bee shelter*. This would often back on to a wall and have solid ends that supported the shelves, and a sloping penthouse-type roof. Occasionally the front was closed in, except for openings through which bees could fly.

FIG. 137 *The set of bee boles in the garden wall at Charity Farm, Lovington, Somerset (131), with Lovington Church in the background.* PHOTO: H. C. TILZEY.

(c) Skeps were placed along one or more of the inside walls of a more elaborate free-standing structure called a *bee house*; the beekeeper could move about inside the bee house and examine his skeps from there. Openings were provided so that the bees could fly out. (In German-speaking countries bee houses have remained in use for modern hives; see Chapter 8.)

(d) Skeps were put singly (or more rarely in twos or threes) in *bee boles*, recesses or niches built into the structure of a thick wall (Fig. 137).

In this book we use the above descriptions to define the terms bee shelter, bee house and bee bole. Bee shelters and bee houses, together with associated structures, are discussed in Chapter 8. This chapter explores the extraordinarily rich archaeological heritage of bee boles, which have been recorded at 690 sites during the past thirty years. By far the greatest number of sites are in Britain and Ireland (678 altogether). We know the number of boles at 618 of these sites (411 in England, 28 in Wales, 160 in Scotland, and 19 in Ireland). Only 12 sites are known in continental Europe. Reasons for this distribution are suggested later. A few sites (i.e. properties) have more than one set of bee boles (a group in the same wall); for example, there are two sets in adjoining walls at Upper Hall, Elton, Gloucestershire (227).*

A register of bee boles has been built up by the International Bee Research Association since 1952, through the interest of hundreds of house owners, archaeologists, beekeepers, travellers and others who reported their finds. It contains a standard record form for each set of bee boles, completed as far as possible, with photographs, drawings, plans, etc., as available. I find it astonishing that any type of functional building structure should have evaded systematic recording until now, but this

*The numbers given here and in Chapter 8 refer to the IBRA Register of bee boles, bee houses and other shelters for skeps, which is summarised in Appendix 1.

FIG. 138 *Painting on wood (c. 1700) at Charity Farm, showing the house, garden, owner and the bee boles in Fig. 137.* INFRA-RED PHOTOGRAPH BY H. C. TILZEY.

chapter is the first comprehensive account and analysis of these bee boles; the IBRA Register is summarised in Appendix 1. Regional reports have been published for England and Wales,[101] Cornwall,[253] Cumbria,[302] Devon,[117,157] Kent,[90,91,92,93] Yorkshire,[122] and three sets, in the West Midlands, Gloucestershire and Staffordshire, have been described in detail.[315,316,317]

There is nothing to suggest that bee boles were ever the usual way of housing skeps, which commonly stood on benches or separate stands in the open. According to Samuel Bagster in *The management of bees* (1834),[23] 'The most *inefficient* [way] is to procure a ledge in a wall . . .'. Apart from this statement, one or two published contemporary references to the use of bee boles are known. William Lawson, in *A new orchard and garden* (1618),[184] ended his instructions on keeping bees with the sentence: 'Some (as that Honourable Lady at Hacknes, whose name doth much grace mine Orchard) use to make seats for them in the stone wall of their orchard, or garden, which is good, but wood is better.' Then in 1822 J. C. Loudon's *Encyclopedia of gardening*[194] described the labourer's cottage and garden, starting with the explanation: 'This may be reckoned too humble a country residence for the consideration of the landscape gardener; but we conceive it to be of very great importance to the general good, that these should be improved, and there [sic] inhabitants ameliorated.' His list of amenities for the cottage (cooking and living room, garret, pigsty and so on) ends with 'A Nitch in the Wall of the south-east front of the house, to hold two or more beehives, with two iron-bars, joined and hinged at one end, and with a staple at the other to lock them up to prevent stealing'. One of Loudon's other designs is shown in Fig. 183.

There is a later reference to bee boles in *The haunted garden* by Handasyde (1907):[140] 'At one time there had been bees in the garden; the deep recess in the enormously thick wall, divided into four compartments by slabs of stone, was where they had been kept. The builder took care to choose the sunniest corner for the hives of plaited straw, and plants of balm, to be bruised and rubbed along the platforms of each hive, still grew along the haunted wall . . .'

There is one early painting of bee boles in use. At Charity Farm, Lovington, Somerset (131), where there are 11 bee boles (Fig. 137), the living room fireplace has a smoke-darkened wooden panel above it; a photograph taken by infra-red light (Fig. 138) shows that it depicts the bee boles, with skeps in them, and (presumably) the owner, walking in his garden. Judging by his clothes, the painting dates to about 1700.

Bee boles often survive as long as the wall itself stands, and the wall had to be substantial to accommodate the recesses, usually at least 20 in. (50 cm) thick. Doubtless many walls that contained bee boles have disappeared, and some bee boles have been lost by incorporation in later building operations. A few in mortared walls have been blocked up, leaving their outlines still visible.

In 1953 Rosamund Duruz and I published a paper[101] on the 74 sets of English bee boles recorded by then, expressing surprise that so many had been found. The following remarks on terminology still stand:

> It seems desirable to define the term bee bole more exactly than has been done in the past. The word bole or boal is defined in Jamieson's *Dictionary of the Scottish language* (1927) as 'a square aperture in the wall of a house for holding small articles'; it is said to be derived from the Welsh word *bolch* or *bwlch* meaning 'a gap or notch, an aperture'. The more recent *Scottish national dictionary* (vol. 2, 1941) gives the meaning simply as 'a recess in the wall'. The term bee bole has been in common use in Scotland, and we have taken it into use for the whole kingdom—there seems to be no corresponding English word—defining it as 'a wall recess made to shelter a (straw) hive'. We include among the bee boles recesses which take two skeps, but not shelters built *on to* a wall, nor separate structures. A bee bole is essentially an integral part of the wall . . .

In the present analysis, however, all structures with separate recesses for the hives are included with the bee boles, except for a few large recesses (alcoves; Chapter 8).

Results given in this chapter are brief summaries of an analysis, made by Penelope Walker and myself, of the records up to June 1981. All original records and details of the analysis are retained at IBRA. In Appendix 1 a summary of records for bee boles is arranged alphabetically under county for England, Wales, Scotland, Ireland, and then for the only other countries concerned, France and Greece.

In this chapter, after a list of some interesting examples open to the public, bee boles are dealt with in relation to: the type of property they are on; their situation within it (in house, garden, orchard wall, etc.); their number; their state of preservation; and evidence that they were in fact used for bees. Finally, characteristics of bee boles are discussed

Bee boles: skep beekeeping in Britain and Ireland

that depend on the wall in which they are built: their date, the direction they face (aspect), the walling material, their shape and size, height above ground and distance apart. Some curious and interesting features of individual sets of bee boles are recounted, with possible reasons for them. Then, from these results, an explanation is offered to account for the building of bee boles and for their distribution in Britain and Ireland. As a postscript the few bee boles outside Britain and Ireland are described and, finally, various recesses are mentioned that had to be discarded from the bee bole list, as they had been built for quite other purposes.

To the best of our knowledge, the details reported here for individual sets of bee boles were correct when they were recorded. Recording has continued from 1952 to 1981, but many sites have not been visited since the record forms were completed, and so information about accessibility and state of repair—or even survival—is not necessarily still valid.

IBRA welcomes records of bee boles, shelters and houses not listed in Appendix 1, and also reports and photographs which may enrich or update present entries in the Register.

BEE BOLES AND OTHER STRUCTURES ACCESSIBLE TO THE PUBLIC

Most bee boles belong to private houses, and it should not be assumed that visitors are welcome, especially if unannounced. There are, however, some splendid examples in a good state of preservation which are on National Trust properties or otherwise accessible to visitors. Accessible bee houses, shelters and alcoves (Chapter 8) are also included below, and indicated as such.

ENGLAND

Berkshire
621 Berkshire College of Agriculture, Hall Place, Burchett's Green (restored bee house, Fig. 189)

Cornwall
627a Godolphin House, Breage, Helston (by appointment)

Cumbria
 79 Appleby Castle Conservation Centre, Appleby-in-Westmorland (bee house; Centre grounds open in summer)

Derbyshire
660 Milford House Hotel, Bakewell (by appointment)

Devon
 37 by Brook Stores, Croyde, Barnstaple
 50 Pack of Cards Hotel, High Street, Combe Martin (ask owner first)
208 Abbey Farm, Buckfast (wall is opposite main (private) entrance to Abbey)

121

FIG. 139 *Bee bole at Tudor House Museum, Southampton, Hampshire (684), dating from the early seventeenth century. The skep (standing on a modern board) protrudes from the shallow recess typical of early brick bee boles. Unlike hives in bee houses (or in the open), those in bee boles, alcoves and bee shelters could be handled only from the front.* PHOTO: F. G. VERNON 1981.

Gloucestershire
242   Gloucestershire College of Agriculture, Hartpury (restored bee shelter; Fig. 199)

Greater London
252   Well Hall, Eltham (in public park; may be blocked in)
268   Church House Gardens, Bromley (reconstructed bee bole)

Hampshire
684   Tudor House Museum, Bugle Street, Southampton (Fig. 139)
709   Titchfield Abbey, Mill Lane, Titchfield (blocked in, but interesting; Dept. Environment)

Kent
 78   Quebec House, Quebec Square, Westerham (NT)
287   Cathedral Close, Canterbury
288   Memorial Gardens, Canterbury
290   Maidstone Museum, St Faith's Street, Maidstone

Leicestershire
592   Bede House, Lyddington, nr Uppingham (Dept. Environment)

Lincolnshire
 68   Gainsborough Old Hall, Parnell Street, Gainsborough

Norfolk
664   Lifeboat Public House, Thornham (Fig. 164)

Northumberland
123   Otterburn Towers, Otterburn (alcove)
125   West Woodburn Filling Station, W. Woodburn, Hexham

Shropshire
 46   Attingham Park, Atcham (restored bee house, NT; Fig. 190)

Warwickshire
 15   Packwood House, nr Lapworth (NT; Fig. 143)

West Midlands
580   Wightwick Manor, Wolverhampton (NT; ? bee boles)

North Yorkshire
  7   Nutwithcote Farm, Masham (near public footpath; Fig. 147)

FIG. 140 *Six of 12 square bee boles facing west in the ground of Tolquhon Castle, Tarves, Grampian (110).* PHOTO: F. M. GLENNIE, 1953 OR EARLIER.

WALES
866   behind Tourist Information Centre, High Street, Haverfordwest, Dyfed

SCOTLAND
13   Weaver's Cottage, The Cross, Kilbarchan, Strathclyde (NT for Scotland)
110   Tolquhon Castle, Tarves, by Ellon (Ancient Monument Dept., Scottish Development Dept.; Fig. 140)
374   Pitmedden House, by Ellon, Grampian (NT for Scotland)
147   Arbroath Museum, Signal Tower, Ladyloan, Arbroath (others in Arbroath; the Museum may organise a bee bole trail)
384   Cliffburn Hotel, Cliffburn Road, Arbroath, Tayside
747   St Salvator's College, North Street, St Andrews, Fife
786   Abbey Street, St Andrews (on road)
801   Cathedral Cloister, St Andrews (in cathedral ruins)
432   Garvald Village Hall, Garvald, nr Haddington, Lothian

IRELAND
697   Dromoland Castle, Newmarket-on-Fergus, Co. Clare (now a hotel)

FRANCE
F3   Autoroute Aix-en-Provence to Salon, Halte de Ventabren, Bouches-du-Rhône (350 m east of rest area; Fig. 172)
F6   Lycée de Valognes, Manche (Fig. 173)

FIG. 141 *Five of 9 bee boles built in the sixteenth century at the Old Rectory, Beaconsfield, Buckinghamshire (333).*

BEE BOLES IN RELATION TO THE PROPERTY

Bee boles are usually closely associated with a dwelling house, and on land belonging to the house. For want of a better word 'property' is used here to refer to the house and land together.

Type of property

Of the two early descriptions of bee boles quoted, one in 1822 refers to a labourer's cottage with 'a Nitch in the Wall' for hives,[194] and the other in 1618 to the orchard of 'that Honourable Lady at Hacknes'.[184] A painting made about 1700 shows bee boles at a farm (Fig. 138). These records are representative of the wide range of properties in which bee boles have been found: cottages, larger houses and farms, a few inns, and country estates, including some castles. One set in Greater London, of which 18 boles remain, may have been the apiary at Eltham Palace (251). In Ireland, Dromoland Castle just north of Limerick (697) has 7 bee boles set high in the Tudor garden walls. Eight sets in Scotland are in castle gardens or outside walls of buildings (Fig. 140).

There are many bee boles in ecclesiastical properties—from vicarages and rectories (Fig. 141) to abbeys—and quite a few report forms state that there used to be an abbey or monastery nearby. Hedderwick House in Tayside (383) has two pairs of bee boles originally of very good workmanship and finish. Hedderwick was 'a new built fyne house' in 1682, and was held in superiority as part of the Abacie of Montrose, feu duty being paid in wax. The house is now roofless, and the bee boles almost inaccessible in the derelict garden and orchard. In Hampshire two of the five properties with bee boles belonged to the church: the Palace

House, Bishop's Waltham (594), and Titchfield Abbey (709)—but a third, St Margaret's Priory (175), is in fact a Tudor house. Detailed beekeeping accounts still exist for the Beaulieu Abbey estate in 1269–70,[306] but Frank Vernon searched the estate and found no evidence of bee boles. Nine sets of bee boles in Scotland and 11 in England are in abbey or priory grounds; 12 are at Scottish manses (Fig. 142) and 9 at English vicarages.

We shall see that the number of bee boles built depended on the type of property, but there is no evidence that only the rich, or only the poor, built them, or that they were particularly a rural phenomenon. There are still at least 12 sets in the town of St Andrews in Fife, and there were more; Sandwich in Kent has 7 sets, and houses in Maidstone and Canterbury have them, also Gainsborough in Lincolnshire, and Kendal in Cumbria.

Situation within the property

Bee boles were usually built in a boundary wall of an enclosure of some sort or, less commonly, in the outside wall of a building. Sometimes the layout is no longer clear, or was not recorded, but the situation of 600 sets of bee boles could be identified (Table 4). Of these sets, 68 per cent were in a garden wall, and most of the rest were in the wall of a house (9 per cent), a farm building (5 per cent), or yard, orchard or, as in Fig. 155, a field. However, Scotland differed from other regions in having almost none in the wall of a building or yard, and 79 per cent in garden walls. Whether this is because more shelter was needed, or because of a greater need for security, we do not know: 27 sets of bee boles in Scotland had been provided with an iron bar across the recesses that could be secured and locked, probably to prevent theft of the skeps from them; another likely use for these bars is discussed later, and an example is shown in Fig. 167.

Less usual situations for bee boles included a roadbank (Cumbria, 689), brewery (Kent, 91), dovecote (W. Midlands, 648) and water mill (Lothian,

FIG. 142 *Three of 6 bee boles at the Manse, Dunino, Fife (596), showing a common use of single stone slabs to form the lintels, and to separate adjacent recesses. The straw ruskie has two caps for the bees to store honey in, one above the other.*
PHOTO: E. C. WILLSHER 1980.

Table 4. Situation of bee boles within a property: number (and percentage) of sets of bee boles in different walls, in England + Wales + Ireland, and in Scotland

	Garden	House	Farm building	Yard	Orchard	Field or other enclosure	Other	Total
England + Wales + Ireland	290 (64%)	51 (11%)	26 (6%)	24 (5%)	24 (5%)	24 (5%)	12 (3%)	451
Scotland	117 (79%)	5 (3%)	2 (1%)	2 (1%)	6 (4%)	6 (4%)	11 (7%)	149
Total	407 (68%)	56 (9%)	28 (5%)	26 (4%)	30 (5%)	30 (5%)	23 (4%)	600

597). It is likely that the few on rough ground, well away from houses, were used as heather sites. Apart from these there is no evidence of out-apiaries; bees were kept where they could be watched and guarded, as the following references show.

Authors of beekeeping books have worded their advice on siting hives in different ways, but their views were very similar. Charles Butler's *The feminine monarchie* (1609)[55] is representative:

> For your bee-garden first choose some plot nigh your home, that the Bees may be in sight & hearing, because of swarming, fighting, or other sodaine happe, wherein they may need your presente helpe. Your garden of herbes & flowers is fit for the purpose. See that it be safe, and surely fenced, not onlie from all cattaile ... but also from the violence of the windes, that when the Bees come laden and weary home, they maie settle quietlie.

John Keys, in *The antient bee-master's farewell* (1796),[171] worded it thus: 'Let the hives be set as near the dwelling-house as conveniently can be, or to rooms most occupied, for the readier discovery of rising swarms, or to be apprized of accidents. Besides, the bees habituated to the sight of the family, will become less ferocious and more tractable.' William Lawson, writing in Yorkshire in 1618, was one who favoured putting the bees in the orchard. In *A new orchard and garden* (1618),[184] which mentions bee boles, he said 'for cleanly and innocent Bees, of all other things, love and become, and thrive in an Orchard'. It seems likely that many of the enclosures with bee boles that are near a house, but not adjoining it, were originally orchards.

A thousand years before any existing bee boles were built, the Ancient Laws of Ireland[168] laid down fines for the theft of bees, according to the place from which the bees were taken. If stolen from the courtyard of a house, its garden, or a 'green' (some other inviolable precinct), the penalty and fine were the same as for household goods. If stolen from elsewhere, only half the full penalty had to be paid. It is interesting that the yard, garden and 'green' were the usual sites for bee boles in succeeding centuries, and it is likely that hive stands in the open were placed in the same enclosures. Nevertheless bees *were* also kept outside these enclosures, although permanent stands for them would not in general have been allowed. Thomas Bewick,[40] writing in 1820 of a Northumberland common as it was about 1780, referred to it as 'the poor man's heritage for ages past, where he kept a few skeps . . .'

Changes in beekeeping practice were rare and slow until the nineteenth century. Chapter 9 will show how movable-frame hives, which were introduced then, enabled beekeepers to base their management methods on swarm *control*, not on encouragement of swarms, and apiaries could thus be maintained away from their owner's house. Transport of colonies to work outlying crops has become much easier, and is now very common. Apiaries can be visited at predetermined times, and the beekeeper can in general prevent what Butler referred to in 1609 as 'swarming, fighting, or other sodaine happe'.

Number on one property

The number of bee boles was recorded for most sets, and was less than the original total only if some had been destroyed or hidden. It would not necessarily indicate the limit of the holding, since other hives could be set on stands in the open (and a bee shelter or bee house was sometimes found at the same site as bee boles). We found no evidence that bee boles were used only at a specific season, e.g. for new swarms, or for wintering, although special winter bee houses have been found, and are discussed in Chapter 8.

Table 5. Numbers of bee boles in different regions, of properties where they were found, and the average number per property

	No. of bee boles	No. of properties	No. bee boles/property Average	Range
Cornwall & Devon	420	86	4.9	1–14
Kent & Sussex	154	37	4.2	1–20
Cumbria & Lancashire	409	102	4.0	1–18
Yorkshire & Northumberland	490	91	5.4	1–36
rest of England	645	95	6.8	1–30
all England	2118	411	5.2	1–36
Wales	139	28	4.9	1–25
all England & Wales	2257	439	5.1	1–36
Fife & Tayside	265	92	2.9	1–8
rest of Scotland	266	68	3.9	1–12
all Scotland	531	160	3.3	1–12
Ireland	120	19	6.3	1–20
Total	2908	618	4.7	1–36

Bee boles: skep beekeeping in Britain and Ireland

Table 5 shows the known numbers of bee boles on properties, with the mean and range of holdings in the various regions. The average for England and Wales (5.1) happens to be close to the average number of colonies per beekeeper calculated from annual statistics published from 1945 to 1971:[74] this was 5.0 in 1945/46 and 4.9 in 1971, and fluctuated up and down in intervening years. The smaller mean for Scotland (3.3) is not reflected in recent statistics: for three years between 1945 and 1971 for which figures are available, the mean number of colonies per beekeeper was 6.0, 5.6 and 5.4, when for England and Wales the average was 4.5, 4.5 and 4.7, respectively.

Nowadays the mean number of colonies per beekeeper in an area can often provide a rough and ready index to the profitability of beekeeping. In areas where this is high, beekeepers tend to keep more colonies. We shall refer to this later, in our discussion on the distribution of bee boles in Britain and Ireland, in the light of all the records available.

In England and Wales, there were more than 8 bee boles in 11.3 per cent of the sets, but this was true for only 4 sets out of 160 (2.5 per cent) in Scotland. In Fife and Tayside (where there were 92 sets), only 3 sets had more than 6 bee boles. In 1795 James Bonner, Bee-Master, published *A new plan for speedily increasing the number of bee-hives in Scotland*.[45] He explained how this should be accomplished, but nevertheless said 'There should not be too many hives in one place. Eight or nine are sufficient for one garden.' Bonner also offered the view that 'the principle reason why bees have not been reared in greater numbers in this country, is the almost total neglect of them, by gentlemen of property'. Much higher limits were set in England, for instance in Edmund Southerne's *A treatise concerning the right use and ordering of bees* (1593):[281] 'In a peece of ground being not past a quarter of an acre, you may not place above fortie stocks.' Also, the records show that in England it was often the gentlemen of property—in the counties favoured for country estates—who built large groups of bee boles. Here are some examples of the number of bee boles built at large country houses:

21 at Olveston Court, Avon (447)
18 at Burnham Abbey, Berkshire (523)
21 at Heathwaite, Cumbria (217)
18 at Elton, Gloucestershire (227)
18 and 15 at Eltham, Greater London (251, 252)
26 at Porter's Mill, Hereford and Worcester (283) (Fig. 144)
18 at Eydon, Northamptonshire (228)
30 at Packwood House, Warwickshire (15) (Fig. 143)
36 at Skyers House, South Yorkshire (499), now demolished
25 at Llanfair-Dyffryn, Clwyd (553)
20 at Ballyneale House, Ballingarry, Co. Limerick (698) (Fig. 145)

ABOVE: FIG. 143 *Part of the garden wall at Packwood House, Warwickshire (15), where the 30 bee boles are a notable feature.* PHOTO BY KIND PERMISSION OF COUNTRY LIFE 1952

BELOW LEFT: FIG. 144 *A brick wall showing some of the 26 bee boles at Porter's Mill, Hereford and Worcester (283); they are 11ft apart.* PHOTO: K. M. SHAW.

BELOW RIGHT: FIG 145 *Row of 20 bee boles at Ballingarry, Co. Limerick (698), which by 1979 had been restored to show their original style.*

Care and preservation

IBRA record forms included an entry 'condition', and this brought to light information of a good many losses within the period of the register. We have no knowledge of the number destroyed earlier, but the remarkable thing is that so many still remain. Reasons for disappearance include:

- being filled in or built over (or, in walls of buildings, converted into windows), which greatly reduces the chance that they will be recognised

destruction of the wall itself, in the course of further building operations or road works
gradual delapidation of the wall through lack of repair
becoming covered with ivy or other vegetation

Sometimes there are several contributory causes.

Some owners of bee boles did not know what they were, or that they had any special significance. Many, however, were interested to have them on their property, kept them in good repair, and found out about their history. In general bee boles on National Trust properties are very well cared for, and are pointed out to visitors. Many others are also well maintained; we know that one, 5 ft wide, at Balkello Smithy, Tayside (547), was repaired in 1963. This is unusual in having space for 3 or 4 skeps (Fig. 146). In North Yorkshire a bee bole wall at Nutwithcote (7) was repointed about 1950 (Fig. 147); the house to which it belongs was a monastic grange of Fountains Abbey in 1453. In the same county a set of 8 at Follifoot (27), in poor condition in 1952, had been restored by 1973. There is a preservation order on a wall with 8 bee boles at Burton Leonard, North Yorkshire (240, Fig. 148), and also on a wall with 6 bee boles at Donegore Hill, near Antrim (812). A 1490 wall in Canterbury, Kent (375) is scheduled (Fig. 152). In Bromley, Greater London, one of the bee boles originally at The Grete House has been reconstructed in Church House

ABOVE: FIG. 146 *A wide undivided bee bole at Balkello Smithy, Tayside (547), which has a single long lintel.* PHOTO: E. C. WILLSHER.

ABOVE RIGHT: FIG. 147 *Details of two of the 6 restored bee boles at Nutwithcote, Masham, North Yorkshire (7), showing the stone facing and brick interior.* PHOTO: E. HAWTHORNTHWAITE.

RIGHT: FIG. 148 *A two-tier set of bee boles in a stone wall at Burton Hall, Burton Leonard, North Yorkshire (240). Note the separate lintels/bases, and the built-up piers between bee boles, instead of single slabs as in Fig. 142.* PHOTO: R. G. WILBY 1979.

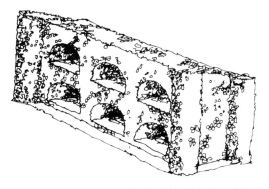 

FIG. 149 *A two-tier set of arched bee boles in a flint wall at Tilty Hill Farm, Duton Hill, Dunmow, Essex (445). The detail on the right shows the construction of the arches, with two rows of hand-made bricks.* AFTER A DRAWING BY H. C. MOSS 1967.

Gardens (268). Five bee boles that were at Cairnie Farm in Angus (147) have been re-erected at the Signal Tower Museum, Arbroath, Tayside; and at Tilty Hill Farm, Essex (445), a two-tier set of brick and flint has been rebuilt (Fig. 149).

Direct evidence of use

Many of the records show clearly that the recesses concerned had been used for skeps. Some of the evidence consisted of personal memories of use, confirmed dates for recent occupation by skeps of bees including:

   1890 West Woodburn Filling Station, Northumberland (125)
   early 1900s Ribby Hall, Kirkham, Lancashire (189)
   1913 (or possibly 1930s) Tree House, Plaxtol, Kent (248)
c. 1920 Cottage, Dacre, North Yorkshire (121)
   1920 Ring House, Woodland, Cumbria (211)
   early 1920s Lune Bank Cottage, Aughton, Lancashire (102)
   1920s Shebster, Reay, Highland (462)
   1929 Garvald, Lothian (432)
c. 1930 Low Colwith Farm, Little Langdale, Cumbria (107)
   1930s Chawleigh Week Farm, Chulmleigh, Devon (165)
   1934 The Daphne, Enniscorthy, Co. Wexford (185)
   1939 Cropten Lane, Pickering, North Yorkshire (762)
   1939 Edenshead, Gateside, Fife (297)
   early 1950s Dale Head Farm, Westerdale, North Yorkshire (83)
   1953 Ballakeighan, Isle of Man (66)
   late 1960s Trehill Farm, nr Okehampton, Devon (789a)
   1980 Pluscarden Abbey, Grampian (114) had skeps in 2 boles still used
          for stray swarms.

Occasionally remains of skeps or skep bases were found in the recesses. The bee boles at Charity Farm in Somerset (131) contained skeps in about 1700 (Fig. 138).

Some names for recesses had been handed down from previous generations. Our own use of the term bee bole may have been instrumental in making it the most common one. But the entry in our record form 'local name for recesses' brought to light a wide variety of alternatives. There were slight variants such as bees' bole (Co. Down, 443), bee boll (Lothian, 463), and bee bowl (Cornwall, 356; Grampian, 501). Bowly hole (Holy Island, Northumberland, 500) and boley hole (Tayside, 292) are not specific to bees, nor is keep hole (Co. Antrim, 528). Others reported are: bee niche (Derbyshire, 97), skep hole (Northumberland, 351), skip hole and skep rest (Devon, 591 and 347), and also bee stack and bee shell (Cumbria, 217 and 622), bee wall (Gloucestershire, 341) and bee- or by-wall (West Yorkshire, 192). In Tayside bee boles were sometimes called bee houses (134); or bee hives (234; also in Devon), and in Highland, skips (488). In Derbyshire (48) and Kent (44) the term bee garth was used, and in Gwynedd (52) *gardd*. Herrod-Hempsall's *Bee-keeping new and old* (1930)[145] used bee garth as synonymous with bee boles, but it rightly describes an enclosure—on to which bee boles might open—being cognate with *yard*. In North Yorkshire (3) garth was used in this sense.

BEE BOLES IN RELATION TO THE WALL

Date of construction

Approximate or exact dates were entered on report forms for 369 sets of bee boles:

	England	Wales	Scotland	Ireland	Total
12th cent.	1				1
13th cent.			1		1
14th cent.	4	1			5
15th cent.	9				9
16th cent.	38	3	10	2	53
17th cent.	66		18	1	85
18th cent.	65	2	57	6	130
19th cent.	57	1	24	3	85
20th cent.					0
Total					369

The period of construction thus spanned more than 600 years, though printed references to bee boles are known only from 1618, the nineteenth century and 1907. The figures suggest that the heyday of bee bole building may have been in the seventeenth and eighteenth centuries in England and in the eighteenth century in Scotland. Perhaps it was earlier,

FIG. 150 *One of the earliest known sets of bee boles still complete, in the cob wall of the Ring o' Bells at Cheriton Fitzpaine, Devon, built in the fourteenth century (450). The two boles on the left have a semicircular arch springing from the base; in the one on the right the arch springs from quite high up the sides. The rectangular recess has a door at the back.*
PHOTO: K. FRENCH 1964.

for one would expect more recent examples to survive (and to be dated) than very old ones.

The following are some of the earliest known bee boles, most of them built before 1400. Two of the early sets are in Devon. Three bee boles in the ruins of the stone medieval abbey buildings of Buckfast (208), probably on the inside wall of an enclosed garden, are likely to have been built in the twelfth century. Bee boles in a cob wall at Cheriton Fitzpaine (450), probably built in the fourteenth century, are shown in Fig. 150.

At Scowles Manor (266) in Dorset (Fig. 151), there are 8 bee boles dated to the late thirteenth or early fourteenth century; the walls are of Portland stone and the bee boles have flat stone lintels. Farther north, in Northamptonshire, at the Manor House of Lower Boddington (407), are 5 arched bee boles in a dry-stone garden wall, probably built in the fourteenth century. Four bee boles of dressed stone (114) in the boundary wall of Pluscarden Abbey, Grampian, may well date from the thirteenth century, when the abbey was built.

The earliest attested brick bee boles are in a scheduled wall, dated 1490, at St Stephen's, Canterbury, Kent (375, Fig. 152). There are 20 small bee boles with pointed arches on one side of the wall and 3 on the other side; later bee boles in brick walls are very similar.

The construction of new bee boles continued well after wooden frame hives had come into use; one set in South Glamorgan (325) was built between 1890 and 1900.

FIG. 151 *The end wall of (probably) a medieval house at Scowles Manor, near Swanage, Dorset (266), with 6 (or 8?) bee boles. The function of the stone arches visible in the construction is to spread the load where the wall is weakened by the presence of the recesses below; the wall behind them is very thin indeed.* PHOTO: M. NIXON 1980.

The earliest bee boles are not notably more primitive, or smaller, or fewer in number, or in any other notable way different from those built later in walls of similar construction. We can, in fact, see little or no chronological development in bee boles; like hives themselves, they seem to be characterised by immutability rather than change.

Recesses similar to bee boles, in neolithic buildings, are referred to on p. 160.

Direction (aspect) and shelter

The aspect was recorded for almost all bee boles, as N, S, E, W or (slightly less frequently) NE, SE, NW, SW, and we therefore have direct evidence as to the direction favoured by beekeepers in past centuries. About half the sets of bee boles faced south, which was consistently preferred all over

Table 6. Direction faced (aspect) by bee boles in Britain and Ireland

	NW	W	SW	S	SE	E	NE	N	Total no. sets
No. sets	$4\frac{1}{2}$	43	58	306	98	78	15	$24\frac{1}{2}$	627
%	0.7	6.9	9.3	48.8	15.6	12.4	2.4	3.9	

The few entries giving intermediate directions (eg SSW) are divided between the two adjacent directions listed in the table the country. Usually south east came next, followed by east and south west or west. Table 6 gives the overall picture.

The direction bee boles face usually determines the amount of sun the hives get during the day, and also the direction from which they are sheltered—by the wall itself, and possibly by other walls or buildings. Both shelter and winter sunshine were considered important by early beekeeping writers. Sunshine requires a southerly aspect anywhere, but the direction that gives shelter from the prevailing wind—or from the prevailing cold wind—varies from one area to another. Such variations as exist in the aspect of bee boles in different regions seem to be linked with the direction from which protection against wind was needed. In Fife and Tayside, which are exposed to east winds, few bee boles face east (and more face south). In Kent, similarly exposed, 7 out of 33 sets faced east, but these were in very sheltered positions. In Devon and Cornwall, fewer than average faced west or south west, the direction of the prevailing winds.

The many bee boles that were close to a house would generally have been well sheltered. A sheltered site was usually chosen for building a house, and trees were often planted to give further shelter. The house and outbuildings would extend the shelter afforded by the wall in which the bee boles were built. Shelter is, however, not easy to quantify, and no analysis for degrees of shelter was undertaken.

The idea that hives of bees should face south dates back at least to Roman times. For instance Columella says in his *De re rustica*, 'A position must be chosen for the bees facing the sun at midday in winter' (9.5.1). Later he states (9.7.4–5):

> The heat of summer is not so harmful to this kind of creature as the cold of winter, and so there should always be a building behind the apiary to intercept the violence of the north wind and provide warmth for the hives. Likewise the bees' dwelling-places, although they are protected by buildings, ought to be so arranged as to face the south-east, in order that the bees may enjoy the sun when they go out in the morning and may be more wide-awake; for cold begets sloth.

FIG. 152 *The earliest dated (1490) brick wall with bee boles, 23 altogether, St Stephen's, Canterbury, Kent (375).* PHOTO: J. BAKER 1959.

Bee boles: skep beekeeping in Britain and Ireland

English authors continued the Roman tradition in this as in many other details. Thomas Tusser in East Anglia (*A hundreth good pointes of husbandrie*, 1557[301]) recommended that hives should be 'set south, good and warme'. A slightly different view was taken by John Evelyn. His *Elysium Britannicum*, written around 1655 but still unpublished except for the beekeeping part,[277] says that hives:

> must be placed against the South sun, a little declining from the East, otherwise the Bees will fly out too early and be subject to the mischief of cold Dews.

John Keys, who wrote *The antient bee-master's farewell*[171] in 1796 from Bee Hall near Pembroke in south-west Wales, was much concerned with shelter from the wind:

> The properest situation for an apiary is one exposed to the wind as little as possible; it being detrimental, and proving often fatal to numbers of bees, by blowing them down, or into the water, or overturning the hives. Trees, high hedges, or fences, on the back and western side of the hives, will be necessary, to screen them from the violence of its force. But they should have a free opening in their front to the south, or rather south-east aspect.

There are 3 bee boles at Bee Hall (793), surprisingly facing north west, but they are not mentioned in his book.

W. C. Cotton in *My bee book* (1842)[66] wanted the hives

> secured from wind, that when the Bees come home laden and weary, they may soon settle at their Hives. . . . If possible, set your Bees on the south side of your house, where they may have most sun in winter.

Evidence of industry continued to be the aim throughout Queen Victoria's reign, and T. W. Cowan wrote in the *British bee-keepers' guide book* (1882):[68]

> Hives may have their entrance in almost any direction, yet to have them face south or east is preferable. The sun shining against the entrance induces the bees to begin their work early.

As an endpiece to this section, I quote an amusing similarity between the direction beekeepers thought best for their beehives to face (as evidenced by bee boles) and the actual preference of a certain stem-nesting bee for the direction its nest entrance should face. A French scientist[64] set out nesting tubes facing in each of the eight main compass directions. The order of popularity of the tubes according to their direction, from highest to lowest, is listed below with the order for bee boles as shown in Table 6.

Bee boles: skep beekeeping in Britain and Ireland

Bee boles	S	SE	E	SW	W	N	NE	NW
Nests of the bee *Heriades truncorum*	S	E	SE	SW	NW	N	NE	W

Walling material

Most of the walls containing bee boles (80 per cent of the 606 for which the material was recorded) were built of stone. In England 72 per cent were of stone, 20 per cent of brick and 8 per cent of cob, which is a mixture of clay, gravel and straw. In the whole area covered, bee boles in cob walls were found only in Devon and Cornwall. In Wales 4 out of 25 walls were brick, in Ireland 4 out of 19, and in Scotland only 3 out of 152—all the rest in Scotland were stone.

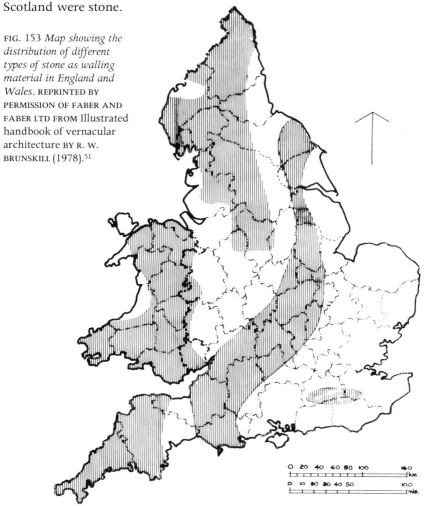

FIG. 153 *Map showing the distribution of different types of stone as walling material in England and Wales.* REPRINTED BY PERMISSION OF FABER AND FABER LTD FROM Illustrated handbook of vernacular architecture BY R. W. BRUNSKILL (1978).[51]

Bee boles: skep beekeeping in Britain and Ireland

On some record forms the type of stone was specified: flagstone, flint, granite, gritstone, limestone, ragstone, sandstone, schist, slate or whinstone. On some others it was referred to as local stone, which would indeed be true of virtually all the walls concerned. Limestone was the commonest specified type of stone in England, and sandstone very much so in Scotland.

Fig. 153 shows where stone is available for walling in England and Wales. The bee bole records conform well with this, although there are some anomalies. Other similar maps[51] show the distribution of walling of cob or other clay-mixtures, and of brick. Brick is the common building material in the eastern countries in general, but there are fewer bee bole walls in this part of England. In Kent and Greater London however, there are 33 of brick, 2 of ragstone, 2 or 4 of flint and 3 unspecified. Bricks can be dated to before or after 1625, because in that year their size was standardised at $9 \times 4\frac{1}{2} \times 3$ in. Of the 22 brick walls with bee boles in Kent, we know that at least 14 were entirely of pre-1625 bricks; Fig. 152 shows one such wall.

No distinction has been made here between walls of dry stone that are common in hill country and other poorer land, and those of mortared stone, freestone, dressed stone, etc. The records were not always clear on this point, and the amount of information varied greatly between recorders. Fig. 154 is an example of what could be done with dressed and shaped stone, and Figs 155 and 156 show an alternative elegance achieved by poorer beekeepers in North Yorkshire and Co. Antrim.

FIG. 154 *Six bee boles at Hutton-le-Hole, North Yorkshire (758). Dressed and shaped stone is used to make arches and piers, and the whole is set into a mortared stone wall.* PHOTO: R. G. WILBY 1980.

FIG. 155 *Two well preserved bee boles in a dry stone wall bounding a field above Muker, North Yorkshire (180b), in 1954.*

Shape and size

The shape of bee boles proved to be closely linked with the material of the wall, and hence with construction techniques. In dry-stone walls and mortared walls of random rubble, almost all recesses are square or rectangular (Figs 155 and 156 are exceptions). The width is limited by the size of stones available to form the lintel, which accounts for some of the irregularity among bee boles in the same wall. The base was often, but not necessarily, also made of one stone. With coursed rubble, dressed stone or freestone, the top was often an arch, semicircular or of a more elaborate shape. Recesses may also be curved with a flat top (Fig. 157), or—very occasionally—triangular (Fig. 158).

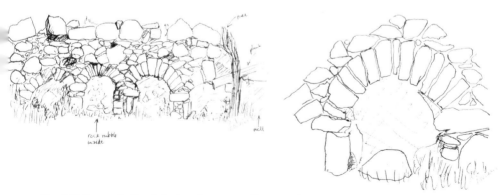

FIG. 156 *Two arched bee boles in a dry stone wall near Red Bay, Co. Antrim (528); detail on the right shows the structure.* DRAWINGS: P. NALDER 1970.

139

ABOVE: FIG. 158 *One of 3 probable bee boles, triangular in shape, in a terrace wall at Lythe House, Grassington, North Yorkshire (644).*
DRAWING: J. H. DEAN 1973.

LEFT: FIG. 157 *A row of Devon bee boles at Gwynant, Georgeham (144). They are rather irregular but all have curved sides and flat tops.* PHOTO: S. HISCOCK 1980.

In brick walls, especially those built in the sixteenth and seventeenth centuries, bee boles were usually straight-sided, topped by a triangular arch (Fig. 136), but a rounded one was sometimes used, springing from the sides quite high up, Fig. 159. Many bee boles in cob walls have the sides and top in a continuous curve springing from the base (Fig. 161; also Fig. 150). In Appendix 1 we describe these as 'domed'.

The above remarks relate to the shape of the opening in the vertical face of the wall. In about 10 per cent of bee boles, particularly in cob walls and some in stone, the back and inside walls are curved in horizontal and/or vertical section, to accommodate the round skep, so to speak (Fig. 157); a pair near Dingle in Co. Kerry (799) are also like this.

Building material was thus the chief factor in determining the shape of the bee boles: rectangular in stone walls, arched in brick walls and with the arch springing from the base in cob walls. But individual fancy and skill still shine through, so that there is no hard and fast rule. Out of 488 sets of bee boles in stone walls 75 per cent were rectangular, but 15 per cent had a rounded arch and 4 per cent a triangular arch. Of 82 sets in brick, 85 per cent were arched (rather more rounded than triangular), but 12 per cent were rectangular. And 88 per cent of the 33 sets in cob walls were arched, all but one bole (in 450) with the arch springing from the base; 3 sets were rectangular.

The size of bee boles varied quite widely, and we were especially interested to establish what, if any, was the main reason for these variations. Were they just chance ones? Were bee boles built larger over the course of centuries? Did some regions favour large or small bee boles? To what extent did the type of wall limit the size of bee boles built in it? In particular, what could the measurements tell us about the size of hives used in different regions and at different times?

FIG. 159 *Seven larger bee boles with rounded arches in the brick wall of a brewery built in the eighteenth century, Boroughs Oak Farm, East Peckham, Kent (91)*. PHOTO: V. R. DESBOROUGH.

Analysis of the measurements of bee boles shows no consistent change in size with succeeding centuries. The size is, however, related to the building material of the wall, and since the latter is a regional factor, bee boles in some areas are larger or smaller than in others.

At first we assessed the size of a bee bole by its width, which would limit the diameter of a skep placed in it. Later, we used the base area, to take into account the depth as well.

The great majority of bee boles were between 14 and 24 in. wide. A few were narrower (down to 9 in.); some were up to 30 in. wide, and then there were scattered examples even up to 72 in. We believe that most bee boles more than 30 to 36 in. wide were meant for more than one skep.

In England the commonest width was 19 in., and Wales and Ireland showed no different pattern. In Scotland 21 in. was most common, and the range was smaller (11 to 29 in. for a single skep, and 36 to 62 in. for more than one).

The mean base area (in square inches) for certain groups of bee boles is as follows; the smallest (165) and the largest (393) correspond to a square with sides 13 and 20 in. respectively.

		Area (sq. in.)	No. of sets
Brick	Kent and Greater London	165	(29)
	rest of south-east England	224	(16)
	rest of England and Wales	304	(23)
	mean for all England and Wales	171	(68)
Cob	Devon and Cornwall	293	(25)
Stone	England and Wales except north (below)	312	(107)
	Cumbria and Lancashire	381	(79)

Table 7. Recommendations for sizes of skeps, Britain, 1593–1890

Year	Author	Diameter	Height	Comments; volume (in litres on right)	
1593	E. Southerne[281]			not more than 2 pecks	>18
1609	C. Butler[55]	15	15	2–4 pecks	18–36
1657	S. Purchas[241]	12+		2–4 pecks	18–36
1768	T. Wildman[318]	10	7	nearly a peck	nearly 9
1780	S. Cooke[62]			$2\frac{1}{2}$–$3\frac{1}{2}$ pecks	23–32
1795	J. Bonner[45] (Scotland)			$2\frac{1}{4}$ pecks Linlithgow measure	23
1796	J. Keys[171]	12	9	inside measurements; 2 pecks	18
1827	E. Bevan[38]	12	9	'in the clear'	18
1838	H. Taylor[291]	14	8 or less	inside measurements	22
1838	J.H. Payne[232]	12	9		18
1847	R. Golding[129]	$11\frac{5}{8}$ top $10\frac{1}{2}$ bottom	9	'in the clear'	15
1851	P.V.M. Filleul[67]	12	9		18
		or 15+	8	inside measurements	18
1870	A. Pettigrew[235]	20	12	$5\frac{1}{2}$ pecks	50
		or 18	12	5 pecks	45
		or 16	12		40
1870	A. Pettigrew also received replies to a questionnaire, from:				
	'man from Hull'			1300 cubic in.	21
	Wycombe, Bucks	12	12		25
	Cornwall	14	11		28
	Lincs.	12	8 or 9		16–18
	Northumberland	15	12		32
	Ayr, Perth, Wigtown and Midlothian: 'similar' to Northumberland				
	Ireland, Wales: 'no answer to our questions'				
1890	BBKA[49]	15	7–9		19–24

Yorkshire and Northumberland	393	(79)
Scotland	393	(135)
mean for England and Wales and Scotland	369	(400)

On average, therefore, bee boles were largest in stone walls, intermediate in cob walls and smallest in brick walls, which are in general thinner than stone walls, allowing only rather shallow bee boles. Apart from this relation with walling material, bee boles tended to be larger in the north than in the south. Were larger skeps perhaps favoured in the north than in the south? If not, why were the northern bee boles larger, even when built of similar materials?

Skeps in existence now do not date back further than a hundred years or so, though they may well continue a tradition from much earlier. A search in British beekeeping books led to the recommendations for skep size in Table 7; Armbruster[11] has discussed hive sizes in general. Lengths are in inches, and volumes in pecks, as usually quoted; 4 pecks = 1 bushel, and 1 peck = 9 litres. For comparison, a ten-frame Langstroth brood box has a volume of about 40 litres, and a hive with 4 such boxes 160 litres.

Measurements in Table 7 show no consistent change in size over the 300 years concerned. Alfred Pettigrew in 1870[235] was the first to promote large skeps, but even his largest was not much bigger than one Langstroth brood box. In the main, beekeepers wanted small skeps, because their system of management was based on crowding the colony in order to make it swarm early. Samuel Purchas said in *A theatre of politicall flying-insects* (1657):[241] 'Let your Hives bee rather too little, than too great, for such are hurtful to the increase and prosperity of Bees.' Pettigrew was exceptional in using large skeps, for a more advanced system of management. His questionnaire suggests that somewhat larger skeps may have been used in Northumberland and Scotland than in England to the south; it is a pity that the Welsh and Irish did not reply.

The majority of bee boles were thus considerably larger than the skeps used; almost all recesses were more than 12 in. wide, and most more than 15 in.; their height would not have been a limiting factor for the low skeps advocated. Many of the bee boles would have been high enough to allow the use of a straw cap or an eke (Chapter 6).

The most likely reason for the larger size of the Scottish bee boles, and possibly for those in northern England, is the use of packing round the skeps in winter. In 1795 Bonner,[45] writing in Edinburgh, had this to say about the preparation for winter of skeps in the open: 'The whole hive should be covered all over with a large quantity of pob tow, or straw, which may be fixed to the hive with ropes made of straw, or hay. A large *divot*, or turf, should be laid upon the top of the tow or straw, to hold it close to the hive, and keep the bees dry and warm.' In Miss K. D. Whyte's

1934 report on the bee boles at West Newton, Tayside (80), she wrote: 'E. Young, Secretary of the Arbroath Beekeepers' Association, had seen the bee boles in his youth. For wintering there was additional protection from a layer of straw 3 to 4 in. thick, fastened across the front by rope or wire, which enclosed the ruskies except at the entrance.' I have myself seen dry bracken used for packing skeps in a wooden bee shelter north of Inverness. The provision of iron locking bars to bee boles probably served partly to keep the packing in place, especially as some of the bars had an ornamental S-shaped or other cross-iron. The same is true where doors or boards are used. These features are discussed on pp. 149–50.

Height above ground

Out of 586 sets of bee boles whose height above the ground is known, 385 are no more than 30 in. from the ground, at a convenient level for inspection and handling, and as recommended by beekeeping writers in past centuries for hives in the open; some of the views are quoted later. Where the ground sloped, the recesses were sometimes built step-wise (Fig. 160). Sets that are more than 30 in. above the ground constitute 39 per cent of the total number in England, but only 19 per cent of those in Scotland. In England, some bee boles were placed very high up, so that people could walk below them without inconvenience from flying bees.

FIG. 160 *Five of six bee boles built step-wise. Spark Bridge, Cumbria (814a).* PHOTO: F. R. ALSTON 1981.

FIG. 161 *Two bee boles, high up in a cob wall and sheltered by the thatch, at the Old Post Office, Longdown, Devon (152).*
PHOTO: H. L. HUSTWAYTE 1953.

This mostly applies to house walls, and especially to cob walls of Devon cottages. Usually only one or two bee boles were placed in these high sites, and the skeps in them must have been handled by someone standing on a ladder. The highest bee boles are a pair in a cob wall at the Old Post Office, Longdown, Exeter, Devon (152). When these were recorded in 1953 the reason for the high site was clear: the skeps were snugly under the eaves of a thatched roof and well protected (Fig. 161). At the next visit the thatch had been replaced by corrugated iron, and there was much less protection.

Bee boles at an intermediate height probably represented a compromise, giving access to the skeps without a ladder, but still allowing a person to work on the ground below without too much interference from flying bees.

Some of the high bee boles are the upper ones of two or three rows. We know of 53 sets of bee boles arranged in two tiers (e.g. Figs 148, 149, 164), which usually seem to have been built in this way so that more skeps could be accommodated in a small space; the rows hold any number up to 9 each. A further 7 sets were in three tiers (Fig. 162), but only one of these held more than 6 skeps altogether. Often there seems to have been some architectural reason for the multi-tier layout—the bee boles were made into a special feature, or their site was unusual or awkward; Fig. 163 provides an example.

English counties that had more than one multiple-tier set were Cornwall, Cumbria (12 sets), Devon, Hampshire, Kent, Northamptonshire, Northumberland and North Yorkshire. Wales had 6 sets, Scotland 5 and Ireland 4.

Among writers who gave advice on the height at which hives should be placed (although they did not refer to bee boles) are the following. Edmund Southerne wrote in 1593:[281] 'Let them stand within two foote of the grounde, whereby the wind shall not have so much power over them as otherwise it would.' Sir Jonas Moore's book *England's interest; or, The gentleman and farmer's friend* (1707)[211] had a chapter on beekeeping, which recommended single wooden stools for the skeps about 1 ft above

145

ABOVE: FIG. 162 *Welsh bee boles in three tiers, 12 in all, at Blackhall Farm, Castle-upon-Alun Mid Glamorgan (837).* PHOTO: P. SHEPPARD 1981.

RIGHT: FIG. 163 *Three bee boles one above the other, built in the early eighteenth century across a corner of the garden at Little Dean, Bramdean, Hampshire (638). In 1980 they faced onto the herbaceous border.* PHOTO: S. PULLINGER 1976.

BELOW: FIG. 164 *Six bee boles in two tiers at the Lifeboat Public House, Thornham, Norfolk (664); the arched bases of the upper row are unusual. We know that skeps were kept in them.* PHOTO: C. CORDNER 1978.

the ground. Thomas Wildman's *A treatise on the management of bees* (1768)[318] favoured $2\frac{1}{2}$ ft. A. Pettigrew (1870),[235] the beekeeper who used extra-large skeps, wrote in *The handy book of bees*:

> Bees are as healthy when placed 2 inches above the ground as when placed 20 inches. If hives are raised 2 and 3 feet, the bees, when heavily burdened, often miss the flight-board on their return from the fields, and thus come unexpectedly to the ground; and, by reason of the sudden and severe shake, do not rise for some time—and some are chilled to death ere they gain nerve and resolution enough to make another attempt. If an elevated position has any advantages at all, we have failed to learn what they are.

Distance apart

Not all records give the distances apart of bee boles, and within many sets these distances varied. Of the sets comprising more than two bee boles each, 359 were sufficiently regularly spaced for the distance apart to have a clear meaning. Of these, 212 sets (59 per cent) had less than 20 in. between adjacent recesses; 19 per cent were separated by 20 to 39 in.; the rest were more widely spaced, many from 5 to 9 ft but some up to twice this distance; Fig. 144 shows an example. In Scotland, many sets had only a single stone slab separating recesses (Figs 142, 165), and 72 per cent came within the first group (less than 20 in.).

In some gardens where the bee boles formed part of the architectural plan, for instance Packwood House, Warwickshire (15, Fig. 143) their distance apart would be determined by the overall design.

The records show a gradual transition between bee boles set separately in a wall, a group of bee boles together in a wall, and a structure built out

FIG. 165 *Three bee boles separated by slabs 4 inches thick; the whole structure, made of sandstone, is set in a brick wall, in front of which it protrudes 7 inches at the base and 12 inches at the top, providing extra shelter. Annacroich House, near Cleish, Tayside (825). The brick wall is unusual for Scotland.* PHOTO: E. C. WILLSHER 1981.

from a wall (or separately), but still with individual recesses for each hive and therefore grouped here with the bee boles. Finally, what I call bee shelters have one or more shelves each holding several skeps, and they may be built against a wall, or as a separate free-standing unit; they are dealt with in Chapter 8.

No writers discussed the distance between hives *in bee boles*, but from Roman times onwards beekeeping authors have been vigorous in preaching ample separation of hives in general from each other, each hive having its own stand or being partitioned off from its neighbours. Columella (9. 7. 3) put it thus: 'If there are no partitions between the hives, they will, nevertheless, have to be so placed as to be at a little distance from one another, so that, when they are being inspected, one which is handled in the course of being attended to may not shake another which is closely joined to it, and alarm the neighbouring bees.'

Later, English beekeeping writers were almost universal in their view that hives in an apiary should be well separated, 'about a yard asunder' or even 'four yards asunder'. There are several probable reasons why bee boles should be more closely bunched together than hives in the open. It would be easier to construct them in a group than to interrupt the wall building at a number of separate points. Also where people would pass to and fro, they might well prefer having to avoid one bee flight area than several. Access (often across cultivated ground) might present less difficulty and cause less damage in one place than in a number.

Of three authors already quoted, Southerne (1593)[281] recommended that hives should be 'set at the least three foot asunder, for feare of infection by standing too nigh together'. Sir Jonas Moore (1707)[211] arranged his hives with military precision 'in straight ranks or rows . . . five feet from one another, measuring from door to door'. Thomas Wildman (1768)[318] said: 'The stands for bees should be four yards asunder; or if the apiary will not admit of so much, as far asunder as may be, that the bees of one hive may not interfere with those of another hive.'

Special features of bee boles

J. C. Loudon, quoted earlier, recommended in 1822[194] that the 'Nitch in the Wall' should have 'two iron-bars, joined and hinged at one end, and with a staple at the other to lock them up and prevent stealing'. Such locking bars were recorded on 28 sets of bee boles, but we were puzzled that so many of these were in the north: 22 in Tayside and 5 elsewhere in the south of Scotland (Figs 167, 168). Enquiry showed, however, that J. C. Loudon (1783–1853) was born in Cambuslang in Lanarkshire, and did not leave the south of Scotland for London until 1803. He was a prolific writer on horticulture with a special interest in architecture. It may thus have been his influence and writings that led to the use of such locking

FIG. 166 *One of two bee boles with rebated edges for receiving a board to close it up for winter; like locking bars, this was a Scottish feature. Kinghorn, Fife (24).* PHOTO: R. BLANCE. *(Compare Fig. 217)*

bars. The bars would help to keep winter packing in place, especially where there was also an ornamental S-shaped cross-iron. We have already seen that Scottish bee boles tend to be larger—especially deeper—than most English ones, and this could well have been to accommodate winter packing material. The bee boles in Northumberland and Yorkshire are similarly large, suggesting that winter packing may have been used there also. We have no direct evidence on this.

Packing for winter did not seem to be a general feature of English beekeeping. The classical book written in 1609, Charles Butler's *The feminine monarchie*,[55] did not mention it: 'At *Scorpio* dresse your hives for winter . . . clean and cloome them close, and mend all the brackes and faults about them: & where the hacles be worne set new in their steades.' Only a few English bee boles have bars or other fittings. In Cumbria there is a section of metal bar at a lower corner (606), and in 1907 and later editions of Cowan's *British bee-keepers' guide book*,[69] fig. 139 shows 6 recesses at Maxstoke, Warwickshire (795), with Scottish-type iron bars.

In 25 Scottish sets of bee boles there are indications that doors were used, or boards fitted into the rebated (i.e. recessed or 'checked') outer edges of the bee bole; see Fig. 166. In England, doors exist or existed on sets

LEFT: FIG. 167 *One of three pairs of recesses with plain locking bars, at Balglassie, Aberlemno, Forfar (239).* PHOTO: R. FORBES 1980. RIGHT: FIG. 168 *Bee boles with ornamented locking bars at East Church Manse, Carmyllie, Arbroath (23).* PHOTO: K. D. WHYTE, *c.* 1934. *Locking bars across bee boles seem to be a speciality of Tayside.*

Bee boles: skep beekeeping in Britain and Ireland

in Northumberland (262, 408), Derbyshire (660a), Avon (314, not certain), and in Ireland, Co. Down (443). One set in Devon (316) has wood and nails above each recess for fitting sacking as protection. A splendid set in Cornwall (356) consists of 14 arched recesses, each with a wooden door pivoted horizontally just below the arch. The bee outlets are cut into the bottom of the doors.

In most of the bee boles with doors, the recesses are oriented in the usual way and the door closes the opening, except where the bees fly out. But in a few cases the bee boles are built back to front, so to speak; a flight hole is provided from the back of the recess, and the bees fly out on the other side of the wall. Examples are: Clevedon, Avon (314), Skyer's House, South Yorkshire (499), Prestwick Hall, Northumberland (262), Ferrybank, Lothian (402).

The small depth of many bee boles (compared with their width) has puzzled us, and it seems likely that the skep often stood on a wooden or stone base that projected in front of the wall; in a few places stone skep stands were found in the recesses. At least 6 records mention an 'alighting area' for the bees, and 3 bee boles at West Farm, Westerhope, Tyne and Wear (263) have prettily shaped projecting bases (Fig. 169). At Sheviock Rectory in Cornwall (318) there is a 'slate alighting slab running along all of the 8 bee boles', and each of the 4 bee boles at Fyfield, Oxfordshire (225) has an 'alighting projection'. The most elegant construction is at Cumrew, Cumbria (136), where a very wide bee bole for four skeps has a shaped stone base with an alighting projection for each skep.

Projections above the bee boles, giving added protection against rain, are more varied. At Broad Hinton, Wiltshire (355), there are 20 bee boles below a thatched roof that projects 21 in. beyond them (Fig. 170). At Charity Farm (131, Fig. 137) and two other sites in Somerset the lintels project above the bee boles.

In Scotland it is not uncommon to find a set of bee boles constructed close together, the whole structure protruding in front of the wall (Fig. 165)— one in Tayside by 7 in. (28) and another by $18\frac{1}{2}$ in. (134). Large projecting

FIG. 169 *Shallow brick bee boles with a shaped projecting stone base; they are in a garden wall at a farm near Newcastle, Tyne and Wear (263).*
PHOTO: C. WEIGHTMAN 1956.

FIG. 170 *Well preserved and unusual curved wall of chalk blocks roofed by thatch, immediately under which is a row of 20 bee boles separated by brick piers. The thatch projects 21 inches beyond the wall, giving generous protection. Uffcott Farm, Broad Hinton, Wiltshire (355).*
PHOTO: P. WALKER 1981.

stone slabs were used to build the 4 bee boles $19\frac{1}{2}$ in. deep shown in Fig. 171, each for two skeps. Alternatively, greater depth may be provided by extending the wall at the back, as at Balbegno Castle, Grampian (422), and in England at Titchfield Abbey in Hampshire (709) and at Pond Hall, Essex (177).

Bee boles associated with large houses that are open to the public (see p. 121) include some of the most interesting architecturally. Among others, one in St Andrews, Fife (745), has a decorative feature above, like an aumbry, which may also have been used for a skep. Eight in a sandstone and brick wall at Coton, Staffordshire (763) have curved sides giving them a bulbous shape, and a small decorative carving above. At Ochiltree, Strathclyde (438), there were two recesses 6 ft above the ground in a house wall, and these were accessible from inside the house, like a cupboard. Beatrix Potter's farm at Far Sawrey, Cumbria (34) has 3 bee boles, one of which appears in an illustration in her *The Tale of Jemima Puddle-duck* (1908).[238]

We often came across evidence of local fashions in bee bole structure, as if one set had been copied from another nearby, and this may well have happened, especially when the beekeeper with the first set was successful.

FIG. 171 *These bee boles would house 2 or possibly 3 skeps each. They are unusually deep for their height, and further protection is provided by protruding flagstones at the top, which slope to shed the rain. Newington House, Cupar, Fife (453).* PHOTO: G. DRAFFEN 1964.

Bee boles: skep beekeeping in Britain and Ireland

THE DISTRIBUTION OF BEE BOLES IN BRITAIN AND IRELAND

Bee boles have been recorded in 36 of the 46 counties in England and in the Isle of Man, in 6 of 8 counties in Wales, 8 of 9 mainland regions of Scotland (but in none of the 3 island regions), and in 11 of 32 counties in Ireland. The counties with an especially large number of properties with bee boles are Cumbria (92), Devon (75), Tayside (64), North Yorkshire (63), Fife (39) and Kent (35). Bee boles stretch from within a few miles of Land's End (Sancreed) to near John O'Groats (Dunnet), and from the most westerly part of Ireland near Dingle to the east Kent coast at Sandwich.

Fig. 271 shows the location in Britain and Ireland of bee boles, according to our records up to June 1981. The other structures providing permanent protection for skeps (alcoves, bee shelters and bee houses) are included in this map because their numbers are insufficient to warrant separate maps.

As the search for bee boles in Britain began to yield results, we were intrigued at their distribution, which seemed to follow no recognisable pattern. There were a great many records from certain areas, a few scattered sets in others, and none at all in quite large parts of the country. The rate of survival would be higher in some areas than in others, and more examples showed up where there was an active searcher. Only after the records were analysed was it possible to work out the likely causes of the distribution, and these are now discussed, with special reference to bee boles, which form the most homogeneous and most numerous type of structure, at 678 sites in Britain and Ireland.

Fig. 271 shows five notable agglomerations, listed below with the areas of highest density first.

(1) Fife and Tayside (east Scotland)
(2) Furness and the part of Cumbria that was Lancashire (north-west England)
(3) north Kent (south-east England)
(4) the Pennine area of North Yorkshire and Lancashire (north central England)
(5) Devon and Cornwall (south-west England)

Certain areas have almost none of these structures including:

(a) the highlands and islands of Scotland (although they exist on lower ground round the main mountain massif)
(b) the highlands of Wales, similarly
(c) the central boglands of Ireland
(d) lowland eastern England, between the Thames and Humber
(e) the Weald of Kent.

The structures found in our survey are those remaining at least 100 (and up to 600) years after they were built. The destruction rate is likely to have been highest in the following areas: where the level of culture was high and where innovations in building styles were therefore frequent, older structures being pulled down or altered; where war or unrest caused damage to property; where there was extreme change in land use, for instance countryside converted into towns or, latterly, into roads that avoid the towns. The enclosures of the eighteenth and nineteenth centuries would not in general have affected walls in the immediate vicinity of houses, where most bee boles were, but materials from existing walls would often have been used for new construction work.

In contrast to most of continental Europe, Britain and Ireland have an oceanic climate, with relatively mild winters but frequent rain, especially in western and hilly regions. Except in areas that were too high or too wet, the equable temperature and adequate rainfall supported many flowering trees and other plants. Beekeeping developed early, and must have been widespread in both Britain and Ireland before A.D. 1000.

There is considerable climatic variation within the islands, and three types of region are relevant here. The best, which became the favoured farming areas, shared some of the benefits of part of continental Europe in having summers that were warm and not too wet; their more equable climate gave relatively mild winters, and strong winds were not a problem. In these areas bees flourished; they could store honey in summer, and wintering was not too difficult. At the other extreme were regions too harsh and inhospitable for bees to survive, where even human populations were sparse and their existence a marginal one, in which bees played little or no part. (Unrest and uncertainty of land or house tenure would also discourage the building of permanent structures in some areas, and this could help to explain the small number of bee boles in Ireland except in the more favoured regions, where they are associated with houses of the richer people.)

I suggest that bee boles were used in a third, intermediate, type of area, where the rainfall was high and winds were often strong and damaging, but the frequent rain and equable temperatures produced a wealth of flowers. Beekeeping could be profitable there, provided the bees were kept alive through the worst of the weather by giving them extra protection—such as bee boles and other shelters supplied. In the west, higher areas might come into this category, and agglomerations 2, 4 and 5 are in areas where the standards of climate, and of living and culture, were lower than in many lowland areas, but still high enough to incorporate beekeeping. The harvests of honey and wax would be specially welcome, and worth an extra effort to procure. In much of these areas the proximity of heather would provide a harvest in late summer after colonies and swarms had built up strong populations.

These factors could account for the large number of bee boles in parts of

Cumbria,[302] the Lancashire and Yorkshire Pennines, Devon east of Dartmoor, and Cornwall east of Bodmin Moor, and those in South Glamorgan. Many of these bee boles belong to small and humble houses.

In more northern regions, and elsewhere in very high areas such as Wales and very wet ones as in Ireland, beekeeping was not possible at all, but it could be carried out in the better conditions beyond them if adequate protection was provided. In the north of Scotland, for instance, the bee boles tend to be in the valleys just beyond the highlands themselves.

On the east coast, Fife and Angus (in Tayside) have the greatest concentration of bee boles found anywhere, except perhaps the Furness area of Cumbria; this is in spite of the fact that they experience damaging cold winds from the north and east. North of the Tay there were many castles—large and small—and good farms, and in Fife the coastal towns have been called 'a golden fringe to a beggar's mantle'. It was in such locations that bee boles were built. Other, smaller concentrations of bee boles occur on the Moray Firth, the Firth of Forth, Norfolk and parts of Kent. Apart from such areas, the concentrations are all in the wetter or higher areas of Britain.

Of the 11 sets of bee boles recorded in France, 8 are to the west of Provence, in a region not unlike those of the western concentrations in Britain.

The main reason for the lack of bee boles in the climatically more favoured areas such as the Midlands and eastern England is, I believe, that bees could be kept there in skeps in the open, without the protection afforded by bee boles. In many of these areas stone was available as

FIG. 172 *Four of 12 bee boles in a dry stone wall near Ventabren, Bouches-de-Rhône, France (F3)*. PHOTO: R. VERHAGEN 1974.

building material (see Fig. 153) and indeed was sometimes used as bases for skeps, so the presence or absence of stone cannot be the ruling factor.

We have no figures for the total number of hives of bees kept in past centuries in different counties, but the Ministry of Agriculture published statistics for England in the 1940s. Ignoring urban counties, where the number of hives is high because of the large human population, the density of hives (number per 1000 acres) in 1944–45 was *highest* in Worcestershire, then Hampshire, Sussex, Oxfordshire, Norfolk and Essex, in that order. According to our calculations, the accommodation for skeps in bee boles that we know of in these counties was very low— between 10 and 33. In contrast, accommodation for skeps was high in the counties listed below, which were those with the *lowest* numbers of hives per 1000 acres in 1944–45.

Yorkshire	511 skeps
Cumbria	377
Devon	369
Cornwall	177
Kent	153
Northumberland	80
Lancashire	64

These include the 6 English counties with the most bee boles, and all are well above the range 10 to 33 for the good beekeeping counties. Counties vary in size, and county boundaries were changed in 1974, but calculations taking these factors into account do not alter the main conclusion: bee boles were not built in the most productive beekeeping areas, where most bees were kept, but in less productive areas where climatic conditions were harder—although good enough for beekeeping to be profitable if adequate shelter was provided for the bees.

BEE BOLES OUTSIDE BRITAIN AND IRELAND

I first learned that there were bee boles in France when an English woman living at Entrecasteaux in Provence read about the English ones in a copy of *The Countryman* and wrote to tell me that there were some in her garden. Several years later I was able to visit them (F8) and a set in Cotignac nearby (F9). Then Raoul Verhagen, also living in Provence, made a search with help from the French beekeeping journals, and 7 more sets were reported;[304] details are included in Appendix 1.

Eight of the 11 sets known to us are in a small area (100 × 60 km) east of the mouth of the Rhône and south of the Alps, in country corresponding roughly to Devon (or Cumbria) but with a warmer climate. There are sets of 3 to 12 bee boles at Entrecasteaux (F8), Cotignac (F9) and Ventabren (F3, Fig. 172), all humble, dry-stone structures virtually indistinguishable from

FIG. 173 *Eight arched bee boles in a stone wall at the Diocesan College, Lycée de Valognes, Manche, France (F6).* PHOTO: M. FRANÇOIS 1974

many in English hill country. At Lurs-en-Provence (F1) a substantial farm, Le Toron, has 6 bee boles in a walled garden. Two other sets are comparable with the large sets on country estates in England: La Chartreuse de Bonpas (F2), built in the seventeenth century, has 30 facing south across a vineyard (Fig. 174), and the Château de Lacoste (F7), built in 1785 and once the home of the Marquis de Sade, has 16 arched recesses in an unmortared wall of dressed stone.

Of the three sets outside Provence, one substantial house at Vicq-Exemplet near the centre of France has 13 bee boles of dressed stone in two walls of the garden (F5). Another at Dormans 110 km east of Paris has 15 square recesses (F4). Finally there is a fine set of 8, with curved arches, at Valognes (F6, Fig. 173), near Cherbourg, built, maybe, in the seventeenth century. The original building was an eleventh-century manor, so these bee boles could be early ones. They are somewhat similar to those at Nutwithcote in North Yorkshire (7, Fig. 147).

Of this small sample of bee boles in France, all are of stone; south is the favoured aspect, with east next. They are rather larger than the English average, and 5 sets are square and 4 arched. A higher proportion of sets are on larger properties than in Britain. The average number of bee boles per property (12) is correspondingly higher, but this may have been partly influenced by the fairly high honey yields obtainable in France. Perhaps more sets on small properties have been destroyed than in Britain; it is almost certain that there are many sets we do not yet know of.

Given the local conditions of climate* and type of property on which bee boles have been found, there seems very little to distinguish those in

* Britanny seemed an obvious location for bee boles, but none could be found. Then *Ouest Apicole* for April 1982 published a photograph of a two-tier set of 10, captioned 'Traditional dwelling house of Morbihan with niches in the wall for skeps'.

FIG. 174 *Thirty bee boles, in 2 tiers, facing across a vineyard at La Chartreuse de Bonpas, Caumont-sur-Durance, Vaucluse, France (F2).* PHOTO: R. VERHAGEN 1974.

France from those in Britain, and the period in which they were built seems to have been similar. No other country is known to have adopted this method of housing bees—except for Greece where we know of one example only, and that now destroyed.

In 1952 Brother Adam of Buckfast Abbey found 98 bee boles in Greece, in a stone wall round a garden a few miles east of Mycenae (Fig. 175). They were tallish flat-topped recesses, wider at the top than at the base, to suit the wicker top-bar hives in them (described in Chapter 9). Their date of construction is not known; by the late 1960s a factory had been built on the site, and the bee boles destroyed. Enquiries have not brought to light

LEFT: FIG. 175 *One of 98 bee boles in a garden wall (now destroyed) near Mycenae, Greece; it contains a top-bar wicker hive (see Chapter 9).* PHOTO: BROTHER ADAM 1952.

ABOVE: FIG. 176 *Nāsr bin Hamid al-Ghaythi, Zahib, Sharqiyah region, Oman, dividing one of the thirty Apis florea colonies he keeps in shallow bee boles in his garden wall. He will move the original colony (now in the recess, with bees covering the comb) to an empty recess. He will leave behind the small piece of comb shown in front, containing brood from which a queen can be reared, and some of the bees.* PHOTO: R. WHITCOMBE 1980.

Bee boles: skep beekeeping in Britain and Ireland

any other bee boles in Greece. In 1972 John Whiston found two in Yugoslavia (about 15 in. high, 10 in. wide). A few structures are known that have similarities to them, and these are discussed at the end of Chapter 8.

In 1982 news was received from Robert Whitcombe of the use of wall recesses in northern Oman for housing colonies of *Apis florea*. This tiny bee builds a single comb in the open, and cannot be hived (Chapter 1), but in parts of Oman it is 'kept' in artificial caves; Fig. 176 shows one of several beekeepers who hive their bees directly into bee boles.

RECESSES FOR OTHER PURPOSES

We kept records of recesses that might be confused with bee boles, in order to help us identify other such finds, but they are not included in Appendix 1. Recesses in a stone wall are a very old type of structure—in Orkney they survive from late neolithic times—and one would expect them to have been put to many uses. Recesses are widely found, often singly or occasionally in pairs, for storing or standing objects that were in everyday use.

Six recesses at ground level in an outside wall facing south at Dudley Arms Hotel, Himley, West Midlands (474), measure 23 × 23 × 16 in. and have a ridge in front 6 in. high; they were probably built for geese. An outbuilding at Temple Sowerby House in Cumbria (655) has 13 recesses that could be for hens or for skeps of bees; a small entrance at ground level suggests the former.

Hawks were provided with mews (the word is derived from Latin *mutare*, to change, because the birds needed special shelter while they were moulting), and these can look superficially like bee boles. They are rather larger, and those we found had a side chamber within the wall, fitted with a perch. The following are examples:

69	Horham Hall, Thaxted, Essex	opening 22 × 18 × 11 in.	facing SSW
483	Sandhouse Farm, German, Isle of Man	opening 12 × 11 × 12 in.	facing SW
521	Patrick, Isle of Man	opening 14 × 10 × 15 in.	facing N
778	East Riddlesden Hall, Keighley, West Yorkshire	opening 14 × 10 × 15 in.	facing N

I have been told that still larger recesses were used for peacocks, but I have not seen any.

Rows of smaller recesses were built close together round the inside walls of dovecotes, as nests for the pigeons; current recommendation is 1 cubic foot of space per nest. They would not normally be confused with bee boles, which are out of doors, but their location is similar to that of

FIG. 177 *The recess on the right is one of two bee boles at 14 Charles Street, Pittenweem, Fife (235). The smaller recess on the left is a charter bole, showing that the wall belongs to the property on this side of it.* PHOTO: E. C. WILLSHER 1979.

recesses in the winter bee houses discussed in Chapter 8. They are, however, smaller and much more numerous (up to several thousand), and dovecotes have flight entrances large enough for the birds.

A recess of about the same size, but quite unrelated, occurs singly in some places in Scotland as a charter bole. It was incorporated in a boundary wall to indicate which of the adjoining properties owned the wall. There are two bee boles in Pittenweem, Fife (235), and a smaller recess in the same wall that is probably a charter bole (Fig. 177). Three well spaced recesses ($13\frac{1}{2} \times 13\frac{1}{2} \times 10\frac{1}{2}$ in.) at South Court, St Andrews, Fife, may well have been built as charter boles, each facing on to a separate garden.

Five or more recesses at ground level near Brandon in Suffolk (603), a flint-knapping area, are very similar to openings to condenser flues constructed to recover lead from furnace gases by condensation,[53] so their purpose was possibly industrial.

Some walled gardens had a series of widely separated recesses in which a fire could be lit (sometimes also flues leading from them); these served to warm the walls, against which fruit trees were grown. At Gowthorpe Manor, Swardeston, Norfolk (348) there are 8 such recesses in a brick wall, $20 \times 28 \times 14\frac{1}{2}$ in., $4\frac{1}{2}$ ft from the ground. Records indicate that charcoal was burnt in them in the nineteenth century. At the Plough Inn, Redford, West Sussex (400) 2 recesses $14 \times 18 \times 9$ in. are said to have been used to heat a greenhouse built against the wall.

Both single and multiple recesses of various sizes and shapes were incorporated inside buildings, and in walls outside them, as part of a structural design or simply as an ornamental feature. There are examples at Harefield, Greater London (738), and at Edzell Castle, Tayside (420). The walled garden at Edzell Castle has 26 such recesses the size of bee boles, which are hollowed out at the bottom to hold earth for plants. There are also twice as many smaller arched recesses, and twelve times as many smaller rectangular ones. These are in two well preserved walls of red sandstone, built in 1604; a third wall (in ruins) may have contained a similar pattern of recesses.

Bee boles: skep beekeeping in Britain and Ireland

It is possible that a few of the alcoves in Chapter 8 (for instance 654 in Wiltshire) belong in this group. Niches for statues are usually tall in relation to their width; there are two at St Cross, Winchester, Hampshire (174).

These, then, are specific and general uses of recesses that we came across in our search for bee boles. If a group of recesses is found with the following charactersitics, they are almost certainly bee boles.

> 3 or more similar recesses in a row, out of doors but near to a house, facing south or south east;
> each recess at least 12 in. high and 15 to 30 in. wide, the depth not more than the width and not less than 10 in., or 8 in. in a brick wall.

THE ORIGIN OF BEE BOLES

In 1954 Professor L. Armbruster[15] suggested that the ancestry of bee boles was to be sought in wild nests of bees in holes in a rock face (shown in rock paintings), via the hiving of bees directly into holes in built walls, and the Roman practice of siting hives in holes that go right through the wall of a house. He saw links with 'beehive' huts and tombs that are shaped like skeps and regarded bee boles in Britain as a result of cultural diffusion from the Mediterranean.

There are several arguments against this view. One is the concentration of bee boles in Britain and Ireland. Ireland and northern Scotland were never part of the Roman Empire, and England and Wales were only an outpost. If bee boles derived from Rome, one would expect more evidence in continental Europe, Asia Minor and north Africa, whereas only a few in France and one set in Greece are known. The Roman influence is much more easily seen in bee walls in Spain and other Mediterranean countries (Chapter 4).

A second argument is that recesses *like* bee boles were built and used for various purposes in Britain, long before Roman times. On the west side of Mainland, the largest island of the Orkneys off the north-west coast of Scotland, are the remains of a neolithic village, Skara Brae. Its date cannot be established with certainty, but 2500–2000 B.C. is usually quoted as likely. The stone buildings were preserved because they were engulfed with windblown sand, as Pompeii was with lava. Beds, shelves and cupboards took the form of recesses in stone walls, and some of them are indistinguishable from recesses built more than 3000 years later as bee boles (Fig. 178). In Harray, the central and once the most remote area of Orkney Mainland, stands Winksetter, a ruinous longhouse long used as a farm building. The first Norwegian settlers came here around A.D. 780. In the room behind the hearth is 'a rich array of stone-built cupboards and two goose-nests close to the floor'.[24]

ABOVE: FIG. 178 *Dwelling at Skara Brae, Orkney, built around 2500–2000* B.C. *Some of the recesses incorporated are similar to the much later bee boles and bee shelters.* PHOTO: DEPARTMENT OF THE ENVIRONMENT (CROWN COPYRIGHT).

RIGHT: FIG. 179 *Recess inside a drystone clochan, the middle one of three probably built in the early Christian period. Cathair-fada-an-dorais, Glenfahan, Co. Kerry, 1982.*

A study of the distribution of the use of wall recesses in general domestic and farm stone building, in different parts of Europe, would be useful in pursuing the subject further. (The funerary use of such recesses does not seem very relevant.) As a start, and while this book was in proof, I searched in an area with a great concentration of early buildings with dry stone walls: west of Dingle in Co. Kerry, which is the most westerly tip of Ireland. It is likely that wood was always extremely scarce here.

At Glenfahan, on the south-west edge of the Dingle peninsula, there are remains of many clochans, both dwelling houses and monastic buildings, corbel-built with dry-stone walls. In one group, Cathair-fada-an-dorais, I found four recesses in the inner walls, presumably used for storage (Fig. 179). Cathair-fada-an-dorais was probably built between the fourth and tenth centuries A.D. A few miles north east, at Kilmalkedar, two medieval stone houses have a number of similar interior wall recesses. The construction and shape of all these recesses, and the size of many, are indistinguishable from those of many bee boles in dry-stone walls. This may

161

perhaps indicate what was possible when walls had to be constructed of large random stones, with no mortar, without reducing the stability or strength.

I am most grateful to the specialists who provided me with clues as to where searches might be profitable, and also with a published reference[110a] to interior recesses in clochans on a monastic site on Skellig Michael, a rocky island 8 miles off the south-west Irish coast. The buildings probably date to between the sixth and the ninth centuries; the monastery was raided by Vikings at least as early as A.D. 812.[183a]

Norsemen did not reach Orkney until 3000 years after Skara Brae was built, so they clearly did not initiate the building and use of recesses in stone walls. Nevertheless if one compares the distribution of bee boles in Britain and Ireland with that of Norse settlements, for example the Isle of Man, Cumbria, Yorkshire, and areas on the east coast, some interesting common features emerge. At the most, the connection would be no more than part of a general cultural diffusion in Britain and Ireland, mediated by, or connected with, Norse seamen. The adaptation for bees of an easily made structure used for many other purposes where stone was the building material, seems a more likely explanation than a transformation of Roman-type bee housing

As already suggested, the oceanic climate of Britain and Ireland may well have been the main reason for the use of bee boles in these islands and almost nowhere else, and also for their geographical distribution within the islands. The discovery of so many examples has provided a unique record of evidence about beekeeping in past centuries.

The IBRA register of bee boles was initiated by Rosamund Duruz in 1952; she soon left to live in Australia, and Joan Harding continued the work. She was followed by James Swarbrick (1954 to 1975) and then by John Whiston. They all contributed many records and photographs, and among others who have made major contributions in this way are: F. R. Alston, Mrs M. Braithwaite, A. E. Cowle, Miss A. S. Cowper, Mrs V. Desborough, Arthur Gaunt, Mrs Janet Glennie, Yeo Jenn, Mrs June Lander, A. R. McLellan, James Riley, Ms Rosemary Robertson, Dr A. Ronald, S. H. Scott, Miss K. D. Whyte, R. G. Wilby, Mrs E. C. Willsher and Mrs Enid Wilson.

At IBRA, the records have been maintained since 1971 by Annette Crownshaw, and Penelope Walker has prepared them and their analysis for publication. To these colleagues, to those mentioned above, and to the many more who have provided information and photographs, I am most grateful.

FIG. 180 *Bee shelter belonging to a farmhouse 300 years old, at Lievelde (Gelderland), Netherlands, near the German border. The house is open as a museum.* PHOTO: C. VAN EDE, ZANDVOORT

8. Bee shelters and houses: Britain, Ireland and continental Europe

This chapter deals with a wide variety of buildings and other structures designed for skeps, some of which are of architectural interest and unusual design. We start with a group of recesses, many quite large and fitted with shelves, that probably held enough skeps to supply a household with honey and wax. The smallest of them might have been classified as bee boles, but several are more than 5 ft high. Almost all are surmounted with a rounded arch, and some are rather elegant; the term *alcove* is used for them here. Britain also has examples of separate buildings known as *bee houses*, in which skeps were placed along one or more sides, and the beekeeper could work behind them, inside the house. The bees flew out through holes suitably placed in the wall, or alternatively the appropriate side of the building was left more or less open. There is clear evidence, also, of bee houses without flight holes, containing internal recesses like bee boles, where bees were wintered in skeps, in the dark and with minimal ventilation. Finally, we shall discuss *bee shelters*: some large and complex, but most of them simple lean-to or free-standing structures with one or more shelves, protected by a roof. These were common in many parts of continental Europe, but the fraction that survive anywhere is small, for reasons that will become clear. Fig. 180 shows a rare archetypal survivor in use.

The alcoves are not known to us from any other country, although they may well exist. Bee houses were until very recently the normal way of housing skeps in German-speaking countries, and some were very large and elaborate. I do not know of winter bee houses, of the type described here, except in Britain and Ireland, but the same purpose was served in Russia by 'winter cellars', where hives were stacked half-underground, below a thick insulating roof.

LEFT: FIG. 181 *Bee alcove at Daresbury, Cheshire (16), ornamented with a design of skep and bees.* PHOTO: C. LINDLEY. RIGHT: FIG. 182 *Triple bee alcove at Cristionydd, Clwyd (51).* PHOTO: G. B. MASON 1952.

BRITAIN AND IRELAND

Alcoves

These large shelved recesses, almost always vaulted, seem to belong to gentlemen's houses rather than to humble cottages. The list in Appendix 1 includes 41 alcoves at 20 sites, 17 scattered over 10 English counties, 1 in Wales and 2 in Scotland.* A good many of them are in richer farmland than the areas of bee bole concentrations, and they have some, if only slight, architectural merit; one feels that they were designed to provide an ornamental setting for the skeps that they contained. Two examples are shown in Figs 181 and 182. Of the 19 alcoves for which dates are known, 16 were built in or before the eighteenth century. Like bee boles, these alcoves were certainly made to house skeps, and they pre-dated wooden hives, for which some of the bee houses were built.

Lack of firm evidence about the number of shelves and whether skeps were placed also on the ground below, makes it difficult to estimate the number housed at each site. At the most conservative, thirty-four of the alcoves may have accommodated only 2, 3 or 4 skeps each, and seven alcoves 5 or 6 skeps each. The (conservative) average is 3.0.

The alcove at Otterburn Towers, Northumberland (123) was used for bees in the 1890s. This was remembered by a local beekeeper, because the landlady at the time wanted to charge rent for its use, on the grounds that the bees gathered nectar from her flowers.

*On six properties with bee boles there were also alcoves, 13 in all, but we do not know whether these were for bees, and they are not included in the analysis below.

Daresbury Hall in Cheshire (16) has a pretty alcove supported by three stone pillars (Fig. 181), which is further ornamented with a carved skep and bees. A brick structure is set into a flint wall at Eynsford in Kent (196) that would hold about 12 skeps; two alcoves are set side by side—each with a solid base and two shelves—and a smaller alcove above, which is partly flint lined. A pair of alcoves at Midmar, Grampian (115b), are surmounted by classical pediments. At Pen-y-cae, Clwyd (51), there is a stone alcove built in the nineteenth century which is divided into three Gothic arches, each with a shelf (Fig. 182); the whole is raised up on a platform. It would take 6 skeps altogether, and was known as a bee *gardd* (garth).

A stone wall in a garden at Mickleton in Co. Durham (475) had an alcove $6\frac{1}{2}$ ft wide that contained round stone stools to take three skeps. The house was built in 1752, and around 1800 John Dent—an ancestor of the owner who told us about the alcove—paid tithes of bees to the Rector of Romaldkirk. In the 1790s the swarms were valued at a penny each, and he gave 1 swarm in 1792, 4 in 1793 and 2 each year from 1794 to 1797. He paid no further tithe of bees until 1802, when inflation (or scarcity?) had increased their value; in 1803 he gave 3 swarms (6s) and in 1804 2 swarms (5s).

In Chapter 7, J. C. Loudon's 1822 recommendation[194] that a labourer's cottage should have 'a Nitch in the Wall . . . to hold two or more beehives . . .' was included among the few contemporary references to bee boles. Loudon subsequently published plans, elevations and detailed descriptions of three designs for a labourer's cottage, and each shows skeps in what he refers to as a bee house. All three fall into the present category of alcoves and, like the cottages themselves, have some elegance. The early dates of some of the surviving alcoves show that Loudon was describing an existing practice, not initiating a new one. Fig. 183 shows Loudon's second cottage, which had a verandah all round supported by trunks of spruce or larch. The alcove for bees, facing north east, had three shelves

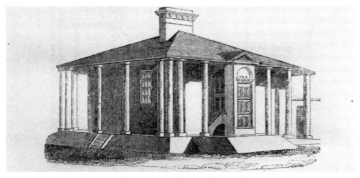

FIG. 183. *J. C. Loudon's second design for a labourer's cottage, with accommodation for bees as well as for pigeons and a dog; fig. 8 from* The cottager's manual *(1840).*[194a]

each for three skeps: above it was the pigeon house and below it was 'a place for a dog or a rabbit, entering from beneath the steps to the porch'. In one of the other designs the skeps faced south east, and the poor dog had to enter his hole directly below the skeps; the 'pigeonry' was over the house porch. Another, with a second floor, incorporated a 'cornice for swallows'; the bees and pigeons were together but faced north east, and the dog was not mentioned.

The cottage walls (and those of the alcove for bees) are shown in the plans as built of brick, but 'where stone is cheaper than brick, that material may be used . . . The walls may also be built of compressed lumps of earth, or in the *pisé* manner, or in the Cambridgeshire or West of England method of building mud walls [clunch, cob, etc.]'.

In his *Encyclopedia of gardening* (1822),[194] Loudon writes of 'a few shelves in a recess of a wall or other building exposed to the south, with or without shutters, to exclude the sun in summer, and, in part, the frost in winter. . . . Bee-houses may always be rendered agreeable and often ornamental objects.' He illustrates a thatched one, with shutters mounted on rustic-work frames.[137]

Bee houses

The bee houses seem to take us still further up the social scale than the alcoves. Nearly all of them probably post-date the alcoves, and their heyday came a century or more later.

In bee houses, as defined here, it was possible to move about in the building, and to stand behind the hives to inspect them or open them, as a beekeeper normally does in an apiary today. This was not very important

FIG. 184 *One of a pair of stone bee houses at Castle Ashby, Northamptonshire (161); one other faces it, on the right.* PHOTO: J. GOFF 1973.

with skeps because inspection and manipulation were minimal. I believe that bee houses were built instead of alcoves as wooden hives came into use. By the early nineteenth century when various types of wooden hives were being used (Chapter 6), and even more towards the end of the century after movable-frame wooden hives were developed (Chapter 9), it would be most unhandy for the beekeeper to have his hives backing on to a wall, as in a bee bole, shelter or alcove. Fig. 210 later in this chapter shows a plan for a bee house published in 1775 in Austria, which was possibly the prototype followed for many English bee houses.

Appendix 1 lists 23 bee houses, 19 in England, 1 in Wales and 3 in Scotland. Five of these are in Hereford and Worcester, but otherwise only 1 in a county, or occasionally 2. As with the alcoves, the counties with bee houses are scattered, and are not those notable for large numbers of bee boles. Of the 10 bee houses that could be dated, most were built after the mid-nineteenth century. The most common size was for 12 hives (7 examples); the largest bee house held 32 hives, at Appleby Hall, Humberside (72), and several others were designed for 20 or more. On average a bee house held 13 hives—far more than the other structures we have discussed so far.

Fourteen of the bee houses were rectangular, two had 6 sides, two had 8, one had 10, one was circular, and one at Llandinam (830) was shaped like a miniature castle (Fig. 185). A few of them were certainly for skeps, not wooden hives. A brick one at Cherry Croft, Herstmonceux, East Sussex (634) is $10\frac{1}{2} \times 6$ ft, with a door at each end, and each long side has a shelf for skeps just under the roof and more than 6 ft from the floor, with 6 quite elaborately made entrance tunnels. There would be plenty of room for working and for storage below the shelves, and the present owners have used the bee house for pigs, hens and dogs, although not for bees. Mr E. R. Miller has provided comprehensive details, plans and photographs of this bee house, which could well be contemporary with the dwelling house, built in the seventeenth century.

An octagonal bee house of ornamental black and white bricks at Gadgarth House, Aunbank, Strathclyde (168) has 12 ornate skep-shaped entrances carved in stone, with a semi-circular wooden alighting board outside. It was built by a firm founded in the mid-nineteenth century.

A bee house of dressed stone at Ecton Lees in Staffordshire (98) had an elegant type of fitment. On the south side there were two rows of 10 flight holes, one above the other (with single alighting boards along the whole length). Inside, the two shelves for the skeps had countersunk bowls. Each skep stood over its bowl, and when it was 'tipped over for emptying' (upturned to cut out the honey combs?), it sat securely and firmly in its bowl. This building, said to be dated before 1850, was 'always' called the honey house.

We have very full details of a pair of matching bee houses constructed

FIG. 185 *Photograph of unknown date and origin showing bee house at Berthddu, Llandinam, Powys (830).*

in limestone ashlar and symmetrically situated facing each other, inside a curved garden wall at Castle Ashby, Northamptonshire (161, Fig. 184). In each there are spaces for two rows of 6 skeps, and although the buildings are only about 5 ft high, the beekeeper could enter through a door in the back wall and (stooping) handle his skeps from the inside. We do not know their date, but the kitchen garden is attributed to Godwin in 1865. The bee houses were in use after 1920.

We know of several interesting bee houses only from drawings or descriptions. Fig. 187 shows a hexagonal bee house (783) for collateral hives. (John Jones wrote a book, *The eclectic hive*,[162] describing the management of his 'Herefordshire collateral bee-boxes', which was published in 1843 by the Hereford Times.) The *Woolhope Transactions* for 1918–20 (p. 76), referring to the design of Baylis, report that Charles Anthony, who founded the Hereford Times, built a bee house (827) at the back of the Mansion House. It could have been like the one Baylis designed, but no description has been found, and it is no longer there.

Dr Edward Bevan used a romantic picture of his own thatched bee house as the frontispiece for his book *The honey-bee*, published in 1827.[38] He commented that he stored potatoes in a cellar sunk into the ground below it. After 1849 Dr Bevan lived at the Old Friars, Hereford, and in 1852 floods swept his bees away down the Wye, so these could hardly have been in a bee house.

Fig. 185, reproduced from a faded photograph marked 'bee house at Berthddu, Llandinam' (830), shows entrances for 6 or 7 hives, in what seems like a doll's castle, the chair also being on a small scale. We should like to know what hives were inside, and what book the owner was reading. An elegant bee house (839) was erected in 1860 by the Apiarian Society of London (Fig. 186); W. B. Tegetmeier, who probably designed it, published *Bees, hives and honey*[292] about the same time. He was a vigorous correspondent and 'expert' on gardening, pigeons and bees.[250]

FIG. 186 *Reproduced from an engraving entitled 'Experimental bee house, Muswell Hill, Hornsey. Erected for exhibiting the working of scientific and improved hives by the Apiarian Society, W. B. Tegetmeier, Hon. Secy.' (839).*

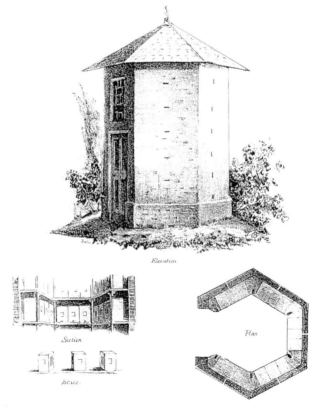

FIG. 187 *Design for a hexagonal bee house (783) by T. Baylis, a builder, issued for Ballards' Lithographic Press, Ledbury, Herefordshire, about 1840. There were shelves running along 5 of the 6 sides. The three 'boxes' that formed each hive are shown; they stood tightly side by side (see 'Section'), and the bees moved from the central brood box to the collateral honey boxes through the slots.*

Bee shelters and houses: Britain, Ireland and continental Europe

The picturesque wooden bee house in Fig. 188 (495) belonged to the Rev. Tickner Edwardes. This photograph, together with the others published in his books *The bee-master of Warrilow* (1907)[105] and *The lore of the honeybee* (1908)[106] (and the original manuscript of the latter), were presented to IBRA in 1952 by the author's widow. To locate the site of his cottage and bee house, turn left down hill just beyond Burpham post office. (The view across the valley to Wepham is still the one shown opposite p. 46 of the earlier book, with the flint boundary wall.) Burgh Cottage stands on the site of the old 'bee-master's cottage'; his bee house was behind it, under an earthwork known as the Saxon Bank. Neither building exists now.

In my view the prettiest of all the bee houses in England is one that still exists, and has recently been rehabilitated, at Hall Place, Burchetts Green, Berkshire (621, Fig. 189). Made of wood, with 10 sides, it has a gracefully curved roof of wood and zinc, and lattice ornamentations on the sides. The bee house was built, probably in the 1880s, round a large tree trunk which formed its centre post. It probably exhibited the 'new' wooden frame hives, each on its stand inside one of the nine windows, which could be shuttered; the door occupies the tenth side. In 1976–77 the bee house was carefully dismantled, rebuilt and tastefully painted, and it is now a showpiece of the Berkshire College of Agriculture that occupies the house and grounds.

FIG. 188 *'An old Sussex bee house' at Burpham, West Sussex (495). The photograph appears as the frontispiece of* The bee-master of Warrilow *by the Rev. Tickner Edwardes (1907).*[105]

FIG. 189 *Restored Victorian bee house for 9 hives at Hall Place, Burchetts Green, Berkshire (621).* PHOTO: G. LEWIS (COPYRIGHT FEDERATION OF BERKSHIRE BEEKEEPERS' ASSOCIATIONS).

The hives in a bee house were not always completely enclosed as in the examples just mentioned. Instead, a lattice of wood or metal might be used. At Bretforton Manor House in Hereford and Worcester (808) there is a wooden bee house with a hipped roof of tiles. Six window spaces fitted with iron grilles are set in one of the long sides (30 ft), and there were semicircular wooden alighting boards outside, two of which remain. The building is $5\frac{1}{2}$ ft wide, with a bench inside to support the hives. This bee house has been used 'within living memory' (1980), and still contains parts of wooden frame hives.

Attingham Park, a National Trust property in Shropshire, has an attractive wooden bee house with a hipped roof of slates (46, Fig. 190). Nearly the whole of the front, and adjacent parts of the end walls, are of wooden lattice work, leaving generous arched openings for flight from twelve hives, in rows of 6. The mansion was built in 1701, and the bee house was in the main orchard, but we do not know its date. After a period of decrepitude, it was lovingly restored in the 1960s, and skeps were kept in it for 8 years afterwards.

The past and present use of bee houses in Germany is described later in this chapter, and their general use in modern beekeeping has been discussed elsewhere.[80]

Winter storage for bees

A set of 11 bee boles at Ballachurry, Rushen, Isle of Man (26), was shown on the cover of the leaflet *English bee boles*.[101] At the same farm was a square windowless outbuilding of two storeys (141, Fig. 191), each storey having its own door and two tiers of recesses, like bee boles, round the

FIG. 190 *Bee house at Attingham Park, Shropshire (46), after its restoration.* PHOTO: N. HODGSON 1974.

inside walls, 31 in all (Fig. 192). This was a winter bee house, where skeps were kept at an equable temperature, protected from vagaries of the weather, and in the dark. Under these conditions colonies of bees are least active and do not fly, and they consume a minimal amount of food. In 1811 the *Agricultural survey for Aberdeenshire*[167] referred to this practice:

> The great objection to the keeping of bees is the expence of feeding them in an unfavourable spring. An ingenious friend of the reporter's has contrived to keep them in an *ice-house*, in a state of insensibility, which is a *saving of their winter provisions.*

Ice houses were also designed to provide a low equable temperature; perhaps this one was used for ice in summer and bees in winter. The only winter bee house we know of in the Aberdeen area is at Midmar (115a), described on p. 174.

In both Russia and North America beekeepers have brought colonies through quite long and severe winters by stacking them together in well insulated spaces half underground, often referred to as cellars. In Värmland in Sweden I found an old potato cellar used as a winter bee house. (Up-to-date versions of these cellars in northern USA and Canada are above-ground buildings with automatic atmospheric control.)

In 1978 I had an opportunity to search out a ruined building (584) in Co. Mayo in the west of Ireland, that we had known about since 1950. It proved to be one of several outbuildings of a substantial house on the eastern shore of Lough Carra—all now roofless and threaded through with brambles. Inside, the two longer walls contained 46 recesses for skeps, typically about $18 \times 18 \times 16$ in. They were in three tiers, or sometimes two. On the same journey I called at Ballingarry House in Co. Tipperary to see the 18 bee boles there (167a), and when I mentioned the winter bee

Bee shelters and houses: Britain, Ireland and continental Europe

house at Castlecarra I was told of 8 similar recesses in a building that is now the pantry at Ballingarry House (167b).

I then started a search for other winter bee houses, and we now have records of 18, of which 6 have outdoor bee boles or alcoves at the same site: in Cornwall (718), Devon (789), Isle of Man (26, 141), Oxfordshire (821), Grampian (115), and Co. Tipperary (167).

I am sure that other winter bee houses exist, unrecognised, for we have only been searching actively for 2 or 3 years. They mostly belong to substantial houses, and some are quite old; 3 are dated to the seventeenth century, 4 to the eighteenth and 2 to the nineteenth. They are very widely scattered—from the west of Ireland through Wales to Bedfordshire, and from Cornwall to Grampian. It seems likely that they must be the relics of a much greater number, some perhaps fallen down, and others put to different uses, maybe with their recesses filled in and forgotten. We do not know of similar buildings outside Britain and Ireland, but they may exist.

Twelve were separate buildings, and these had the most recesses. The other 6 were presumably used for other purposes, and the recesses were often in only one wall. Of course additional skeps could have been stood on the floor, or on stands or benches, inside a bee house; see Lady Anne's bee house (79, below). I do not know of any description of their use, or whether the skeps were packed, or the bees fed, or when they were put in or taken out. There is, however, a reference to bees in *Yn y lhyvyr hwnn*, the earliest book to be printed in Welsh, in about 1546.[155] It gives

LEFT: FIG. 191 *Square two-storey winter bee house at Ballachurry, Isle of Man (141), showing both entrance doors.* ABOVE: FIG. 192 *Some of the recesses in the upper room of the building.*

FIG. 193 *Interior of winter bee house at Butterhill Farm, Dyfed (817), showing some of the 21 recesses.* PHOTO: ROYAL COMMISSION ON ANCIENT AND HISTORICAL MONUMENTS IN WALES.

gardening instructions month by month, and for January it says: 'Dig your garden and spread it with dung, move your bees . . . make use of your plough . . .'. Why should bees be moved? One possible reason is that they had been kept indoors and were to be moved out, although January would seem early for doing this.

Fig. 193 shows a small winter bee house near Milford Haven, Dyfed (817), probably typical of many; it is similar to the much larger one at Castlecarra (584). In 1981 the owner promised to cut down the elder tree that pushed the slates off the roof, and to protect the interior with polythene. Another building in which recesses occupy most of the wall area stands behind a farmstead at Midmar, Grampian (115a). In the garden wall in view of the castle are 2 elegant alcoves for skeps (115b) mentioned earlier in this chapter. The back (north) wall of the bee house has two rows of 10 recesses, and each side wall has three rows of 3, plus 1 above at the front (south) end of this wall, the roof sloping up towards the front. Francis Glennie's photographs in 1953 show 8 openings in two tiers of 4 in the front (south) wall, and occupying a good deal of it. They are about 4 ft high, and their purpose is not clear. In 1980 Bernard Möbus found that the front and the roof had fallen down, exposing the recesses, which are about $24 \times 17 \times 15$ in.

At Radland Mill, St Dominick, Cornwall (307), there is a building with four rows of recesses inside one gable end wall, 5 in the bottom row, then

Bee shelters and houses: Britain, Ireland and continental Europe

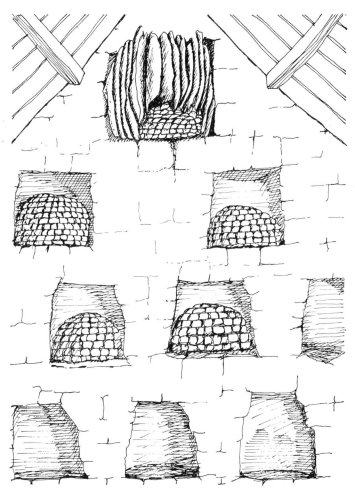

FIG. 194 *Five skeps (and, at the top, much comb built outside the skep) in a winter bee house at Radland Mill, St Dominick, Cornwall (307).* DRAWING AFTER A PHOTOGRAPH BY C. D. CRADOCK.

4, 2, and 1 at the top (Fig. 194). Each recess now has a flight hole through the wall with a slate alighting shelf outside, so the skeps could be kept there all the year round. But evidence suggests that the flight holes may have been added after the recesses were constructed. A photograph published by C. D. Cradock in *The Field* for 26 November 1959 shows 5 skeps in position, the top one full to overflowing with combs built above and around the

175

Bee shelters and houses: Britain, Ireland and continental Europe

skep—14 combs in all. The building would have been erected before 1800; in 1980, sadly, it was roofless, exposed to the weather and vandals, and in imminent danger of demolition, as the land had been sold off for leisure plots.

At Appleby Castle in Cumbria is a square, two-storey building, very much like the winter bee house at Ballachurry (Fig. 191). Traditionally said to have been built by Lady Anne Clifford between 1650 and 1676, it is known as Lady Anne's bee house (79). Her diaries mention stables and almshouses that she built, but not the bee house. Neither exploration of the building, nor a search through Lady Anne's diaries and other records, shows any link with bees, but (like the ice house mentioned earlier) it could well have been used for wintering skeps, stood on the floor or on the stands used outside. The upper storey was probably windowless, although it was later converted into a chapel, and windows were inserted. The lower storey has its own door and a small original window. It may have been a guard house; the position of the building above the River Eden makes it suitable for a look-out post.

In Chapter 7 we saw that in the north of Britain skeps were sometimes packed with straw and bracken for the winter. In Cumbria one garden (754) has a large 'winter bee bole' as well as summer ones, and at a house where there are 5 bee boles (532), there are also two pairs of shelters (632) described as 'winter bee butts'. Fig. 202 shows skeps in a shelter packed for winter.

Bee shelters

We use the term bee shelter for any roofed structure, without individual recesses, that protects hives without allowing space for the beekeeper to move around in it. Although some were free-standing, often of wood (Fig. 195), many were built against an existing wall, usually (Fig. 198) by constructing two solid ends for the shelter, which supported a roof of thatch, tiles, or other material, and one or more shelves beneath.

Bee shelters were undoubtedly the most usual type of structure specially built for housing bees in Britain and in many other European countries—though we found no evidence of any surviving in Ireland. The bee shelters we recorded in Britain are likely to be the relics of thousands that have existed over the centuries.

Fig. 254 is taken from a painting made in the sixteenth century, but shelters like this were almost certainly used much earlier. Fig. 196 is the first illustration of a shelter in a published book,[184] and it is described thus:

> A Frame standing on posts with one floor (if you would have it hold more Hives, two floores) boarded. . . . In this frame may your Bees be dry and warm, especially if you make doores like doores of windows to shroud them in winter, as in an house: provided you leave the hives mouth open.

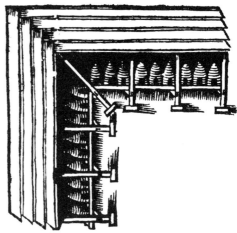

FIG. 195 *Free-standing wooden shelter described in Edward Bevan's* The honey-bee *(1827)*[38] *as a 'shed'. It is viewed from the back; the front is closed 'with the exception of such openings as may be necessary for the entrance of the bees and for their accommodation in bad weather'.*

FIG. 196 *Earliest known representation of an English bee shelter from William Lawson's* The country housewife's garden *(1618).*[184]

Not all authors approved of bee shelters, or indeed of a number of hives sharing the same shelf. John Keys (1796)[171] said:

> It is very *wrong* to place *hives* on benches, which is always the source of mistakes, quarrels, and often slaughter, by their interference with one another. A still worse contrivance is that of little *cots*, or sheds, with shelves therein, one above another; affording a greater harbour for their enemies, very inconvenient for the management, and indeed *impracticable* in the story method.

Shelters, particularly those simply constructed, may collapse or be removed more easily than walls with bee boles, and many of the shelters once used in Britain and Ireland must have disappeared without trace. We are fortunate that 60 have been identified and recorded (47 in England, 4 in Wales and 9 in Scotland—5 at separate sites and 2 pairs elsewhere). This is a sufficient number to make it worth noting their aspect. Of 43 for which this was recorded, 23 faced south, 10 south east, 6 south west and 4 east, so the preference seems to have been similar to that for bee boles. Like the alcoves and bee houses, surviving shelters are well scattered over the country—in 13 counties from Devon to Norfolk, and from Somerset to Cumbria. All 9 Scottish shelters are on the eastern side. Cumbria springs the greatest surprise: 29 of the shelters are there, more than half the total number in England. They are in gardens of cottages as well as farms and large country houses.

Of shelters that can be dated, 10 were built in the nineteenth century, 14 in the eighteenth, 4 in the seventeenth and 1 probably before the

LEFT: FIG. 197 *Typical simple Cumbrian bee shelter, at Grizebeck, Kirkby in Furness (210).*

ABOVE: FIG. 198 *Bee shelter in use in the 1920s, standing against a house in Herefordshire (744).* PHOTO: A. WATKINS (HEREFORD CITY LIBRARY, NO. 3111).

sixteenth. This early shelter (242) was built at Nailsworth in Gloucestershire (Fig. 199). Other bee shelters known to be early were built in the 1600s (Church House, Warton, Lancashire, 89), 1670 (Oaks Farm, Langdale, Cumbria, 195) and 1702 (Nab Cottage, Grasmere, Cumbria, 155, where De Quincey and, later, Hartley Coleridge lived). Also in Grasmere at Sykeside next to Dove Cottage, once the home of William and Dorothy Wordsworth, there is a shelter (513b) as well as 2 boles (513a). Dorothy Wordsworth entered in her diary on 27 April 1802:[320] 'When I came back I found that . . . John Fisher . . . had sodded about the Bee-stand.' On 27 January in the same year she had noted: 'The bees were humming about the hive. William raked a few stones . . . I cut the shrubs.'

The 29 bee shelters in Cumbria whose size we know were relatively small and modest: 10 took 3 skeps, 7 took 4, and only 3 took more than 6. However, the largest, at Well Know, Cartmel (156), had space for 20. In other counties a shelter usually housed from 2 to 10 skeps, but the one in Gloucestershire (242) could take 28 or more. The average number overall is 6.3 skeps per shelter, or 5.8 if 242 is excluded. At least 43 of the shelters were built of stone, but 4 were of brick and 9 of wood.

Several properties with bee boles also have a shelter. The garden of a house near Ulverston in Cumbria has 5 bee boles (532) and a shelter with two compartments (632a); there is a somewhat similar double shelter in a nearby field (632b). Altogether there would have been covered accommodation for between 28 and 38 skeps, depending on how they were arranged.

In the following account of some interesting individual shelters, we start with the simplest and then move on to more complex and less usual ones. Fig. 197 shows one of the 29 Cumbrian shelters (210), which is typical of

Bee shelters and houses: Britain, Ireland and continental Europe

many. It backed on to a wall, faced south, and would have held 3 or possibly 4 skeps. In 1953 I found a similar, longer, shelter in High Farndale, North Yorkshire (128), still in use and being reroofed with tiles. It would have held perhaps 10 skeps, but at the time contained 6 skeps, 1 frame hive and 2 boxes with a skep above, all standing on a low stone shelf. A remote hill farm in Gwent had a simple shelter (323) whose roof sloped down to the front; 210 had a flat roof, and 128 one that sloped down to the back.

Fig. 198 (744) shows a common position—against the sunny side of a house. Another photograph from the same area shows a single-storey cottage (741) with skeps in a double shelf tucked under the eaves, further sheltered by projecting slates. This is the only example I know of in Britain of a type of shelter (like a hanging bookcase) that was very common in Switzerland.[280]

Other shelters were recessed into a wall. Constantine Abraham built one at Great Fryup, Lealholm, North Yorkshire (197), with a tiled roof; like many others it rose from the ground and the skeps would have stood on flat stone stands. (In one of the alcoves (475), the skeps were similarly raised on stone stands.) Abraham planted trees for his bees all round, and these were well grown in 1954. A similar shelter was built in Danby Dale nearby, at a house called Honey Bee Nest.

The most spectacular, and the largest, bee shelter in Britain used to be at Hive House, Nailsworth, Gloucestershire (242), which later became a police station. In 1968 the shelter was threatened with demolition, and Mr W. J. Robinson, Secretary of the Gloucestershire Beekeepers' Association, obtained permission to dismantle it and re-erect it at the Gloucestershire College of Agriculture, Hartpury, where it now stands—a unique historical monument (Fig. 199). This quite elaborately carved stone shelter has two rows of 14 niches for skeps, and a further 5 or 10 skeps could have

FIG. 199 *Elaborate shelter for 28–38 skeps built at Nailsworth, Gloucestershire (242), with stone from Caen.* PHOTO: GLOUCESTERSHIRE COLLEGE OF AGRICULTURE, HARTPURY *(where the shelter can now be seen.)*

FIG. 200 *Wooden shelter with 8 skeps at Larling in Norfolk (828).*
PHOTO: I. J. UNDERWOOD.

FIG. 201 *Enclosed wooden shelter with brick ends built in 1861–62 at Fairfield, Lorton, Cumbria (73); compare Fig. 195.*

stood in the recesses below, possibly up to 38 in all. Mr E. G. Burtt has provided the following information. The shelter is made of stone from Caen in Normandy, where an Abbaye aux Dames had been endowed by Queen Matilda, wife of William I. The Lady Abbess owned the Manor of Minchinhampton (of which Nailsworth was part), and the Steward of the Manor took the rents to the Abbey in Caen each autumn. This link ceased with the Reformation, so the bee shelter was probably built before about 1500. In Chapter 7 we reported bee boles at Valognes in Normandy (F6).

We found *a pair* of shelters at Gordonstoun School in Scotland (113). They are sited symmetrically with respect to the house, free-standing, and well built, each with two single-stone shelves, and with recesses and sockets for shutters. The shelves were over 8 ft long and $2\frac{1}{2}$ ft deep, so together these shelters might have held up to 20 skeps, with plenty of room for winter packing. It seemed to be common knowledge locally that these buildings were for bees.

A shelter built about 1750 at Auchterhouse, Tayside (280b), still has door hinges at the ends (although the doors are gone), and also a sturdy iron locking bar across the front. The roof of this shelter, as of many others, is a single stone slab, and its shelf another slab. It backs on to a garden wall, which also has 6 bee boles (280a). A 'shelter for bee butts' near

Bee shelters and houses: Britain, Ireland and continental Europe

FIG. 202 *Wooden shelter packed for winter, Muir of Ord, Highland Region (836).*
PHOTO: M. LOGAN 1954.

Dunnabridge Pound in Devon (230) had an enclosing door hinged at the bottom.

Some shelters were found with doors or shutters intact. A free-standing shelter was built in 1861–62 in Cumbria, at Fairfield, Lorton (73, Fig. 201). with brick ends and a roof of local green slate. It had separate compartments for 5 skeps, which were almost completely enclosed by hinged doors at both back and front. In 1953 there were still colony records written in pencil inside the doors, although only one was still legible: 'June 24th 1866. Swarm left.' Nearby, at Oakhill, Lorton (74), was an almost identical shelter built between 1855 and 1861 by the same beekeeper.

Wood was used in making some of the bee shelters that survive, and is likely to have been used for many more that have disappeared. Fig. 200 shows a free-standing shelter in Norfolk (828), built entirely of wood. Each row of skeps is protected by a half-door hinged at the top. A somewhat similar wooden shelter in Somerset (829) had doors at the back to give access to the skeps. Fig. 202 shows a smaller wooden shelter in Scotland, with 3 skeps packed with dry bracken and sacking for winter (836).

In Herrod-Hempsall's book,[145] figs 389 and 390 are photographs taken in the 1930s of an elegant little shelter '300 years old', at Holmbury, Surrey (832). It has a thatched roof and a carved wooden front that makes 3 places for skeps on each of its 3 shelves.

There is evidence of recent use of some bee shelters in our records. Shelters in use since 1900 include: West Ings, Muker, North Yorkshire (182a), about 1900; near Ulverston, Cumbria (608), 1906; at Dunnabridge Pound, Devon (230), said in 1953 to have been used 'within living memory'; Chatley House, Great Leighs, Essex (18), late 1920s. Skeps were kept in a shelter at Warton, Lancashire (89) regularly until 1933 and occasionally until 1946.

181

Bee shelters and houses: Britain, Ireland and continental Europe

We have classified as bee shelters many structures that were known locally as bee houses, for instance in Cumbria (608, 646) and Essex (18). In Cumbria (632a) the term 'bee butt' was also used, and in Devon (230) 'shelter for bee butts'. 'Bee holes' was used in North Yorkshire (182a), and 'hive shelter' in Gloucestershire (242).

Various structures for housing hives

Beekeepers are often inventive, and sometimes devise unusual places for keeping their bees. In 1788 Richard Hoy[150a] advertised from his Honey-Warehouse at 173 Piccadilly '. . .Bee-hives contrived so as Ladies may have them on their dressing Tables without the least danger of being stung'.

In at least two buildings in England, arrangements were made for keeping bees under a window-seat. Benthall church, near Ironbridge in Shropshire, has a special housing for two hives of bees (826). In 1893, when a new door was built, two churchwardens who were beekeepers contrived a window-seat in the gallery just above the old door, under which the hives were placed, and lead pipes led from each hive to the two sides of the open mouth of a stone lion's head, which can still be seen on the outside wall. Sir Paul Benthall remembers one of the hives still *in situ* around 1920; it was an earthenware jar, possibly made in the pottery nearby.

The Nag's Head Inn at Avening, Gloucestershire (273), has a rather similar arrangement for four hives to which access was gained through a window-seat in a bedroom. The flight entrances are about 10 ft up, and form an integral part of the carved stonework above the ground-floor window.

I was shown an oddity in the garden of 69 Belmellie Street, Turriff, Grampian (369): a hollow red sandstone pillar, perhaps 2 ft across and 6 ft high. This massive structure was built to house a skep, and has an opening at the side of the appropriate size and shape.

Another unusual find was a stone building at the side of the Cottage, Cowbridge, South Glamorgan (324), fitted out to house a Nutt collateral hive, and with accommodation for two skeps below (Fig. 204). Nutt published a description of his hive in 1832 (see Chapter 9), and it was astonishing to find one still in first-class condition and in use in 1959.

Fig. 203 shows one of two skeps at Beehive Cottage, Bladon, Oxfordshire (36), on a stone base resting on wooden supports jutting out from the wall, level with the first floor windows. A beekeeper lived in the cottage about 1880, so perhaps he erected them. They were presumably reached by a ladder.

Stone or slate stands were sometimes discovered during searches for bee boles, shelters and houses, and in adjoining gardens and fields. They were

FIG. 203 *Skep on a stone stand resting on wooden supports on a cottage wall, and covered with an inverted cream pan, at Bladon, Oxfordshire (36) (see Appendix 1 under 'Other structures').*
PHOTO: J. D. ROBINSON 1952.

FIG. 204 *Nutt collateral hive in its own 'house', 1959. The Cottage, Cowbridge, South Glamorgan (324), (see Appendix 1 under 'Other structures').*

usually round, sometimes with a projection like a short pan handle, that served as an alighting board. Skep stands in general lie outside the scope of this chapter, and only one large group of stone stands is mentioned.

For several years my husband and I took our bees to the heather at Old Fold Farm, Carlton, on the North Yorkshire moors near Helmsley. On the first occasion, about 1950, the farmer directed us to a row of shaped stones set in the shelter of a wall near the house. Our modern hives sat on them well enough, and subsequent explorations brought to light other rows in the farm enclosures, and two long rows in the moor above. We counted about 150 skep stones altogether, and were told that many others had been incorporated into walls. In 1952 a beekeeper in Helmsley remembered his father taking his skeps in carts to the same farm. These skeps were moved on their stands (presumably of wood), and not—as in Germany and Holland—inverted and covered with a square of sacking pinned at each corner into the thickness of the straw. By 1976 the area round the farm had been planted by the Forestry Commission, and none of the stones could then be found *in situ*.

FIG. 205 *Woodcut by Johann Grüninger in Sebastian Brant's* Georgica, *Strassburg, 1502, showing skeps in two simple bee shelters.*[189]

The National Trust has a property, Erddig Hall near Wrexham, for which garden plans still exist from 1740, drawn by Badeslade. In 1976 when the grounds were being replanted John Sales, the Trust's Garden Adviser, drew my attention to these plans because of certain recesses shown in the hedges. The plans include two parallel broad hedges with semi-circular bays scooped out, 3 backing on to 3 in one hedge and 3 backing on to 2 in the other; the bays face north and south and are about 15 ft across. All 11 bays were contained in a compact area, like an apiary, and they may have been designed for hives: a single longer avenue would have been more effective for ornamental urns or statues. We may well never know what their purpose was. The hedges are now replanted with yew.

One type of structure about which we should like more direct evidence is the Irish bee tower. These towers, also known as honey towers or honey pots, must have been rather like pigeon lofts. They are said to have been introduced into Ireland by Anglo-Normans in the twelfth century, but I do not know of similar structures elsewhere, and evidence about the Irish ones is slight. Descriptions of three are available.[148,309] Around 1230 Nicholas de Verdon erected one north of Dublin at Clonmore, Co. Louth, which no longer exists. It was 10 ft square and not more than 50 ft high with flight entrances for the bees, one above the other, on the east, west and south sides. Inside, the several floors were reached by a central ladder, and there were racks to take the honey combs. The inside of each loft was lined with straw matted together to keep the hives warm. Only the ruins of de Verdon's Castle remain today, and the bee tower has gone.

The Cistercian Monastery at Mellifont, also in Co. Louth, had a bee

FIG. 206 *German bee shelter in a woodcut of 1583.*[9]

tower 40 ft high; no more seems to be known except that Lord Drogheda had it demolished in the early eighteenth century to provide building stone for his mansion.

Farther north, just outside Belfast, Sir John Rawdon built a bee tower in 1746 at Moira Castle, Co. Down. This was 30 ft high, with a spiral staircase to the different floors, each of which had a window. The bee lofts 'were lined with hay covered with brown linen, and there were comb racks made of wood on which the honey combs were built. . . . In winter the tower was heated by a charcoal stove at the bottom of the spiral stairs'.

Baylis's bee house (Fig. 187), with 6 tiers of shelves, is only 10 ft high below the roof.

STRUCTURES FOR HOUSING HIVES OUTSIDE BRITAIN AND IRELAND

Bee shelters

In the section on bee shelters that survive in Britain and Ireland, it was stated that such shelters were widely used in Europe in past centuries. They were probably used at least as early as A.D. 1000,[115] although the examples shown in Fig. 205 and Fig. 206 are from the sixteenth century. Some bee shelters have had very long histories. In 1951 I visited Sebastian Eckstein in Fischbach near Nürnberg in Germany, whose family had kept bees on the same site for 400 years, and the large bee shelter I saw was the successor of earlier less elaborate shelters erected from 1550 onwards. In Chapter 10, Figs 253, 254, 258 and 268 show bee shelters.

Skeps kept in the shelters were not normally handled except to turn

185

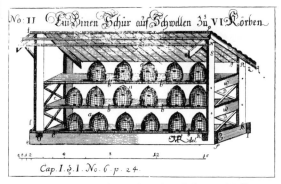 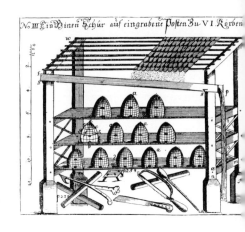

German design for a bee shelter, 1719:[52]
LEFT: FIG. 207, *viewed from the front;*
RIGHT: FIG. 208, *from the back.*

them upside down at appropriate times to check that the bees were in good order, to lift them to judge the weight—and therefore the amount of honey—in the autumn and, of course, to remove the honey.

A book published in Berlin in 1719[52] gives clear details of a well made shelter (Figs 207, 208) 15 ft long, 10 ft high at the front and somewhat less at the back. It has three shelves each holding six skeps, 19 in. high and 16 in. internal diameter at the base; the shelves occupy the whole depth of the bee shelter from front to back, which is only 2 ft.

In Carniola, a region of Slovenia in Yugoslavia, the traditional hives were horizontal ones made of wooden boards,[22] probably a survival of Roman horizontal hives discussed in Chapter 4. By the eighteenth century, if not before, these hives were customarily kept in shelters (Fig. 209) which had a roof sloping down to the back and extending above an overhang at the

FIG. 209 *Carniolan bee shelter near Kranjska, 1953.*

front that shelters the hives. The hives were placed tightly together, their front ends (containing the flight entrance) thus forming the front wall of the shelter. Ferula hives in Italy have sometimes been kept in such shelters.[7] Fig. 36 shows some of the wooden hive fronts.

German bee houses

We have seen that in England bee houses were a later development than bee shelters. The same is true in continental Europe, and a book published in Vienna in 1775[214] probably provides the key to the transition from the bee shelter to the bee house, which started in the Austrian Empire, and reached other countries—including Britain—in due course.

Anton Janscha was born at Breznica in Carinthia (now in Carniola, Yugoslavia) in 1734. In 1769 he was appointed by the Empress Maria Theresa as Imperial and Royal Beekeeper, to lecture and advise on

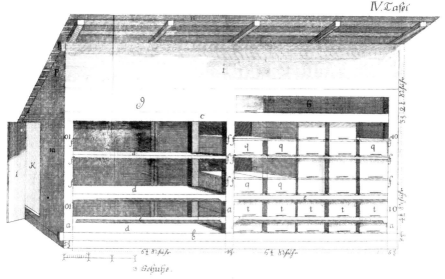

FIG. 210 *Anton Janscha's design for a bee house, published by J. Münzberg (1775).*[214]

beekeeping in Vienna. In 1771 he published a book in German on the swarming of bees (a shortened version is available in English, with an account of his life[114]) but he died from typhus in 1773, before he could publish any comprehensive beekeeping book. He taught many of the facts that were rediscovered by the blind Swiss scientist François Huber several decades later. A book setting out Janscha's teaching was published in German in 1775 by his successor and devoted disciple Joseph Münzberg.[214] As a result of these two books, Janscha's ideas and practices exerted a great influence over the German-speaking world. In the second book we find the prototype of our bee house (Fig. 210), which allows the

187

beekeeper to enter and to work his hives from inside. The following is a free rendering from its description on p. 27:[214]

It is good if the back of the bee house is against a wall, fence, trees or similar protection. Its sloping roof should extend sufficiently to prevent rain reaching the hives, but not to shade them too much . . . [In the drawing the roof slopes up right to the front edge.] The various inspections and manipulations of the hives require a building that is convenient and not too small. If it is planned to leave the hives in the bee house over winter, the front of it must be fitted with doors [to enclose the hives]. If there is danger from thieves, the hives should be protected in such a way that they cannot be removed.

The length of the bee house can be according to choice. In my drawing it is 13 ft, and there are 10 hives in each row. It is $4\frac{1}{2}$ ft high at the back and a little over 7 ft at the front. There are three rows of hives each 16 inches high, including the upper box. Above is a shelf for one-storey hives, or for ventilation when the hut is closed, by a door (g) hinged horizontally at the top. It is not convenient to work with hives higher than this.

The bee house is $6\frac{1}{2}$ ft from front to back, $2\frac{1}{2}$ ft for the hives plus 3 ft or more to provide a working space. At the left-hand end m is the door l (entrance k) $2\frac{1}{2}$ ft wide. [The door cannot be more than $4\frac{1}{2}$ ft high.] The roof extends $1\frac{1}{2}$ ft beyond the front wall. *01* on the left and *10* on the right show the position of hinges for winter doors that close across the three lowest rows of hives, and similarly (g) for the upper rows (except that these doors drop down, or are kept open by a wedge when ventilation is needed). The remaining space above can be used for storing equipment.

The hives shown in Janscha's bee house (Fig. 210) were a development of the Carniolan hives discussed in Chapter 4. Janscha used several of these boxes superimposed as the boxes of a hive are today, and probably because of the consequent need to handle separate hive boxes with bees in them, he made a bee house with room for the beekeeper behind the hives. The access door (on the left in Fig. 210) was very low, and the passage it led to was only 3 ft wide, but it sufficed. It is interesting to compare Fig. 207 and Fig. 210 with Fig. 211 published in Germany by J. L. Christ in 1802.[59] Many details of the latter bee house seem to be copied from the 1719 one[52] (Fig. 208)—even the water-chute and specimen of roof; the hives are still made of superimposed wooden boxes, though of a different style from Janscha's. There is also space behind the hives for the beekeeper, although it is even narrower and more awkward of access than Janscha's.

It is noteworthy that all the bee houses found in Britain that have a working space behind the hives post-date Janscha's description,[214] but bee houses of this type have never been common anywhere except the German-speaking areas of Europe. There they blossomed into elaborate buildings, often erected near to the human dwelling house and sometimes looking nearly as large. Some of the existing bee houses are most attractive, with

Bee shelters and houses: Britain, Ireland and continental Europe

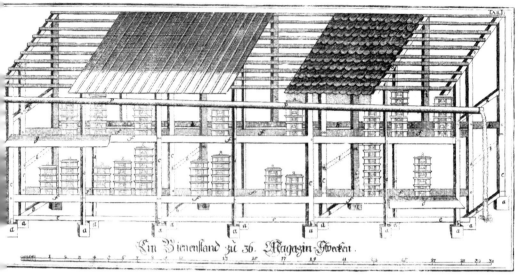

FIG. 211 *Design for a bee house published by J. L. Christ (1802).*[59]

bright patches of colour on the side the bees fly from, which help returning bees to locate the entrance to their own hive. Inside, they are likely to include a workbench, honey extractor, and beekeeping equipment. If they are away from the beekeeper's house, there will probably be facilities for refreshments, and sometimes beds, so that they can be used at weekends.

Fig. 212 shows the geographical distribution of bee houses as far as I have been able to ascertain it from my own travels and from enquiries in the countries concerned. Bee houses are common in areas in which the German language has been well known, although it may not now be the official language. Examples can be found in present-day Denmark, Poland, Czechoslovakia, Hungary and Yugoslavia, as well as in Germany and Austria. They are in Switzerland too, but only in the German-speaking part; in the French-speaking areas separate hives in the open are usual

Another thread was woven into the story of 'German' bee houses in the nineteenth century. In Chapter 6 I showed how some of the early hives in the north-European forests were logs with a door cut in the side, simulating the access door to a nest of bees up in a tree. When, later, upright hives were made of wooden boards, they too opened at the back. Movable-frame hives introduced after the mid-nineteenth century in the German-speaking forest areas followed the same style: the frames were inserted from the back, not from the top as everywhere else.[259] These back-opening frame hives could be operated conveniently in a bee house but not in the open, and each helped to perpetuate the other. Again, books published in German by beekeeping leaders, for instance Dzierzon

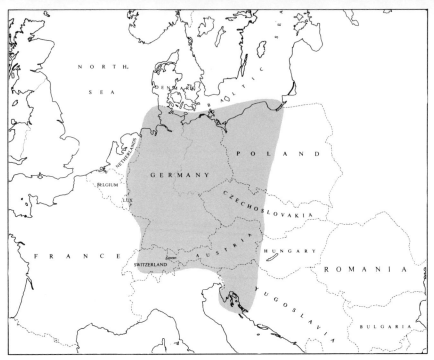

FIG. 212 *Central area of Europe (shaded) in which German-type bee houses are (or were) the most common way of keeping hives, including modern movable-frame hives.*

(1848),[102] Baron von Berlepsch (1860)[35] and Dathe (1870)[87] spread and perpetuated this style of beekeeping.

Photographs and diagrams in Gerstung's book *Der Bien und seine Zucht* (1926)[126] show the whole spectrum of German bee houses: with skeps on the outside wall; shelters that have space only for the hives, and are completely enclosed except for flight holes; and bee houses accommodating a large number of wooden hives, arranged in one or two tiers and worked from the back, with access space behind (Figs 213, 214). Zander's *Die Zucht der Biene* (1923)[321] gives other illustrations.

Since the 1960s the characteristic German bee houses have been going out of fashion, largely because of the increasing cost of timber. German beekeepers are at last putting their hives out into the open, each standing separately, as beekeepers in most other countries have done for the past hundred years. So existing bee houses will in due course become part of beekeeping archaeology, as older ones are now. But they are still in common use (Fig. 215).

Various structures for housing hives

Examples are given here of structures that have been reported in various countries, starting with hive recesses in Europe other than bee boles, which are the subject of Chapter 7.

There is a priory chapel at Arènes near Saintes, north of Bordeaux in

Bee shelters and houses: Britain, Ireland and continental Europe

France, built in the twelfth century. On the interior of the south side, and well spaced out, are 11 rectangular recesses set in threes one above the other, the highest being 7 m above the floor. They are 22 to 30 cm high, 16 to 20 cm wide and 72 to 75 cm deep, taking up nearly the whole thickness of the wall; the closure is pierced by a group of three flight holes 12 to 15 mm in diameter. It is believed that the recesses were made when the abbey was built.[48] Presumably a skep was set in each recess, but it is unusual to find recesses so spaced out and so high, and in a church. There are records from 1483 and 1582 of wall recesses in Cyprus that housed colonies of bees; 'they have little holes, to go in and out, and the wax and honey are thus inside the houses'.[60]

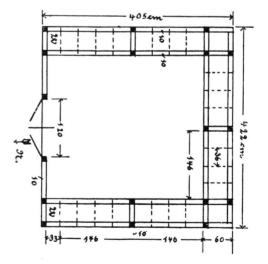

TOP RIGHT: FIG. 213 *A German bee house for 48 modern hives and 6 skeps, from F. Gerstung's* Der Bien und seine Zucht *(1926 ed).*[126] RIGHT: FIG. 214 *Ground plan showing the dimensions in cm.*[126] ABOVE: FIG. 215 *Modern German bee house, on the same pattern, with coloured decorations on the hive fronts to help bees find their own hive entrance.* PHOTO: E. WOHLGEMUTH 1951.

RIGHT: FIG. 216 *A complex for 2800 to 3000 hives used by Berber beekeepers at Tinzerki, Morocco.* PHOTO: P. HACCOUR 1961.

BELOW: FIG. 217 *Wall of volcanic tuff near Brindisi, Italy, with hives cut in it.*

At Szomolya, Komitat Borsod, Hungary, a great many recesses were cut out of a cliff face of the volcanic rock known as tuff, probably between the eleventh and fourteenth centuries. They are 60 to 70 cm high, 25 to 40 cm wide (slightly narrower at the top) and 20 to 40 cm deep (24 to 28, 10 to 16, 8 to 16 in. respectively). The edges are set in, so that a board can be fitted over the recess. The recesses have been variously described as Celtic remains, and as niches for urns or other cult objects. But a report on them by Andor Saád, a Hungarian archaeologist,[20,294] follows local folklore in referring to them as stone hives, and beekeeping was certainly developed in Hungary by the time of their construction. Farther south, on the island of Brać off the Dalmatian coast of Yugoslavia, hives constructed of stone slabs actually in the ground (Fig. 131) were described in 1776 and were still in use until the 1940s.[251] It is likely that bees were also housed directly into the tuff recesses, as in the hives of Brać; this was done in Brindisi (Fig. 217). There are other examples in the Balkan region of beekeeping in the earth or rock itself,[10] for instance in Albania,[257] and searches might well bring to light still more. Our

knowledge of them is not systematic enough to warrant a further discussion here.

An unusual cluster of nesting places for honeybees was reported in 1921 from Macedonia in Greece.[216] Near Lake Katlanovo, the local people had made many cavities about the size of a melon in the face of a cliff of loess. They coated the cavities inside with cow dung, and over each entrance fitted a wooden door provided with a flight hole. Swarms would settle in these cavities to nest there, and the beekeepers could remove them to stock their own hives.

Bee shelters similar to those used for skeps have also been used for horizontal hives, and Fig. 216 shows what is probably the largest in the world.[136] It is really a complex of bee shelters, constructed about 1850 at Tinzerki at the westward end of the Atlas mountains in Morocco, in an area rich in bee forage. There is accommodation for 2800 to 3000 hives, brought there by Berber beekeepers each year for the honey flow. The shelters are still operative, although they now contain only a few hundred

hives. The hives are horizontal cylinders of woven reeds coated with cow dung, such as are common in North Africa. In Malta the entrance to a cave may be more or less 'filled with standstone, cut into a series of openings, behind which the ceramic jars [used as hives] would be laid. Then an iron door would be inserted and padlocked'.[283a]

An interesting apiary in Central Anatolia, Turkey, has been reported to me by Dr D. P. Erdbrink. He came across a beekeeper at Yarǐ near Pazarören whose garden, containing many fruit trees, consisted of a number of compartments separated by brick walls. Some 30 to 40 large unglazed earthenware jugs, like amphorae, had been incorporated into these walls, lying horizontally, at places where the trees shaded them. All the jugs were occupied by bees. The beekeeper had knocked the bottom out of each, and replaced it with a wooden plug so that he could reach into the jug from behind. The bees flew from the mouth of the amphora, always facing the garden. Dr Erdbrink comments that a number of agricultural practices from Ancient Rome linger on in the area, and these hives are probably one such survival.

One part of the world that is likely to be well worth exploring for methods of sheltering hives is in the Himalayas and associated mountain ranges. The bees here are strains of *Apis cerana* that can withstand the harsh climate, but about which rather little is known. Winters are cold but summers may give fairly good honey yields. Many of the valleys are isolated, and traditional ways survive. In Kashmir I have seen many houses with internal shelves in the thickness of the mud walls, containing rows of horizontal hives of mud, clay, wicker or wood (Chapter 4), a flight

FAR LEFT: FIG. 218 *Beekeeper showing his hives built into the house wall, and pointing to one of two flight entrances. Agru, Nuristan, Afghanistan.* PHOTO: R. VERHAGEN 1970.

LEFT: FIG. 219 *A beam newly hollowed out for a colony of bees, ready to be incorporated into the structure of the house.* PHOTO: R. VERHAGEN 1970.

BELOW: FIG. 220 *Hive built into a wall at the Saharan oasis of Erfoud, Morocco, 1963; after taking the honey the beekeeper plasters the wall up again with mud.*

entrance being provided for each to the outside. The wall warmed by the cooking hearth was always well stocked with hives. (In England bee boles were occasionally found built into the wall of a chimney, e.g. 49, 175, 245.) Honey is taken from the end of the hive inside the house, the shutters being thrown open to let flying bees escape; there are no glazed windows. I also saw water pots embedded in walls, on a similar principle to those mentioned in Anatolia, except that alternate hives faced in opposite directions, perhaps to reduce drifting.

In Ladakh, a higher and more remote valley, a row of wooden doors was found to be fitted between the upright beams of a house wall, with a flight entrance for bees in the middle of each at the lower edge. I have not been able to discover if the bees were in hives inside the doors, or hived directly into the cavity.

Figs 218 and 219 show how bees are kept in one valley in the high mountains farther north in the Hindu Kush. The houses are constructed of wooden beams interspersed with courses of stone, and one of the beams is hollowed out to make a hive for the bees (*Apis cerana*). As in Kashmir, access to them is obtained from inside the house; here, a board closing the rectangular opening shown in Fig. 219 is removed to take the honey. In at least one oasis in the Sahara, bees are kept directly in cavities in the mud walls, where they are insulated against the heat (Fig. 220).

FIG. 221 *Sir George Wheler's drawing of a Greek top-bar hive.*[314]

9. Towards movable-frame beekeeping

This chapter is concerned with material remains from past centuries —and their equivalents still in use—that help to show us the sequence of events that led up to the watershed between fixed-comb and movable-frame beekeeping in 1851. Speculation will be necessary about dates of certain developments, but we will start with what is *known*, although this takes us back only to 1682.

The hives described in preceding chapters are fixed-comb hives: the bees built their combs downwards from the top of the interior surface of the hive, and generally attached them to the sides as well. The beekeeper could remove a comb only by cutting it out. This was not so difficult with horizontal hives as with upright hives where the only access was at the bottom, well away from where the comb was attached.

THE EARLIEST KNOWN EXISTENCE OF MOVABLE-COMB HIVES

The use of movable-comb hives was first described in a book published in London in 1682 by Sir George Wheler: *A journey into Greece.*[314] The author was very impressed with the hives he had seen at St Cyriacus's monastery on Mount Hymettus in Attica, and he published a drawing to show their special features (Fig. 221).

> The Hives they keep their Bees in, are made of Willows, or Osiers, fashioned like our common Dust-Baskets, wide at the Top, and narrow at the Bottom,

and plaister'd with Clay, or Loam, within and without. They are set the wide end upwards, as you see here. The Tops being covered with broad flat Sticks, (as at C.C.C.) are also plaistered with Clay on the Top; and to secure them from the Weather, they cover them with a Tuft of Straw, as we do. Along each of those Sticks, the Bees fasten their Combs; so that a Combe may be taken out whole, without the least bruising, and with the greatest ease imaginable. To increase them in Spring-time, that is, in *March* or *April*, until the beginning of *May*, they divide them; first separating the Sticks, on which the Combs and Bees, are fastened, from one another with a Knife: so taking out the first Combs and Bees together, on each side, they put them into another Basket, in the same Order as they were taken out, until they have equally divided them. After this, when they are both again accommodated with Sticks and Plaister, they set the new Basket in the Place of the old one, and the old one in some new Place. And all this they do in the middle of the day, at such a time as the greatest part of the Bees are abroad; who, at their coming home, without much difficulty, by this means divide themselves equally. This Device hinders them from swarming, and flying away. In *August* they take out their Honey; which they do in the day-time also, while they are abroad, the Bees being thereby, they say, disturbed least. At which time they take out the Combs laden with Honey, as before; that is, beginning at each out-side, and so taking away, until they have left only such a quantity of Combs in the middle, as they judge will be sufficient to maintain the Bees in Winter; sweeping those Bees, that are on the Combs they take out, into the Basket again, and again covering it with new Sticks and Plaister.

The stone hives on the Dalmatian island of Brač (see Fig. 131) also had top-bars, since the roof-slab could be 'removed at will'.[251] Twigs of olive or mulberry wood were pushed in between the side plates and the roof, and the bees built their combs downwards from these.[128] However, it was the Greek hive with the sloping sides that made the greatest contribution to hive development, and experience in using it shows why this was so. The essential features of a top-bar movable-comb hive are: (a) that if the bees build their combs down from each bar, they are separated by the bees' own natural spacing, and (b) that the sides of the hive slope in from the top as bees' natural combs do (Fig. 1).

EVIDENCE FROM HIVES STILL IN USE

Wicker hives of the type described by Wheler are still used in parts of Attica and elsewhere, though they are fast disappearing. In Crete,[324] top-bar hives of clay, like giant flower pots, are used similarly (Fig. 222), and in the west of the island the wicker hives are used as well; they are 'plaister'd' (cloomed), but if the hives are moved from one place to another the clooming falls off (Fig. 223).

In 1958 one of the cloomed wicker hives from Attica was sent to us in

ABOVE: FIG. 222 *Pottery top-bar hive from Crete, with an earthenware cover above an insulating layer of oak leaves (1979). Height 37 cm, diameters 36 cm (top), 26 cm (base), internally.*

RIGHT: FIG. 223 *An apiary of about 100 wicker top-bar hives taken to Kambani (Akrotiri) in northern Crete for the thyme flow. Each rests on a stone slab to which the wicker is plastered.* PHOTO: P. PAPADOPOULO 1939.

LEFT: FIG. 224 *Comb with bees and brood, removed from a top-bar hive; the gap left by the top bar can be seen just below it.* PHOTO: P. PAPADOPOULO 1965 IN ZIMBABWE.

BELOW: FIG. 225 *Greek top-bar wicker hive used experimentally in England in 1958. To the right of the hive, one of the combs removed from it stands upside down, on its top-bar. Height 53 cm, diameter 43 cm (top), 28 cm (base).* IBRA COLLECTION B58/1.

FIG. 226 *Top-bar wicker hives with hackles, at Phyli, Attica, Greece, 1979.*

Towards movable-frame beekeeping

England, and I kept bees in it for several years (Fig. 225). As Wheler said, the combs were removable (Fig. 224): one could divide a colony into two parts in spring, and the part without a queen would rear its own; in autumn, combs full of honey could be lifted out and harvested, the adhering bees being swept off. Any comb could be removed for inspection at any time and replaced.

In 1979 I visited an apiary of several hundred of the hives only a few miles from Athens (Fig. 226), although not more than a hundred still contained bees. Each hive was covered with a 'Tuft of Straw' (a hackle) as in Wheler's day.

WERE MOVABLE-COMB HIVES USED IN THE ANCIENT WORLD?

There is no evidence to suggest that such top-bar hives existed in Ancient Egypt. Although it is sometimes assumed that they were used in Ancient Crete and Greece, we have no direct evidence before 1682. The most interesting questions are how and when they originated—and, in particular, whether such hives did exist in the ancient world—and why, with their (to us) manifest advantage over fixed-comb hives, their use apparently did not spread beyond a small part of the Mediterranean region.

The following argument may be presented in favour of their existence in classical times. Top-bar hives of wicker are used today in Crete, except that pottery is used in the mountains where willows do not grow. The wicker hives are also used in Attica in Greece (Fig. 226). It is likely that pottery was used for the hives first and the change to the lighter and more convenient wicker construction made later. It is also likely that top-bar hives were used in Crete earlier than in Greece,[259,260] and it is easier to envisage the transmission of the use of these hives from Crete to Greece in classical times than in subsequent centuries.

Arguments against the existence of the top-bar hives in Ancient Greece include the following:[131]

(a) They are not mentioned in Roman writings (Chapter 4)—but we know that other advances were made and lost again.
(b) Roman authors referred to difficulties in feeding bees and in cleaning hives, which would not exist with such hives.
(c) Aristotle made clear his desire for more observed information about bees than he could get, yet on a comb removed from a top-bar hive he could have seen the queen laying eggs and many other activities in the hive. On the other hand, the correctness of many of Aristotle's observations[293] has been considered by the Greek beekeeper N. J. Nicolaidis[219] as an indication that he did have top-bar hives.

FIG. 227 *Drawing of Orestada vessel in the museum at Isthmia.*

It is much more likely that a pottery hive would survive from antiquity than a wicker one, and a number of horizontal pottery hives from Ancient Greece are now known. The discovery and dating of a top-bar hive from antiquity would be most significant.

In 1961, Professor C. Kardara published an account of the excavation of a dyeing and weaving works at Rachi, Isthmia, near Corinth. Her report[166] included a photograph of a vessel with the name Orestada incised on it (Fig. 227), which Dr O. Broneer had earlier[50] described as probably a wine press. She said that Mr Pallas had identified the vessel as a hive; it had certain similarities with the hives found in the Justinian fortress in Corinth (Chapter 4). I was able to go to the Isthmia Museum in 1979, and inspect the vessel. It had what appeared to be a flight entrance, and looked as though it could be an earlier, smaller, version of the top-bar pottery hive now used in Crete.

During and since my visit Miss Penelope Papadopoulo in Athens has rendered a very great service by discussing our problems with the archaeologists concerned, and by pursuing questions that arose about this and other possible hives. Close examination by Professor Kardara and Miss Papadopoulo showed that the entrance slit, which looked similar to that in a Cretan pottery top-bar hive, was in fact made during the reconstruction of the excavated vessel: there was no evidence from the fragments to show what shape the hole was, or indeed if there was a hole at all. Moreover the vessel is very small for a hive: its (outcurved) mouth diameter is 34 cm and its height 28.5 cm. Its volume would not be more than 17 litres, less than half that of the traditional top-bar hives in use in Crete today.

In 1980 Professor Kardara noticed two more vessels that resembled the Orestada vessel. These had been found at the Sanctuary of Poseidon close

Towards movable-frame beekeeping

to Isthmia, and have since been catalogued (IP 2512, IP 2215). The former has a round hole 2 cm in diameter near the base, and the latter a slit 4.5 × 0.8 cm just beyond the bottom edge, in the base itself. These Greek vessels are even more tantalising than the Iron Age and Roman English pots (Chapter 6) that have groups of holes, as if for bees, but are too small to be hives. Other comments must be postponed until further studies have been made. I could not obtain permission to publish photographs of these vessels.

There still remains the question of how the top-bar hives were derived from the earlier long horizontal, end-opening hives of the classical world (Fig. 26)—the top-bar hives never completely *replaced* them, and even today in Crete and in Attica, long horizontal hives are used (Figs 29, 24). Dr Ifantidis in Thessolonika[153] has suggested that the movable-comb hive might have been preceded by a 'movable-nest' hive. Suppose that a swarm had taken possession of an Ancient Greek thimble-shaped hive (as Fig. 26) or a sloping cylinder hive in Crete (Fig. 29), standing empty with its wide open mouth uppermost and covered with a lid. The beekeeper could well have discovered that, when he lifted the lid, the combs came out of the hive with it, being attached to its underside. It seems most unlikely that, at this period, the top-bar hive was devised as a result of a logical argument, as the movable-frame hive was in the nineteenth century. But it could well have existed, and been used, without the users understanding or fully exploiting its merits.

CABINET-MAKER'S HIVES

We now turn to a quite different approach, which concentrated on providing separate honey chambers in a hive, that could be removed with the combs of honey firmly fixed inside them. With the development of cabinet-making in Europe in the seventeenth century, it was possible to improve greatly on a skep enlarged by a cap or bell jar. In his *History of beekeeping in Britain*[116] and elsewhere,[113] Dr H. M. Fraser has described a number of these cabinet-maker's hives, of which there were two main types: the tiered hive consisting of a pile of similar boxes superimposed, and the collateral hive with several boxes arranged side by side.

Tiered hives

The first of these hives was designed by the Rev. William Mew in Gloucestershire in 1649, and Christopher Wren made a drawing of it (Fig. 228), which was published in 1655 in Samuel Hartlib's book *The reformed commonwealth of bees*.[143] This hive was used by a number of educated and influential people, including Samuel Pepys, John Evelyn (who described it in his diary) and Robert Hooke. It had a number of successors,[116]

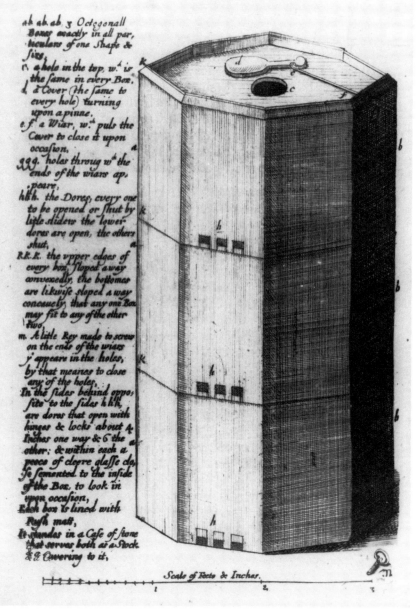

FIG. 228 *Mew's octagonal tiered hive, drawn by Christopher Wren and published by Samuel Hartlib in 1655.*[143] *The 'hole in the top (c) is the same in every box', as is the cover (d), 'turning upon a pinne' to open or close this hole.*

FIG. 229 *Stewarton hives in use, with William Walker, probably the son of an experienced beekeeper of the same name who helped Ker, and who died in 1840.* PHOTO: WEST OF SCOTLAND COLLEGE OF AGRICULTURE.

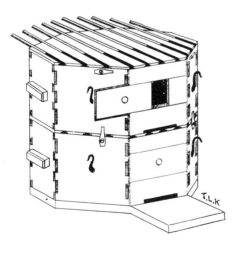

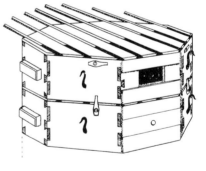

FIG. 230 *Drawings of a Stewarton hive, showing (left) two deeper boxes that together form the brood nest, and (above) two shallow honey supers. Each box has two inspection windows, which can also be seen in Fig. 229.* REPRODUCED FROM J. STRUTHERS.[285]

Towards movable-frame beekeeping

following broadly the same principle as a skep with several caps, but by precision in construction, and the use of movable slides, the bees could be given access to each box as and when the beekeeper wished.

The earliest type of which examples survive (Fig. 230) is a hive devised in 1819 by Robin Ker (or Kerr) who lived in Stewarton, near Ayr (now in Strathclyde), and the hive was named after his village. Fig. 229 shows some of the hives in use in the nineteenth century.

The top-bars of each box (Fig. 230) were *fixed across them*, each with a thin strip of wood underneath at the correct distance apart for the bees, from which they built their combs. The deep brood boxes had 9 of these fixed top-bars and the shallow honey supers 7, more widely spaced (see p. 13). Supers full of honey were separated with cheese wire, and after taking them away from the hive, the bees were allowed to engorge themselves for a while, after which they flew off to their hive. Suitable management, which included adjustment of the slides, kept the queen out of the honey supers, and these boxes full of honey (20 pounds each) won many prizes at shows. Many other tiered hives were invented, in Britain and in other countries, but there is no space to deal with them here.

Collateral hives

Meanwhile, hives on the collateral system were being developed from 1750 onwards. The first were devised by the Rev. Stephen White in 1756—three 9-in. (23-cm) cubical boxes set side by side, with adjustable passage-ways between; the central box was the brood chamber and the side boxes were for honey. When the honey in one box was to be removed, access for the bees was adjusted so that they flew out of this box and returned to the central box. Fraser[116] described other collateral hives which were designed, but no examples survive. Fig. 231 shows a near-survivor: John Jones's 'Herefordshire eclectic hive' (1843),[162] which is mentioned in Chapter 8.

FIG. 231 *Two of John Jones' eclectic hives at Weobley, in gothic-style outer structures. The thatched building is a summer house.* PHOTO: A. WATKINS (HEREFORD CITY LIBRARY, NO. 1660)

FIG. 232 *A medley of hives and hive shelters in 1881, all supported well off the ground. The woman is presumably about to put some syrup in the feeding drawer of the Nutt hive, which stands unprotected in the open.* REPRODUCED FROM ISSUE 246 OF The cottager and artisan, PUBLISHED BY THE RELIGIOUS TRACT SOCIETY.

Nutt's collateral hive:
BELOW: FIG. 233 *Photograph of one of the hives in the IBRA Collection (B65/5), from Bradford City Art Gallery and Museums.*
RIGHT: FIG. 234 *Drawing from Nutt's 1832 book.*[222]

Towards movable-frame beekeeping

There is, happily, one hive of which at least four examples have come to light in England in the last 16 years (IBRA Collection B65/5, B70/142, B70/143, B74/23). It was devised by Thomas Nutt, whose 1832 book *Humanity to honey bees*[222] made it widely known, especially as translations in German, Portuguese and Hungarian were published (1834, 1835, 1836). I saw a fifth Nutt hive in the Beekeeping Institute near Budapest in 1963.

In Fig. 233 the photograph shows the Nutt hive set up for use, with the central brood box, and two side boxes and a central upper one for honey. In the drawing (Fig. 234), the hive is spread apart to show the slots that communicate with the side boxes, and a bell jar full of comb honey stands above the central box. The middle drawer below was used for feeding, and the 'block-fronts' at the sides were pivoted and 'so contrived that ten thousand Bees can with ease leave their prison and their sweets in the possession of the humane apiarian, without the possible chance of a single intruder forcing its entrance to rob the magazine or to annoy the apiarian'.[222] The Nutt hives that survive are of superb craftsmanship, and were probably designed for a gentleman's parlour. The hive shown in Fig. 204 was, however, kept under cover out of doors, and Nutt hives were also used in the open (Fig. 232). A hive devised by Mme Vicat in Switzerland[23,152] was made on a modular principle, with units holding perhaps two frames each, held together by long locking screws.

Leaf hive

This must be mentioned here among the cabinet-maker's hives, because it was the first hive in which the combs were supported (and bounded) by a rectangular frame. I do not know that any early examples survive. In 1792

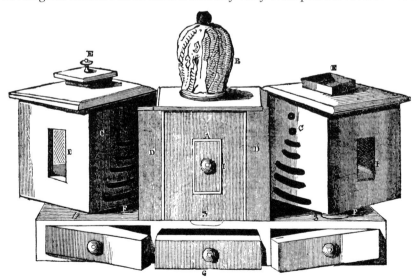

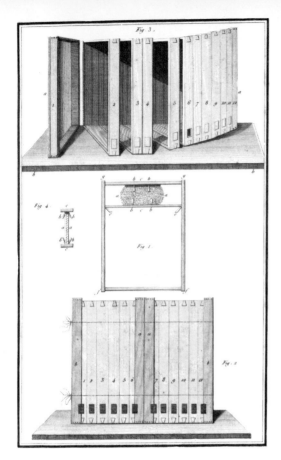

LEFT: FIG. 235 *Leaf observation hive, from* Nouvelles observations sur les abeilles *by F. Huber (1792).*[151]

BELOW: FIG. 236 *Replica of Huber's leaf hive (made by R. R. Greenwood) occupied by bees.* IBRA COLLECTION B53/115.

François Huber, the blind Swiss naturalist, published a detailed drawing (Fig. 235) of the hive,[151] which he devised for the purpose of observing the bees' activities; an assistant made the observations under Huber's direction. I have kept bees in a replica of his hive, and Fig. 236 shows it occupied. The 'leaves' are at the correct bee spacing (which Huber understood), and a small piece of comb is fixed in the top part of each frame; the bees then build their combs within the whole of the frame. By opening the 'book', the bees on the comb surface can be observed, and Huber's many discoveries about bees were made with this hive. It is difficult to close the hive again without squashing bees and it is not practicable for general beekeeping.

Early box hives with frames

Other frame hives were invented, but the bees attached the frames to the hive walls: the frames were not 'movable'. The earliest type of which I have located an example is Peter Prokopovich's (Fig. 237). Prokopovich lived in the area of forest beekeeping (Chapter 5), and his hive had a door at the back, like the upright log hives he knew (Fig. 103). It was also like a

FIG. 237 *Prokopovich's frame hive, invented about 1806, in the Central Beekeeping Research Institute, Rybnoe, Ryazan, USSR, 1962. The example here is not the earliest type; the middle compartment has frames, and the top one contains wooden honey 'sections'. In the brood space below, the bees built comb as they liked, without bars or frames.*

log hive in that it was a single unit, not a pile of tiered boxes; instead, there were three compartments, the top one having wooden frames, with notches in the end bars for the bees to pass through. The frames came close to the walls of the hive, and the bees attached the two surfaces together with comb or propolis. It was necessary to cut them out. This was also true of frame hives devised by Baron August von Berlepsch,[35] J. Dzierzon (1848)[102] and others in Germany,[36] by W. A. Munn and others in England,[39,116] and by beekeepers in other countries.

THE FINAL ACHIEVEMENT OF MOVABLE-FRAME HIVES

We have now examined the innovations in hives that, between them, contributed the advantageous features of the modern hive: movable combs, frames round the combs, tiered boxes that could be handled individually, and separate honey chambers (Table 8).

In 1790 L'Abbé Della Rocca, Vicar-General on the island of Syra (Syros) in the Cyclades, published his *Traité complet sur les abeilles*,[89] and it is clear from a passage in vol. 2 (pp. 467–9) that he understood the importance of the bee space between the combs in top-bar hives he had

Towards movable-frame beekeeping

Table 8. Significant steps towards the development of the modern movable-frame hive

Hive type	Where first developed	First known date	Representative illustration	Movable combs/ frames	With frames	Tiered boxes	Separate honey chambers
Horizontal mud cylinders	Egypt	?	Fig. 16	(yes)	—	—	—
Top-bar pottery/ wicker	Crete/Greece	?	Fig. 221	yes	—	—	—
Superimposed boxes	Europe	1649	Fig. 210	—	—	yes	yes
Collateral boxes	England/Europe	1756	Fig. 234	—	—	—	yes
Leaf hive	Switzerland	1792	Fig. 235	(yes)	yes	—	—
Early box hives with frames	Europe	e.g. 1806	Fig. 237	(yes)	yes	yes	yes
Practical movable-frame hive	USA	1851	Fig. 239	yes	yes	yes	yes

All hives except the first two types were of wood; plastics are sometimes used for modern hives. In column 5, brackets indicate that combs/frames had to be cut out (but could be replaced) or had other deficiencies

seen on 'l'île de Candie' (Crete). François Huber understood the spacing too; in 1792 he wrote of 'the equal distance uniformly preserved between the combs', and (see Dunbar, 1840) for his single-comb observation hive 4 lines ($\frac{1}{3}$ in., 8 mm) is specified between the comb and the glass; if the space is smaller, 'the bees will work against the glass'.[99]

Meanwhile in England and elsewhere, Wheler's account of the Greek top-bar hives in 1682[314] had attracted attention, but progress was slow. Robert Golding, who lived at Hunton in Kent, not far from George Wheler's native village of Charing, described an 'improved Grecian hive' in 1847 (Fig. 238) in *The shilling bee book*.[129] It was a small hive, being intended for use with a honey chamber above, and was made of coiled straw. Like Della Rocca, Golding understood the need to use the bees' spacing between combs:

> The bars should be half an inch thick, $1\frac{1}{8}$ inch wide [29 mm], and seven in number. If properly adjusted, there will be interspaces between them of about half an inch.... The general fault of those who adopted bars, was, that they made them too wide, so that the bees—of course adhering to their own rules—building regularly throughout, often attached two combs to one bar; or built across the intervening spaces.

The final advance was made by the Rev. L. L. Langstroth in Philadelphia, USA, in 1851.[160,217] He had started beekeeping with two colonies of bees in box hives, and he later acquired a Huber leaf hive. He

ABOVE: FIG. 238
Robert Golding's
Improved Grecian hive,
'9 inches high by 11⅝
inches wide at the top,
gradually tapering
down to 10½ inches
diameter, in the clear'.
The shilling bee book
(1847).[129]

RIGHT: FIG. 239 *Within a decade Langstroth's original movable-frame hive was illustrated in* A practical treatise on the hive and honey-bee *(1857)*.[182a]

built up a useful library of books on bees,[161] including Huber's *Letters*, the books by Della Rocca and Golding, and Edward Bevan's *The honey-bee*, including the second (1838) edition[38] that referred much to Golding's hive, and the 1870 edition revised by Munn.[39] This last book described Munn's own bar hive with a shallow super. Langstroth used Munn's hive but improved it by extending the top-bar down at its two ends to make a frame. If the ends of the frame were kept at a bee-space distance from the inner hive walls, the bees did not build comb across the gap, and the frame was removable. He also deepened the grooves on which the bars rested, leaving about ⅜ in. (9 mm) between the cover and the bars; this facilitated the removal of the cover board—in earlier frame hives without this gap, the bees fixed the frames to the cover. The key development, which cuts the history of beekeeping into two halves, was made in the autumn of 1851, and we have Langstroth's own words to describe it (Fig. 239):[182]

> Pondering as I had so often done before, how I could get rid of the disagreeable necessity of cutting the attachments of the combs from the walls of the hives . . . the almost self-evident idea of using the same bee space as in the shallow chambers came into my mind, and in a moment the suspended movable frames, kept at a suitable distance from each other and the case containing them, came into being. Seeing by intuition, as it were, the end from the beginning, I could scarcely refrain from shouting out my 'Eureka!' in the open streets.

Langstroth's intuition was justified: the bees did in fact 'respect' the bee

Towards movable-frame beekeeping

space left between the vertical hive walls and the frames in which the combs were built; they did not build comb across the space, and the frames were, therefore, truly movable. The earlier movable-comb hives without frames had to have sloping sides and could not be more than one storey high.

The movable-frame hive itself (Fig. 240) was being increasingly used in the United States by 1861. It was introduced into England by 1862, and from 1869 Charles Dadant described it in the French and Italian journals. It soon spread to other countries, each of which used its own variants built on the same basic principle.

After 1853[182] other inventions—such as the queen excluder, and sheets of beeswax embossed with the pattern of cells to serve as foundation for comb—soon made beekeeping more efficient and productive.[76] Today, 130 years later, the world's beekeeping industry produces 800000 tonnes of honey a year, mostly from hives similar in principle to Langstroth's.

Beekeeping is now in another critical phase. The great majority of hives are still made of wood, although this material has become increasingly expensive, as has the labour of making the hives, frames, and other components. The consequent increase in the cost of beekeeping must initiate basic reconsideration of hive design and materials. This is also necessitated by the development of beekeeping in countries of the tropics where improvements are sought to the traditional hives still used there. Intermediate hives are being explored; for instance, a long top-bar hive on the principle of the Greek hive (Fig. 225) but with bars all of the same length and therefore interchangeable (Fig. 255).[41,173] The first such hive was made by E. J. Tredwell and P. Paterson in 1965. It may well be that other features of early hives described in this book will play a new part in modern beekeeping,[73,77] and we should keep our minds open to such innovations.

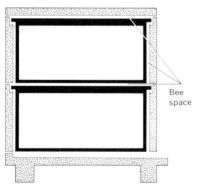

FIG. 240 *Section through a movable-frame hive showing suspended frames in place, and the 'bee space' between frames and also between frames and top, bottom, and sides.*

FIG. 241 *Top-bar hive in Kenya, like the Greek hive (Fig. 225), but rectangular and with all bars of the same length and therefore interchangeable.*
PHOTO: G. F. TOWNSEND.

FIG. 242 *A beekeeping scene by Pieter Breughel the Elder (1565); original in Berlin.*

10. Bees in art and in everyday life

This chapter provides an introduction to artefacts connected with bees that relate to art and everyday life rather than to the practice of beekeeping. Examples are mainly taken from Britain, but those from elsewhere serve to indicate that many other parts of Europe could yield an equally rich harvest.[36,111,198,243,280] Printed material is omitted here, together with woodcuts and engravings, on which information is available elsewhere.[18,111,155,198] Illuminated manuscripts showing bees and hives have not yet been systematically sought out and studied, except for the Exultet Rolls in Italy[21] and the collections in the Bodleian Library, Oxford. Postage stamps[240] are likewise omitted, and banknotes—except to say that the skep did not disappear from English bank notes until 1961.[81]

We shall deal first with sculptures, stained glass, well known paintings that include bees or hives, needlework, and jewellery and other small ornaments. Then we turn our attention to bees and the beehive used as emblems: on coins, tokens, medals and gems, on inn signs and in heraldry. The chapter finishes with a short section on honey in food and drink from the Bronze Age onwards, and another on artefacts made of beeswax, or cast by using beeswax in the *cire-perdue* (lost-wax) method—from world-famous bronze statues to delicate ornaments contrived from the gold of El Dorado.

213

FIG. 243 *François Rude's bronze statue of Aristaeus mourning the loss of his bees (1830).*
PHOTO: MUSÉE DES BEAUX-ARTS DE DIJON.

SCULPTURES

According to Greek legend, beekeeping was invented by Aristaeus. In 1821 François Rude made a bronze statue showing Aristaeus, with a tall stylised woven skep, mourning the loss of his bees; the theme, taken from Virgil's *Georgics*, Book 4, was prescribed in a competition for the Prix de Rome—which Rude won. The statue (Fig. 243) is in the Musée des Beaux-Arts in Dijon, France.

There are many representations of skeps in wood and stone, but I know of only two that show beekeeping scenes, both in Burgundy in France. In Cluny Abbey, built about 1120, the capital of one of the columns carries a stone carving which shows a beekeeper inspecting an upturned skep (Fig. 244). The Madeleine Chapel at Vézelay, some 150 km to the south east, was built at the same time. The capital of one of the columns in the nave shows a similar but fuller scene of the same type, and one is tempted to think that the same mason carved both. There are four men, one operating giant bellows, and each of the others holding an upturned conical wicker skep.[29] Two of the men are shown in Fig. 245.

In England, outside the south door of St James Church in Taunton, is a stone corbel built in the sixteenth century, with a swarm of bees running up into a (rounded, presumably straw) skep, and two sheep, against a background of foliage. When last seen, the carving was much more decayed than it appears to be in a photograph published in 1949.[282] In the Doge's Palace in Venice in Italy, the capital of the third column from the corner facing the Piazza includes a carving of a bear with a mouthful of honey comb, which is covered with bees.

St Ambrose is venerated as the patron saint of beekeepers,[295] and

FIG. 244 *Carving in stone of a beekeeper examining an upturned skep. Cluny Abbey, France, c. 1120.* PHOTO: SERVICE COMMERCIAL MONUMENTS HISTORIQUE

FIG. 245 *Two beekeepers examining upturned skeps, carved in the Madeleine Chapel, Vézelay, France c. 1120.* PHOTO: ALPHABET AND IMAGE

a good many churches are dedicated to him in the western part of continental Europe. He is frequently portrayed in statues and paintings with a skep, but no collective study of them seems to have been made. A few examples are given here. Venray church in the Netherlands (fifteenth/sixteenth century) has an attractive statue. There is another in the church at Gorsem near Truiden, where St Ambrose worked. In the Bavarian Alps in Germany, the Wieskirche is a splendid baroque church built in the eighteenth century, with both a statue and a painting of St Ambrose with his skep; in the church at Ebersberg he is the subject of another painting in the baroque style.

There is a limewood statue of St Ambrose in England. It was made in the seventeenth century, probably in Flanders or Germany, and is known to have been moved from an old house in Lincoln's Inn Fields, London, to Winchester City Museum in 1856. In 1975 it was lent to the headquarters of the International Bee Research Association, where it now commands the

FIG. 246 *Limewood statue of St Ambrose (seventeenth century) now at IBRA, Buckinghamshire, England.* PHOTO: M. P. DAVEY.

215

Bees in art and in everyday life

main staircase (Fig. 246). Ombersley near Kidderminster has been described as the only church in England dedicated to St Ambrose,[146] but I suspect that this is an error. It is local belief that the name derives from Aurelius Ambrosius, son of Constantine, King of Britain, who pitched camp there with his army; the name Ambresley was used until 1721. The present church is dedicated to St Andrew.

St Ambrose was Bishop of Milan in Italy from 374 to 397; the altar of his church (San Ambrogio), rebuilt in the tenth century, shows the infant Ambrose in his cradle with a swarm of bees flying round his head; they were supposedly the source of his eloquence, and of his appellation 'honey-tongued'. I do not know of any representation of St Ambrose with a traditional *Italian* hive. In fact his association with the hive—instead of the bees from it—seems to have evolved north of the Alps, and it may be much later than his association with bees, and hence with honey—sweetness—eloquence.

One of the few saints who kept bees was St Gobnet, who was in charge of a convent at Ballyvourney, Co. Cork, Ireland, around A.D. 500. She used her bees to repel a band of raiders who were stealing the local people's cattle, presumably upsetting some skeps and shaking the bees out. In one version she miraculously changed the bees into soldiers, and a skep into a brass helmet, which she presented to the defending chieftain O'Herlihy. His family is said to have kept and treasured the helmet until the penal days (when Roman Catholicism was proscribed), and then it was lost. In another version the skep was turned into a bell, and up to the nineteenth century St Gobnet's bronze bell was shown, with two holes at the top, where the clapper would have been fixed, through which the bees were said to have come out to sting the marauders.[169] The parish church owns a rare medieval wooden statue of St Gobnet, but the one on view is modern, and shows her with a skep at her feet, like St Ambrose. Near the ruined church on the site of her convent is a 1951 stone statue by Séamus Murphy, with a much larger skep; her grave there is easily located by the discarded crutches and votive offerings left by her devotees. There are many other remains.[223] There is also a strong devotion to St Gobnet at Dunquin, west of Dingle in Co. Kerry, where she is the patron saint. Nualla ni Dhomhnaill told me that in Dunquin St Gobnet is a late accretion of Mor, daughter of the sun and wife of the sea. Mor was a matriarchal goddess linked with Aphrodite, the goddess of love, to whom Ancient Greeks sacrificed honey. St Gobnet's well and the ruins of Kilmore (the church of Mor) lie within a few yards of each other.

A putto (cherub) holding a straw skep can be found in many baroque churches; he is usually tasting the honey which symbolises the sweetness of the words of the patron saint of the church. The best known (Fig. 247) decorates the side altar of St Bernhard in the chapel at Kloster Birnau, Überlingen, Baden, West Germany; it dates from c. 1750. Another is in the

ABOVE: FIG. 247 Putto *known as the* Honigschlecker *(honey licker) by J. A. Feuchtmayr, at Kloster Birnau, German Federal Republic* (c. 1750). PHOTO: JULIUS KITT.

RIGHT: FIG. 248 *Wood carving on one of the panels (doors to cupboards behind) of the Friedenssaal in Münster, German Federal Republic.* PHOTO: WILHELM SCHRÖDER.

monastery church of St Trudbert in Untermünstertal, near Freiburg/Breisgau. But putti with their skeps are not confined to churches. There was an especially lovely one in the city of Dresden, but I think it was destroyed in the Second World War. Three putti with a skep adorn a fountain in Haguenau in France, and there is one at Chartreuse d'Astheim near Volkach-am-Main in West Germany.

Skeps without putti have been used in more serious vein. As an emblem of thrift they have been sculpted on civic buildings, banks and trade centres. A skep is at the centre of a group of statuary that represented industry at the Old Crystal Palace, London; it was used as a model for a corner piece of the Albert Memorial in Kensington Gardens, west London. A skep adorns the facade of the Royal Exchange in London, the Rauhenhaus (home for rescued boys) in Hamburg, and the town hall of Venray in the Netherlands (a town whose arms also include a skep and flying bees).

In Münster in Westphalia the Friedenssaal (Peace Hall) has a splendid panelled wall made in 1540, one scene in which shows a skep being robbed by a bear, which in turn is being belaboured by a woman holding a distaff (Fig. 248). This represents folklore more than symbolism, as does an attractive domestic piece that was in the morning room of Wiggenthorpe Hall near Malton, Yorkshire; the centre panel of the marble fireplace showed a bear attacking one of four skeps on shelves below a thatched roof.

There are skep embellishments on houses, gravestones and memorials in many European countries; sometimes their origin is still clear, but sometimes it is lost beyond recall.

Bees in art and in everyday life

Sculptures and other representations of hives and bees show an interesting shift in symbolism over the centuries. In most of Europe, bees were at first linked with the sweetness of honey—as for example St Ambrose's 'honeyed words'—and the bees' wax provided the light for the Mass candles: the symbolism became Christian. Then, as I have suggested elsewhere,[78] the bees were secularised, so to speak, and were treated as part of the family (for example, the bees were told of a death). They were also used in political imagery, and in the—not specifically Christian—ethic of industry and thrift. So the early concept of sweetness gave way to sterner ideals derived in part from the industrial revolution: work, toil, labour, industry. The skep, rather than the bee, has often been used as the symbol possibly because it was easier to portray.

There is, however, a motto as early as 1596 'Alle vereint die Arbeit' (Work unites all), exemplified by three skeps.[189] And Ely Cathedral in eastern England has a misericord with a wood carving showing a man drinking, a pair of gamblers, and—close by them—the wife holding an empty skep, said to symbolise the fact that all her savings have gone in drink and gambling. An upturned empty skep is also shown on a plaque from a house wall in The Hague, Netherlands, and is now in the Gemeentemuseum there. It is held by a smiling man: an empty skep can perhaps presage honey in the future.

A skep, as the beehive, was adopted as a motif in 1849 by the Mormon church leaders in the Great Basin that they named 'Deseret', meaning a honeybee. A 1980 catalogue of Utah folk art[56] shows 24 recent examples of the symbolic use of a skep in and around Salt Lake City. Elsewhere the bee had ancient symbolisms quite separate from that of the colony, or of the skep or hive that represents it. The earliest of these was kingship. The hieroglyph of the bee was incorporated in the titulary of the Pharaoh of Egypt from about 3000 B.C., when Menes unified Upper and Lower Egypt, and it continued in an unbroken sequence for some 2650 years. The sedge plant (*Scirpus*) symbolised Upper Egypt and the bee Lower Egypt.[185] The two appear together in countless ancient stone and other monuments throughout Egypt (e.g. Fig. 22), and examples can be seen in any museum that has Egyptian antiquities.

Bees also abound in Rome, especially in St Peter's Basilica. There they have a papal symbolism, although this came about almost by chance. The present church was started in 1506 but was not consecrated until 1626, when the Pope was Urban VIII—Maffeo Barberini—whose family arms consisted of three bees. He brought three bees to the papal coat of arms, and they were, so to speak, built into St Peter's; they have settled on Urban VIII's tomb at the side of the high altar, on the elaborate bronze baldacchino which stands above St Peter's tomb, as well as on the altar itself. This canopy was made for Urban VIII by the artist Bernini in 1633 (who also made the beautiful bee fountain in Piazza Barberini, where

FIG. 249 *Part of a sculpture in the Vatican showing the three Barberini bees.* PHOTO: VATICAN MUSEUM.

three stone Barberini bees guard the streams of water flowing into the pool). On the baldacchino there seem to be bees everywhere—foraging on the plants which climb up the twisted pillars, flying against the roof, and in formal triple array on the pelmet which surrounds it. In the Vatican Museum the Barberini bees decorate maps, tapestries and many other treasures, as well as the structure of the building itself. Fig. 249 shows a very elaborate sculpture of the three Barberini bees.

Here in Rome I shall break my general rule of excluding written records from archaeological material. In the Accademia dei Lincei in Rome I found the draft of what was to have been the first scientific book on bees; it was prepared by Prince Cesi by cutting and pasting up the text of a broadsheet which had been presented to Urban VIII at Christmas 1625. The broadsheet showed the first anatomical drawings of insects ever to be made under a microscope, which Galileo had recently brought from Holland. The insects shown were the three Barberini bees, presented 'as a token of everlasting devotion' in an attempt to divert the interests of the Inquisition from the scientists, Members of the Academy, who were being regarded with increased suspicion. Sadly, the attempt failed. With the draft book is a handwritten report, by the Secretary of the Academy, of the Inquisition's examination of Galileo—whose scientific theories were ideologically unacceptable. Here are the speeches of one Member after another, all testifying in his favour. But the Establishment was not won over, and Galileo was sentenced on 22 June 1633, to incarceration at the pleasure of the tribunal, being 'vehemently suspected of heresy'.[72]

STAINED GLASS

Stained glass is not a medium that lends itself to portraying bees and hives, but a few windows have been designed with these as their themes. In England, the only medieval one I know is a piece of hand-painted glass made in the fourteenth century that was in the Prior's Lodging at Castle Acre Priory, Norfolk (Fig. 250). It shows a stylised (straw) skep with a

FIG. 250 *Skep portrayed on glass, Castle Acre Priory, Norfolk.* REPRODUCTION BY ANN RICHARDS.

flight entrance looking like a small door. Castle Acre Priory was founded by monks from Cluny in France, but the rounded shape of the skep on the window is quite unlike the conical wicker one carved in stone in Cluny Abbey itself (Fig. 244). At least one window in the Netherlands shows St Ambrose with his skep.

There is a legend that St David of Wales introduced bees to Ireland, sending them there in the charge of St Modomnoc (or Medoc). The event is commemorated in a window of the parish church of St Mary the Virgin at Swanage, Dorset, with which St Medoc had connections. St Medoc is shown taking a skep of bees on board ship in a wheelbarrow, blessed by St David. I am indebted to Dr Daphne D. C. Pochin Mould for knowledge of a gloss on the entry for St Finan Cam (of Kinnitty, Co. Offaly) in the *Martyrology of Oengus*: 'Finan Cam brought wheat into Ireland, i.e. the full of his shoe he brought. Declan brought the rye, i.e. the full of his shoe. Modomnoc brought bees, i.e. the full of his bell; and in one ship they were brought.' The similar shape (and the mouth-down position) link a skep and a bell together, as happened also in the story of St Gobnet (p. 216).

St George's Chapel, Windsor, has a stained glass window embellished with the Napoleon bee emblem, made to commemorate the Prince Imperial (son of Napoleon III), after his death in 1879. In London, the south side of Gray's Inn Hall has a window that shows the arms of Sir Miles Mattinson, KC, which include bees. Three silver skeps on a red ground feature in the arms of Ernst von Büren, which are displayed on a window in Bern Cathedral, Switzerland. And the three Barberini bees adorn a window in the Church of Santa Maria d'Arocoeli in Rome.

Charles Butler, known as the father of English beekeeping, who published *The feminine monarchie* in 1609,[55] was vicar of Wootton St Lawrence in Hampshire from 1600 to 1647. To mark the coronation of Queen Elizabeth II, a stained glass window, designed by G. E. R. Smith, was placed in the north aisle of the church in 1954; it is illustrated in *Bee World* 1955 (p. 2). It shows Butler and, behind him, a drawing from his book with queen, workers and drone on a honeycomb. There are other scenes, including some skeps under a thatched shelter. The window, and the church, are well worth a visit.

Another designer of stained glass windows, Walter Camm of Smethwick, has introduced skeps and bees into a number of windows: a memorial to Lady Burrell in Shipley church in Sussex; St Francis d'Assisi preaching to his little brethren the bees and birds (in Leamington); and a memorial to Sir John Mitchell in the old chapel of Harborne Church, Smethwick.

PICTURES IN WHICH BEES WERE USED AS A THEME

Woodcuts and engravings are beyond the scope of this book and, unfortunately, so are medieval illuminated manuscripts, some of which have entrancing pictures of hives and bees, and a few even of beekeeping scenes. The Bodleian Library in Oxford has taken the trouble to record all such paintings in its own collection, and has prepared colour slides of them; the set of 45, which includes some of other insects besides bees, can be purchased.[156]

In the sixteenth century a number of great artists produced paintings in which bees were used as a theme — an event which has not happened before or since.

The Florentine artist Piero di Cosimo (1462–1521) painted a pair of pictures in tempera on wood, now in the Worcester Art Museum, Worcester, Massachusetts, USA. Both are reproduced in B. F. Beck's *Honey and health*.[31] One is 'The discovery of honey' by Bacchus and his companions, as told by Ovid. The other is 'The misfortunes of Silenus': Silenus (the foster-father of Bacchus) stood on a donkey's back to reach for honey combs in a wild nest, got stung by the bees, and was shown how to ease the pain of the sting with mud; this is the scene shown in the painting. Dulwich College in south London has a painting by Nicolas Poussin (1594–1665), 'The nurture of Jupiter', with a related theme: nymphs are hunting wild honey for feeding to the infant Jupiter (Zeus).

In Germany, between 1527 and about 1537, Lucas Cranach the Elder made eleven paintings that show Cupid stealing a honey comb, being stung, and seeking consolation from Venus. These are in the National Gallery, London; the Statens Museum for Kunst, Copenhagen (Fig. 251); Villa Borghese, Rome; and elsewhere.[103] According to the legend Venus said to Cupid: 'Art not thou like the bees, that are so small yet dealest wounds so cruel?', and it has recently been suggested that Cranach was using Cupid as an allegory for syphilis,[103] which indeed dealt cruel wounds at that time.

Also in what is now Germany, Albrecht Dürer (1471–1528) drew a picture of Cupid running towards a (well robed) Venus, having upturned one of three skeps of a north-European type, and with a cloud of bees above his head (Fig. 252). In 1519 Mathias Grünewald, another German artist, painted quite a different type of picture in the 'Stuppacher Madonna' (Fig. 253). In the foreground Mary holds the infant Christ,

LEFT: FIG. 251 *Cupid complaining to Venus after being stung by a bee, by Lucas Cranach (1530). Original in Statens Museum for Kunst, Copenhagen.*

RIGHT: FIG. 253 Stuppacher Madonna *painted by Mathias Grünewald between 1517 and 1519.* FROM THE ARCHIVES OF THE STUPPACHER MADONNA (KAPELLENPFLEGE), 6990 STUPPACH-BAD MERGENTHEIM, GERMAN FEDERAL REPUBLIC.

BELOW: FIG. 252 *Cupid complaining to Venus after being stung by a bee, by Albrecht Dürer (1471–1528). Original in the Kunsthistorisches Museum, Vienna.*

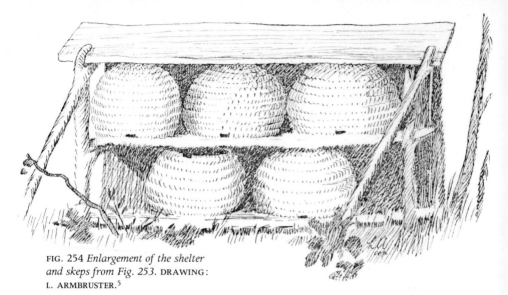

FIG. 254 *Enlargement of the shelter and skeps from Fig. 253.* DRAWING: L. ARMBRUSTER.[5]

with lilies growing at her feet; behind her on the right is a church, and on the left a shelter with two rows of three flattish skeps. The painting exemplifies the revelation to St Brigitta of Sweden, that the three symbols of the presence of the Son of God are the hive, the church, and Mary the Mother of Christ. Armbruster's book *Der Bienenstand in völkerkundliches Denkmal* (1926),[5] which has a great deal of information on representations of hives and apiaries, includes a drawing of the skeps in their shelter (Fig. 254).

The focus then shifted to Flanders. Pieter Breughel the Elder (c. 1520–1569) gave us a detailed and splendid beekeeping scene with tall skeps, and beekeepers protected against stings by coveralls and masks (Fig. 242). Hans Bol (1543–1593) drew a somewhat similar scene, with five beekeepers in masks, a swarm being hived, and another being tanged. Frans Floris (1520–1570) portrayed Aristaeus as the inventor of the hive, with 12 tall woven skeps (again north-European), one being in his hand; the body of Aristaeus is fully exposed to the bees, but none seem to be troubling him. In 1620 Peter Paul Rubens (1577–1640) painted St Ambrose who, as we have seen, was the patron saint of beekeepers. He was shown on a huge canvas, confronting the Emperor Theodosius. Rubens also painted various pictures that incorporated the three bees in the arms of Pope Urban VIII (see Fig. 249).

I have not been successful in finding any representation of the scene after Christ's resurrection when he ate 'a piece of broiled fish, and of an honeycomb' (Luke 24:42, in translations before 1885). A likely reproduction[197] proved to be from a seventeenth-century painting of the Dutch school, illustrating the exhortation in Proverbs 24:13, 'My son, eat thou honey, because it is good . . .'. The catacombs in Rome seemed a possible place, in view of the frequent representation there of the fish

Bees in art and in everyday life

symbol, but the only bee-oriented painting I know of shows Samson wrestling with the lion (Judges 14:5–6); it was discovered near the Via Latina in 1955 and dated to probably A.D. 320–360. Below the live lion a dead lion is shown, with a swarm of bees flying round its head and appearing to be leaving (or entering?) its mouth. This presumably illustrates verse 8, 'there was a swarm of bees and honey in the carcass of the lion', although there is no implication in the painting that the bees are nesting in the carcass; the body appears to be intact. The painting is reproduced as fig. 49 in *The catacombs* by J. Stevenson.[284]

Many people have wondered about the origin of the lion and the bees on tins of Lyle's Golden Syrup, with the caption 'Out of the strong came forth sweetness', which has been used as a motif since 1885. Abram Lyle's son says in *The Plaistow story*:[196] 'My father told me they had great difficulty in choosing a suitable trade mark. The association of Samson's lion with honey seemed to them appropriate and its Scriptural origin appealed to their austere religious upbringing.'

A good many paintings show hives in the background of a pastoral scene, but these are usually come upon only by chance. Two works of Adriaen van Ostade in Mauritshuis, The Hague, Netherlands, show hives. The 'Violinist' (1673) has two skeps on a shelf by the house, and 'Peasant in an inn' (1662) shows a single skep. A row of skeps forms part of the background of 'Le départ du conscrit' by S. Freudenberger (1745–1801), a farewell scene in a peasant family, in the Musée Ariana just above the Palais des Nations in Geneva, Switzerland. A similar row is shown in a water-colour drawing by George Robertson (1742–1788), in the Cecil Higgins Art Gallery, Bedford, England. Entitled 'A swarm of bees', it is a romanticised farmyard scene, in which a swarm on the ground has excited the curiosity of the farm workers and passers by. Fig. 255 illustrates an unusual find: a flat dish showing a variant of this scene. Curt Liebig (born

FIG. 255 *Blue and white dish, part of a 'Beemaster' dinner service made around 1825–1830 by Adams.*[70a] PHOTO: J. CRUNDWELL.

Bees in art and in everyday life

1868) painted 'Mädchen bei der Bienen', now in the Augustinermuseum, Freiburg, German Federal Republic; the little girl is looking at a shelter with three rows of flattish straw skeps (Fig. 258), reminiscent of those in Grünewald's 'Madonna'. Another shelter with skeps in it appears in a painting from the eighteenth century, showing a farm in England after enclosure; it is reproduced in *The Age* (Melbourne) for 3 May 1958. The National Gallery of Victoria in Melbourne has Clara Southern's 'The bee farm', a more recent painting, which was shown in *The Age* for 1 June 1957.

NEEDLEWORK

In my experience it is rare to find a tapestry or embroidery showing a beekeeping scene. However, in one house owned by the National Trust, at Wallington, Northumberland, there is a screen worked in petit point by Julia, Lady Calverly, in 1727 (Fig. 256); it shows a scene taken from Hollar's engraving for Virgil's poems, with swarms in the air and beekeepers trying to bring them down, and the skeps whence they came.

Scattered bees have been widely used in ornamental design, for instance in the stump work of the Stuart period. They adorn tapestries in the Vatican Museum in Rome, from the Barberini connection already mentioned. Their most splendid application has been in robes and hangings of Napoleon Bonaparte (1769–1821) and subsequently those of Napoleon III (1808–1873) and his son the Prince Imperial (1856–1879).

LEFT: FIG. 256 *Needlework screen made in 1727 by Lady Calverly at Wallington, Northumberland.* PHOTO: NATIONAL TRUST. RIGHT: FIG. 257 *Detail from coronation portrait of the Emperor Napoleon Bonaparte.* REPRODUCED FROM K. A. FORSTER.[111]

Fig. 258 Mädchen bei der Bienen, by Curt Liebich (born 1868). Compare the skeps (one inverted) with those in Fig. 254. Original in Augustiner Museum, Freiburg/Breisgau, German Federal Republic.

These can be seen in most collections of Napoleon memorabilia, in Paris (Fig. 257), Corsica, Elba, and England in Farnborough, Hampshire, and they are portrayed in many paintings of Napoleon and his family. The painting by J. A. D. Ingres in the Musée de l'Armée in Paris, shows Napoleon enthroned as Emperor of the French in 1804. It has been said that Napoleon inverted the French kings' fleur-de-lys symbol (symbolising idleness to him) to make a bee (symbolising industry). There are other interpretations; one that is commonly accepted is given in the next section, on jewellery. Four bees, elaborately embroidered in gold thread on a piece of red velvet from hangings in the church of Notre-Dame in Paris, were among the collection of Prince Napoleon (no. 193); they are now in the Musée National du Château de Malmaison. A lovely miniature in *Voyage de Gênes* by Jean Marot shows Louis XII riding off to make war on the Genoese in 1507, and both he and his horse are accoutred in clothes richly embroidered with wicker skeps and hundreds of bees.[245a]

In Utah, the Beehive State in USA, bees are used as a motif in folk art of various kinds, including gloves ornamented with spot-stitched bead-

Bees in art and in everyday life

work, and many bed quilts.[56] I am sure that homes (especially of beekeepers) in various countries have pieces of domestic needlework that incorporate bees into the pattern. An English beekeeper, Reginald E. Dickson, has told me of a sampler in his possession; it is dated 3 May 1790 and was made by Sarah Ann Hitch, who embroidered the following lines:

> Th' Industrious bee extracts from ev'ry flow'r
> Its fragrant sweets, and mild balsamic pow'r
> Learn thence, with greatest care & nicest skill
> To take the good, and to reject the ill:
>
> By her example taught, enrich thy mind,
> Improve kind nature's gifts, by sense refin'd;
> Be thou the honey-comb—in whom may dwell
> Each mental sweet, nor leave one vacant cell.

The hexagonal shape of the bee's cell has provided one of the most common modules for patchwork quilts. Finally, turning to the more delicate medium of lace, Bill Gibb, a modern London designer, uses a stylised bee as his trade mark, and his products may become treasured finds for future collectors.

JEWELLERY AND SMALL ORNAMENTS

The magico-religious significance of bees is extremely old. In *The gods and goddesses of Old Europe*,[127] Marija Gimbutas explores the myths, legends and cult images during the period 7000 to 3500 B.C., and shows that bees already figured among the sacred objects in this period. Snakes were even more important: like bees, snakes live in a dark hole, and are 'venomous'. Swarms of bees miraculously issue forth at a certain season of the year, and so do snakes. Cobras are still a cult object in parts of southern India, and

FIG. 259 *Gold pendant from Mallia, Crete, 2000 to 1700* B.C.
PHOTO: ARCHAEOLOGICAL MUSEUM, HERAKLION.

Bees in art and in everyday life

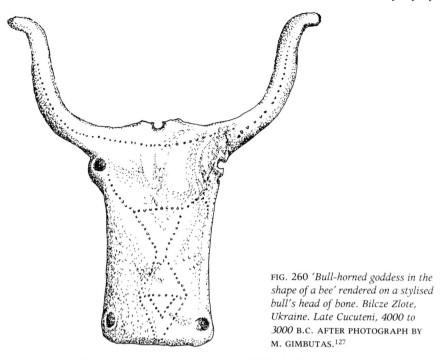

FIG. 260 *'Bull-horned goddess in the shape of a bee' rendered on a stylised bull's head of bone. Bilcze Zlote, Ukraine. Late Cucuteni, 4000 to 3000* B.C. AFTER PHOTOGRAPH BY M. GIMBUTAS.[127]

are kept for ceremonial purposes in a lidded clay pot that would serve equally well as a beehive. These facts suggest links right back to the days of honey hunters living in rock shelters.

The earliest representations of a goddess in the shape of a bee occur in the neolithic complexes of Proto-Sesklo (Greece) and Starcevo (Hungary). The bee goddess appears on the stylised head of a bull carved out of bone, from north-west Ukraine, in or soon after late neolithic times (Fig. 260). Similar representations, but less schematised, are known from Minoan, Mycenaean, Geometric and Archaic Greek art. Many gold rings of Minoan workmanship from Crete and Greece portray her, and gold plaques from Rhodes and Thera dated to 700–480 B.C.; some of those from Rhodes can be seen in the British Museum.

A lovely gold pendant was found in 1930 at Mallia, a burial site in Crete; it is now in the Museum at Heraklion. In this pendant (Fig. 259), which is dated to about 2000–1700 B.C., two insects are shown face to face. They have been much written about and variously interpreted. In 1976 Sinclair Hood[150] cited 85 references and concluded that they were bees rather than wasps, an earlier identification. F. Ruttner[261] was certain that they were queen honeybees, and they were used on the badge of the 27th International Beekeeping Congress in Athens, Greece, in 1979. Meanwhile O. W. Richards[248,249] had presented arguments for their identification as wasps of the genus *Polistes*, which were elaborated and supported by

FIG. 261 *Two golden pinheads with lions and bees. Peloponnesus, Greece. 500–400* B.C. *96.717 is 77 mm long including the pin, and 96.718 is 118 mm.* PERKINS COLLECTION. COURTESY MUSEUM OF FINE ARTS, BOSTON.

another detailed analysis in 1979;[180] a 1981 study veered to bees again.[175]

The technical arguments in favour of *Polistes* wasps are persuasive, but the overwhelming number of artefacts showing the honeybee as a cult object makes it seem to me more probable that the insects in the pendant are honeybees.

Unquestioned golden bees can be seen in the Museum of Fine Arts Boston, USA, on two golden pin heads, found in a grave from 500–400 B.C. near Patras, Peloponnesus, Greece (Fig. 261). Both have four bees, each taking nectar from a flower, together with four lions rampant. One (96.718) also shows three sphinxes, and the other (96.717) a snake. Each part of the pin is made separately from sheet gold, and is of exquisite workmanship. The bees are modelled with complete realism, even to the veining of the wings.

Childeric, the first King of the Franks, died in A.D. 481 and was buried at Tournai in France. In 1652 his tomb was opened and among the treasures found were more than three hundred golden bees. The special symbolism of kingship attributed to these golden bees was probably the reason for Napoleon's adoption of them, mentioned above.

On a lighter note, the Wallace collection in London has a set of charming mother-of-pearl counters dating from the time of Napoleon, which are

Bees in art and in everyday life

ornamented with inlaid designs of bees (Fig. 262). In early watches, a watch cock was used to protect the balance wheel. It was made of brass and often engraved with minute and intricate designs. One in the collection of Colonel A. S. Bates of Manydown Park, Hampshire, includes a house and garden, 3 pigs, 4 hens, 3 ducks, and 5 bees and their skep—all in an area less than 2 × 2 cm (Fig. 263).

Victorian jewellers frequently used insects as motifs for brooches and pendants. Bees, flies and dragonflies in gold and precious stones were popular, and examples can still be found on sale, as well as their modern counterparts.

NUMISMATICS AND GEMS

Coins, trade tokens and medals all carry designs of interest to us. Ancient gems are also included here; in these the design is hollowed instead of in relief.

A splendid catalogue was published by Münz Zentrum for an auction held in Cologne in 1980,[213] which illustrates and describes 473 different coins and medals that show bees or hives. This gives some idea of the richness of the material. However, there are fuller sources for the ancient world: Armbruster described 111 items,[14] as well as other more recent ones,[16] and Nivaille[220] has made a detailed study of Greek coins.

The bee appears on Ephesian coins from the time they were first issued, before 545 B.C., fairly certainly because the bee was a cult object, perhaps

ABOVE: FIG. 262 *Mother-of-pearl counter*. PHOTO REPRODUCED BY PERMISSION OF THE TRUSTEES, THE WALLACE COLLECTION, LONDON. NO. XXA 56.

RIGHT: FIG. 263 *A watch cock (total height 36.5 mm) showing a farmyard scene, with a skep and five bees.* PHOTO: A. S. BATES.

representing a bee-goddess.[243] Most early coins show a single bee, head uppermost and viewed dorsally (Fig. 264). Bee coins were also issued in Delphi, Athens, Ionia, Crete, and Phoenicia, and many other centres in the Greek world. They were not used in Roman times and do not reappear in numismatics until the late sixteenth century, when medals produced in several European countries show them or—more often—a skep or skeps, and occasionally a notional swarm in flight (Fig. 265). The symbolism varies, and sometimes bees seem to have been used almost by chance. Many early designs have a delightful freedom of movement, but this had largely gone by the nineteenth century. The skep then took pride of place over the bee(s), and the symbolism usually had a single focus—industry—with thrift coming a good second. Many of the medals were in fact awarded 'for industry'.

As more and more people in England bought and sold things, the supply of small change became inadequate, and traders issued tokens to substitute for the coins most needed. These trade tokens were issued in the eighteenth and, especially, the nineteenth century, and many show bees or skeps. They were mostly used by the poor, and it must have seemed appropriate that they should carry a design that symbolised both thrift and industry.

In 1896 Fürtwängler[118] published a description of the ancient engraved gems in the Royal Museum in Berlin, and Armbruster[14] reproduced descriptions and illustrations of those that relate to bees, together with others in collections in Paris, London and New York. Many of the gems, like the coins, show a single bee, and several others show a lion and a bee. Hilda Ransome's book *The sacred bee*[243] illustrates a red jasper cameo 'which shows a bee, as a symbol of the soul, in the mouth of a lion, which beast was intimately connected with Mithra', and two gems—one in the British Museum—of a grasshopper ploughing with a team of two bees. Hilda Ransome gives some discussion of the likely symbolisms involved, but now, over 40 years later, the time seems ripe for a new assessment in the light of more recent scholarship. I hope that this paragraph will serve to whet someone's appetite to delve into this interesting byway.

INN SIGNS

The 'sign' outside an inn or public house had a very old origin; Larwood and Hotten's *The history of signboards* (1866)[183] gives an entertaining history of them. In Britain I have found evidence of 126 examples of 'The Beehive' (mostly public houses). The earliest on record[183] was in St Mary's Hill; it was named after its owner, John Hive, in 1667. A 1758 billhead of Eliza Blick, a wax chandler referred to in the discussion on beeswax below, says that she operated 'at the Golden Beehive, opposite the Mansion House, London'. The most famous Beehive Inn is the one in

FIG. 264 *Coins and tokens from Ancient Greece, as depicted by Claudio Menetreio in* Symbolica Dianae Ephesiae *(1657). I and VI are from Delphi, II from Dyrrachium Obrimi and III from Dyrrachium Daminus, IV from M. Plaetorius, and V from Boetium.*

FIG. 265 *One of the earliest medals after classical times to show bees, in honour of Philippe de Croy, Prinz von Chimay, 1567. On the reverse side a hand reaches out through clouds to grip the top of a skep.*
PHOTO: MÜNZ ZENTRUM,[213] NO. 2301.

Bees in art and in everyday life

Grantham, which started to use a hive of bees in a tree for its sign in 1798, after a swarm settled on the inn's current signpost. The tall crocketed spire of St Wulfram's Church can be seen from the road by the inn, and a board outside it proclaims:

> Stop, Traveller, this wondrous sign explore,
> And say when thou has viewed it o'er and o'er,
> 'Now, Grantham, now, two rarities are thine,
> A lofty Steeple and a living Sign.'

Lists of the inns and taverns of London, from records made in James I's time, in the next reign, and at the period of the Restoration, do not include a Beehive.[183] But in 1864 there were '. . . 5 Artichokes, 13 Barley Mows, 9 Beehives, 31 Bells . . .'. Most Beehive Inns were probably founded in the nineteenth century, and their existing sign boards are of varying degrees of interest. Outside London, where there are now at least 45 (and the Queen Bee in East Peckham), 82 have been identified in England, 3 in Wales, 2 in Scotland and 1 in Ireland.

England

Avon: Bath, Westbury-on-Trym
Berkshire: Reading, Upper Basildon, White Waltham
Cambridgeshire: Kirtling, Whittlesford (Bees in the Wall)
Cheshire: Harthill, Macclesfield
Cornwall: Helston
Cumbria: Carlisle, Deanscales, Eamont Bridge
Derbyshire: Combs, Glossop, New Mills, Ripley
Dorset: Parkstone
Durham: Consett, Fishburn, Holmside
Essex: Chelmsford, Colchester, Lambourne End, Little Horkesley
Gloucestershire: Cheltenham, Prestbury
Greater Manchester: Bury, Clifton, Horwich, Hyde, Whitefield
Hampshire: Aldershot, Cosham
Hereford and Worcester: Worcester
Hertfordshire: Barley, Epping Green, Hatfield Hyde, Hemel Hempstead, St Albans, Watford
Kent: Chipstead, Riverhead
Lancashire: Bacup
Leicestershire: Leicester
Lincolnshire: Grantham
Merseyside: Halebank, Sutton
Norfolk: Norwich (2)
Northamptonshire: Deanshanger
Northumberland: Seghill
Oxfordshire: Abingdon, Carterton, Oxford, Russell's Water

Shropshire: Shifnal, Shrewsbury
Somerset: Frome
Staffordshire: Hanley
Suffolk: East Bergholt
Surrey: Egham, Reigate
East Sussex: Brighton
Tyne and Wear: Newcastle, South Shields, Sunderland (2)
West Midlands: Birmingham (2), Halesowen, Sutton Coldfield, Tipton, West Bromwich
Wiltshire: Bradford-on-Avon
North Yorkshire: Newholm, York (Beeswing)
South Yorkshire: Sheffield (2)
West Yorkshire: Buttershaw, Halifax (Beehive and Cross Keys), Ripponden, Thorner, Wakefield

Wales

Dyfed: Pencader
Powys: Brecon, Crickhowell

Scotland

Fife: Anstruther
Lothian: Edinburgh

Ireland

none in the Irish Republic; one in Belfast

A number are thus in large towns, as well as the 45 in Greater London. Many of the urban houses with the sign of the Beehive are not in the town centres but among streets of small houses, and I believe that the symbolism was again thrift and industry (for working people) rather than any link with living bees in a rural setting. Very few other names associated with bees seem to have been used, but there is a Honey Pot in Stanmore (Greater London), Beeswing in York, and Bees in the Wall at Whittlesford (Cambridgeshire), where there is actually a bees' nest inside the wall of the public house.

I am sure that an interesting holiday could be planned, by devising a route through rural England that would take in some of the Beehive Inns, and enquiries made on the spot might well throw light on the history of the inns and their past and present signboards.

HERALDRY

Some heraldic bees and hives are mentioned in different sections of this chapter, according to the type of artefact they decorate. The best hunting ground for readers interested in finding other examples would seem to be towns and villages which include them in their arms. The following is my list to date, but I am sure it is not complete. Where the blazon (technical description of the arms) is not available to me, I have given a straight description. Family arms are excluded, as being less easily found on accessible artefacts.

England

The years in which the arms were granted is interesting. As with the names of inns and public houses, the popularity of the bee and the beehive seems to have followed the industrial revolution rather than to have been linked with the medieval church or rural life.

Aylesbury (Buckinghamshire) a silver cross charged at the centre with a bee; in the first quarter a golden beehive. Use ceased in 1957.
Bacup (Lancashire) 1883 . . . black fleece between 2 bees in proper colours
Barrow-in-Furness (Lancashire) 1867, red on a gold band an arrow pointing upwards towards a bee
Bishopton (Warwickshire) azure 3 bee-hives argent a canton ermine
Blackburn (Lancashire) 1852, 3 bees volant
Burnley (Lancashire) 1862 . . . 2 gold bees . . .
Bury (Lancashire) 1877 . . . a bee between two cotton flowers
Kingswood (Gloucestershire) . . . 3 bees . . .
Kirkby (Merseyside) includes 2 bees
Kirkham (Lancashire)
Luton (Bedfordshire) includes a skep and flying bees
Manchester 1842 . . . terrestrial globe with 7 bees on it
Salford (Greater Manchester) 1844 . . . field strewn with flying gold bees
Widnes (Cheshire) 1893 . . . in the 2nd and 3rd quarters a gold beehive with 4 gold bees

France

Morbihan (Département in Brittany) includes a skep with hackle, surrounded by many flying bees (Fig. 266).

German Federal Republic

Binau (Baden-Wurttemburg) 2 skeps, flying bees, with a fisherman
Peissenberg (Bayern) includes 3 bees

Italy

Elba (island) Napoleon's flag has 3 golden flying bees on a diagonal red band

Luxembourg

Wiltz, skep with hackle, and 6 flying bees (Fig. 266)

FIG. 266 *Examples of arms that include skeps and bees: Morbihan, France (*Gazette Apicole, *1970), and Wiltz, Luxembourg (*Letzeburger Beien-Zeitung, *1973).*

Netherlands
Borne (Overijssel) skep surrounded with 3 large bees
Hengelo (Overijssel) sheaf of corn and skep with swarm of bees
Hoogeveen (Drenthe) includes a skep and bees
Venray (Limburg) includes skep with 10 flying bees

Romania
Mehedinti, a bee
Vaslui, wicker skep and 3 bees

Switzerland
Aclens (Waadt) includes a golden skep
Jentes (Fribourg) golden skep on blue, with green below
La Chaux-de-Fonds (Neuenburg) includes a golden skep surrounded by 7 golden flying bees on silver
Mumpf (Aargau) includes a brown skep
Sarzens (Wallis) 3 golden bees on red

USSR[279]
The towns mentioned are in various parts of European USSR; the district (*guberniya, oblast'*) is given in brackets.

Cherven', earlier Igumen (Minsk) 1796, includes a flowering bush surrounded by 5 bees; the area was famous for its beekeeping.
Klimovichi (Mogilev) 1781, includes a golden bee below the imperial arms.
Nakhichevan' on Don (between Azov and Rostov) 1888, an array of 6 golden bees on silver; below, a golden skep. Here the bees symbolise industry; the town was an Armenian settlement.

Novoe Mesto (Chernigov) 1782, a pyramid of 6 golden skeps; on either side a group of 7 bees on green.
Osa (Perm) includes, on silver, a tree with a log hive in it, and many bees flying, indicating that there is much honey; also a bear climbing the tree. Perm was a strong area of tree beekeeping (see Chapter 5) and Osa had a famous mead factory.
Roslavl' (Smolensk) 1668, includes 2 golden skeps on azure
Sosnitsa (Chernigov) pine tree (*sosna*) with a golden log hive in it, a black bear climbing up to the hive, and many golden bees flying.

In addition, the arms of Tambov district, about 400 km south east of Moscow, which were granted in 1878, include a silver skep on an azure shield, with 3 bees above. The arms of all towns in Tambov include this device, together with some other individual one; an illustrated account of 15 of them has been published.[30] The skeps portrayed there (and in the arms listed above[279]) appear to be built of bricks and, in fact, look rather like a Russian heating stove.[120]

FOOD AND DRINK

Honey is more perishable than beeswax, but some of it nevertheless survives from ancient times. Honey found in Egyptian tombs[322] is on display in the Agricultural Museum in Cairo, and honey from Ancient Rome was found recently in excavations at Paestum in Italy. It would, however, be good if the identifications could be re-examined by modern methods.

Honey pots have been described elsewhere;[78] general museums do not often have examples, but some of the beekeeping museums listed in Appendix 2 do. Almost all are from the past 200 years, although IBRA has a small Romano-British two-handled jar that was probably used for honey (Fig. 267); children still play a 'honey-pot' game in some of the countries that were once within the Roman Empire, in which one child bends her arms to make the two handles of a honeypot and is then 'weighed' by other children.[100,224]

There are archaeological finds in connection with two honey products: mead, made by fermenting honey, and gingerbread. A deposit of plant material was found in a Bronze Age burial from about 1000 B.C. at Ashgrove in Fife, Scotland, and when it was examined by identifying the pollen grains in it,[94] the main plant constituents were found to be: small-leaved lime (*Tilia cordata*) 54 per cent, meadowsweet (*Filipendula*) 15 per cent, ling heather (*Calluna*) 8 per cent, a plantain (*Plantago*) 7 per cent, and a mint (Labiatae) 5 per cent. Material remaining in a beaker in the grave contained pollen from the same main contributing plants. Leaving aside some technical details that do not affect the main issue, *Tilia cordata* is a

Bees in art and in everyday life

FIG. 267 *Romano-British two-handled jar; larger similar jars have been found stamped with the weight of honey they contained.* IBRA COLLECTION B61/1.

good honey source, but it is a tree of the southern part of Britain. It was concluded that the beaker had contained a honey-based drink (possibly mead), which had spilt out, and that imported honey had been used for it, perhaps from England or Denmark. Other graves are now giving similar finds, but these are not yet published.

Mead was drunk from various vessels, one special type, the mazer, being a hardwood bowl that was often silver mounted and ornamented.[124] In 1382 the refectory at Christchurch, Canterbury, possessed 182 mazers, and an excellent example from about 1307 is still to be found at St Nicholas Hospital, an almshouse at Harbledown near Canterbury.[42] Others can be seen in museums; for instance the British Museum has the Rochester mazer (1532), and the national Museum in Dublin has a late medieval mether, an Irish wooden drinking cup for mead. In 1963 a silver mead jug carrying a 1690–92 Chester mark was sold at Sotheby's in London for £1200.

The other honey product that has survived from antiquity was much more important then and in the Middle Ages than it is today: gingerbread or spiced bread (*pain d'épice*).[78, 263] It was traditionally made with honey, flour and spices—not necessarily including ginger—but without raising agent or fat. The stiff dough was shaped into figures; even today shops sell 'gingerbread men', although most of these probably contain no honey. I found examples looking like the archetype of the 'gingerbread man' in the Agricultural Museum in Dokki, Cairo. They have superimposed features of a face and a body, and were catalogued as 'feast cakes for children', dating from 1400 B.C. They were found at Dier el Medinah on the west bank of the Nile at Luxor.

I have not heard of any other gingerbreads surviving from early times, but in northern and eastern Europe gingerbread moulds can be found that were in use one or more centuries ago; I know examples from England and Wales, Belgium (especially Dinant), Netherlands, Germany, Poland,

Bees in art and in everyday life

Czechoslovakia, Hungary and Switzerland. The moulds are wooden boards into which a design is carved in reverse. The prepared dough is pressed into the mould and thus shaped, before baking. The Welsh Folk Museum at St Fagans has 12 such moulds. Birmingham City Museum houses the Pinto Collection of Bygones, and Edward Pinto's book *Treen and other wooden bygones*[236] includes descriptions and illustrations of many gingerbread moulds in the Collection. There is a fine collection in Nürnberg Museum, Germany. An illustrated account has been published of Czechoslovak moulds and other equipment used in making gingerbread.[263] It would be worth enquiring for gingerbread moulds at any folk museum in the countries listed above. It has been suggested[149] that the shaping of the gingerbreads into human (or god-like) form must date back to pre-Christian days, and the examples from Ancient Egypt show that this was so.

The earls buried in honey

In the classical world it was not uncommon for a corpse to be put in a sealed coffin filled up with honey. It might be the only way in which the body of a general dying on a distant battle-field could be preserved for transport to his native land. Strabo tells us that, on his own order, the body of Alexander the Great was placed in white honey in a golden coffin. Agesilaus, King of Sparta, was transported home in honey when he died in Libya in 360 B.C.

In England, with its cooler weather, temporary preservation was not so difficult, but there is one well substantiated case of long-term burial in honey, which is little known, and the story is therefore set out here.[307] In the parish church of St Peter, Titchfield, Hampshire, is a magnificent monument to the first four Earls of Southampton, the Wriothesleys, constructed in 1594. The bodies of the Earls were interred in sealed lead coffins in a vault below the monument. The story was current that they had been buried in honey, and when a subsidence slightly crushed one coffin early this century, the workmen saw a liquid trickling out of the damaged seam. One of them, determined to test the story, ran his finger along the seam and, from tasting the liquid, confirmed that it was indeed honey. The first Earl died in 1550 and the fourth in 1667; the third Earl was Shakespeare's patron, and possibly the 'onlie begetter' of his sonnets.

BEESWAX

Beeswax is secreted by worker honeybees from wax glands on the underside of the abdomen. The Ancient Greeks, however, believed that the wax combs 'were made from flowers', and Aristotle linked the bees' spring comb building (and brood rearing) with the flowering of the olive

Bees in art and in everyday life

trees. In Greece, Penelope Papadopoulo made the interesting suggestion to me that this belief may have arisen because pollen from olive trees is nearly white. In spring, when the olives are in flower, the bees build white comb, and the whitish pollen they then bring in could have been mistaken for this wax; later in the year the wax is darker.

Although honey is usually the primary harvest from keeping bees, beeswax is far more interesting in an archaeological context. It has been put to a multitude of uses, including the production of many decorative objects, and it could have been used for casting even before the Bronze Age. Beeswax is fairly inert and retains its shape at most atmospheric temperatures, thus many beeswax artefacts survive from earlier times, and a separate book could well be written on beeswax archaeology. I shall use the limited space here to indicate the type of object that readers can seek out and enjoy, giving a few examples of museums and other sites that have them, and showing where more information can be found about particular interests. The most important source—and the most magnificent work of all—is a 12-part volume in German published by the firm Farbwerke Hoechst AG at Frankfurt-am-Main.[54]

One of the earliest uses for beeswax was to provide illumination. Mosaics at Enna in Sicily and Carthage in Tunisia show candles in cult scenes from early centuries A.D., and there are paintings from pre-Christian times too. Actual candles survive from the first century A.D. at Vaison near Orange in France (now in the British Museum, which also has cakes of beeswax from a Late Bronze Age founder's hoard), from a Viking grave in Mammen in Denmark, and various other sites. Beeswax candles played an important part in the services of the Roman Catholic Church,[74,78] and splendid examples can still be found in use. In Bavaria, the candles are richly ornamented, and small versions of them can be purchased. Moulds and other equipment for making your own candles are available from Candle Makers Supplies, 28 Blythe Road, London W14; the same shop sells materials for batik (see below).

The Wax Chandlers of London[98] and an earlier article[71] give much information on the history of this Worshipful Company and its influence on the beeswax trade and the making of candles in England; Fig. 268 shows the designs on its silver cup, made in 1683. The guild was tireless in its attempts to prevent malpractices, as witness Ordinance 20 of 1664:

> Wares and commodities of the Art, *viz.* torches, tapers, prickets, flamboys, *etc.*, shall be of good and perfect wax and good wick, not mixed or corrupted with turpentine, resin, tallow, *etc.*, except the casting of torch staves which cannot be done without; every small torch to be a yard long in the wax besides a convenient snuff, every large torch an ell [$1\frac{1}{4}$ yards] long, with like snuff; every yellow link to weigh $1\frac{1}{2}$ lbs., every black link $1\frac{1}{4}$ lbs., both yellow and black to be a yard and a half-quarter long; all book candles, searing candles and soft wax to be made good and clean, ...

Bees in art and in everyday life

FIG. 268 *Engraved designs on a silver cup made in 1683 and given to the Worshipful Company of Wax Chandlers by Richard Normansell. The beekeeper on the left is tanging a swarm of bees, and the one on the right is hiving a swarm.* REPRODUCED FROM T. W. COWAN.[69a]

Searing or cering candles were used to impregnate the cerecloths (winding sheets) that were wrapped round a body preparatory to burial.

Ceroplastics is the term used for the art of wax modelling, and the First International Congress on Ceroplastics in Science and the Arts in 1975[61] opened our eyes to the wide range of examples of these products that survive from earlier centuries, in many parts of Europe. The following list of historical ceroplastics is compiled from the Congress Proceedings and other sources:

>   anatomical models, especially at La Specola in Florence, the Josephinum
>   in Vienna and Musée d'Histoire Naturelle in Rouen
>   plants modelled in wax
>   models in the Army Museum in Leiden
>   wax models in Madame Tussaud's, London
>   many artistic works in wax in Florence (e.g. The scenes of the plague;
>   The descent from the cross)
>   the Medici Venus in Florence
>   portrait waxes in at least 30 European museums[246]
>   wax effigies in the Undercroft of Westminster Abbey, London[289]
>   folk art in many countries, including votive offerings mentioned below.

The British Museum in London has a number of 'magical' beeswax figures from Ancient Egypt. Many others can be seen in Egypt itself; the Agricultural Museum in Cairo has several from 1400 B.C. One statue of extreme beauty was found as recently as 1980 in the tomb of Rameses XI in the Valley of the Kings on the west bank of the Nile opposite Luxor. There are two figures of golden beeswax about 8 in. high: the king, and the goddess Ma'at in front of whom he stands in adoration.

Beeswax is produced by stingless bees too, and in the South Australian

Bees in art and in everyday life

Museum in Adelaide I found sacred human and animal figures modelled from it by Australian Stone Age aborigines.

Beeswax was much used for votive offerings—thanksgivings for recovery from an illness or accident. Many were realistic and sometimes grisly models of the afflicted or cured part of the anatomy, but others were entirely ornamental and richly coloured, constituting a delightful form of folk art. Many German museums have a collection (not always on display); that in the Heimatmuseum at Rosenheim, south east of Munich, has been described in a lovely book, *Viel köstlich Wachsgebild* (Treasures in wax), adorned with coloured illustrations.[141]

Wax seals have been used for important documents from Roman times if not earlier; sometimes an additive was used to harden the wax, and verdigris or vermilion could be used to colour it. In England natural wax was used for routine business, a green colour for documents sent to sheriffs, and scarlet for diplomatic documents. A great many museums have seals to show, and excellent plastic replicas of some interesting and historic English seals can be purchased from Museum Casts, 4 Church Street, Cottingham, Market Harborough, Leicestershire, OE16 8XG.

Bavaria and some other parts of Germany have been noted for their beeswax plaques depicting richly apparelled human figures and spirited animals. The plaques are made in wooden moulds, which can be found in museums; the plaques themselves—often made in the old moulds—can be purchased in shops.

It is easy to inscribe marks on a beeswax surface, and in Ancient Rome a wooden board coated with beeswax was commonly used as a renewable writing pad. The British Museum has a pair that are still coated with wax; the London Museum has others. Writing was done with the pointed end of a metal stylus, and erasing (and smoothing the surface for re-use) with the flattened other end.

FIG. 269 *Sacred figurines made by Australian aborigines from* Trigona *wax; now in South Australian Museum, Adelaide. Others portray kangaroo, emu, echidna, etc.*

Bees in art and in everyday life

Beeswax was used as a binder for the charcoal black pigment on the famous head of Nefertiti, now in the Egyptian Museum in East Berlin. It was used for filling the incised hieroglyphs on the red granite coffin of Rameses II, now in the Louvre Museum in Paris. There are many similar examples.[185]

Beeswax was used very effectively as an adhesive in several ancient civilisations. (The first experiment, in mythological times, was unsuccessful: Icarus managed to fly, using wings of feathers fixed to his body with wax, but he rose too near to the sun, the wax melted and he fell into the sea and drowned.) In the British Museum[32] beeswax can be found attaching alabaster lids to vases, and alabaster vases to their pedestals; fixing the flint teeth of a sickle; together with limestone powder, cementing a handle on to a razor. Beeswax is used—successfully—by Australian aborigines to stick feathers to sacred staves carried in dances, and in making didgeridoos. They also modelled cult objects from it (Fig. 269).

The term encaustic (Greek = burnt in) is now applied to paintings made with wax as the chief ingredient and applied with the pigment, whether or not this was actually heated. The method was used in Ancient Egypt, although so many examples from Roman times still exist that it is often thought of as a Roman technique. Encaustic paintings decorated many of the houses in Pompeii and Herculaneum. The British Museum has a number on display from various sources.[32] Encaustic techniques are still used, and they have also been applied to the colouring of marble statues and wooden icons.

Batik is another decorative art based on the use of beeswax. It is a method of printing coloured designs on textiles by waxing the areas not to be dyed with a particular pigment, before applying the pigment to the fabric. Many colours can be used in sequential waxings and dyeings, with the most splendid results. One need hardly look for museum specimens, for batik is a living craft in many countries of south-east Asia (it originated in Java), and beautiful examples can be bought inexpensively.

One curious find of beeswax itself is worth including here. In 1813 Alexander Henry travelled down the Columbia River to the Pacific coast of the USA. Writing of the local Clatsop Indians in Oregon, he said: 'They bring us frequently lumps of beeswax fresh out of the sand which they collect on the coast to the south where the Spanish ship was cast away some years ago ...'. There is other evidence of the fact that large quantities of beeswax were found and traded, some with markings (which might be merchants' marks like those recorded from Morocco in the seventeenth century[204]). Chemical tests on samples available—it is still washed up—have shown that it is indeed beeswax, dated somewhere around 1680, and more probably from Asiatic than from European honeybees. It is estimated that 10 or 11 tons of this beeswax were the cargo of a ship, possibly sailing from Manila in the Philippines, that was

Bees in art and in everyday life

FIG. 270 *The beeswax model for Benvenuto Cellini's bronze statue of Perseus with the head of Medusa.*
PROPERTY OF MUSEO NAZIONALE DEL BARGELLO DI FIRENZE.

wrecked off the Oregon coast. The *San Francisco Zavier*, which vanished after leaving Manila in 1705, is one possibility.[190]

I have left to the end one of the most important and widely used ancient technologies involving beeswax: *cire-perdue* or lost-wax casting.[207,262] The statue or other object to be cast in metal is first made in beeswax; this is coated with a fire-proof material and the whole heated until the wax has melted and run out. The molten metal is finally poured into the cavity left by the wax. In principle the method is simple, and beeswax was a very suitable medium, available from the earliest times. The method was known to the Sumerians, and in the Indus Valley and Egypt, before 1000 B.C., and probably also in China. In another method the wax model is made on a refractory core; a third method, first used by the Greeks, enabled the original model to be preserved; a layer of wax was formed over it, which was later melted out and replaced by metal.

Many of the world's best known statues were made by the *cire-perdue* process. The wax model for Benvenuto Cellini's famous Perseus can still be seen in the National Museum in Florence (Fig. 270). Cellini's detailed description of the process of casting can be read in his autobiography, which he started in 1558, and it shows vividly some of the problems encountered in casting large amounts of molten metal. Pouring channels and air vents had to be provided, and iron rods used for strengthening. A letter written by Michelangelo in 1507 shows that he had met similar difficulties. The books referred to above[207,262] describe and illustrate many of the bronze statues made by this use of beeswax.

Bees in art and in everyday life

In West Africa the same method was used for casting brass by the Ashanti and related peoples of the former Gold Coast, now Ghana. These people traded among themselves and with colonials using gold dust as currency. The dust was carefully weighed against brass weights, which took the form of tiny figures—often figures of fun—related to situations of everyday life. The Museum of Mankind in London has many hundreds of them, and in 1978 mounted an exhibition, 'Ashanti gold-weights'. The descriptive leaflet says: 'Nearly every senior male engaged in trade among the Ashanti and related peoples ... owned or had access to a set of weights, scales and small gold-dust storage boxes. The hundreds of thousands of weights produced by large numbers of craftsmen up to the end of the last century encouraged a wide range of styles and different representations of the same subject matter, which give a witty and varied portrayal of many aspects of African life. The court and its monarchs, farming, hunting, warfare and other domestic scenes are all vividly illustrated on the weights.' In 1979 the industry was endangered by a shortage of beeswax.[210]

In the Americas there were no honeybees until settlers from Europe took them there, from the seventeenth century onwards. But beeswax existed in tropical America even in prehistoric times, produced by several species of stingless bees (Meliponini). It was harvested from wild nests and from hives, and put to some of the same uses as in the Old World; application of the *cire-perdue* process for gold casting was described by a priest, Sáhagun, in the sixteenth century.[266] In 1979 a Royal Academy exhibition in London displayed 'The gold of El Dorado', and the catalogue[46] describes in detail how the magnificent Inca gold jewellery was made by the *cire-perdue* method, and illustrates hundreds of pieces from the Colombian Andes. The minute filigree work is in complete contrast to the large bronze statues for which similar methods and materials were used in the Old World.

CONCLUSION

The previous chapters of the book provide an introduction to archaeological material surviving from honey hunters and beekeepers. In all parts of the world, and from the end of the Ice Age until the present day, these people have—in spite of stings—worked in direct contact with the bees and looked after them, in order to harvest their honey and wax. Chapters 7 and 8 record for the first time the exceptionally rich archaeological heritage of beekeeping in Britain and Ireland, and I hope that the findings will spur others on to explore and record, more vigorously than before, the archaeology of beekeeping in other countries. The astonishing variety of artefacts discussed in this final chapter bears witness to the fascination that bees have held for mankind through the centuries—and more especially their mysterious and apparently disciplined community within the hive.

The relationship between man and bees—and their honey and wax—has been a long one, and in many parts of the world it has had considerable cultural and economic significance. We should preserve the archaeological remains that will enable this relationship to be explored and understood by future generations.

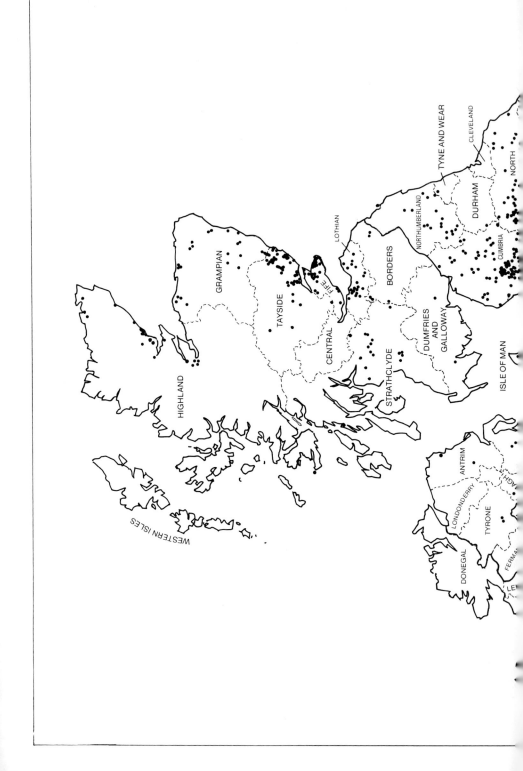

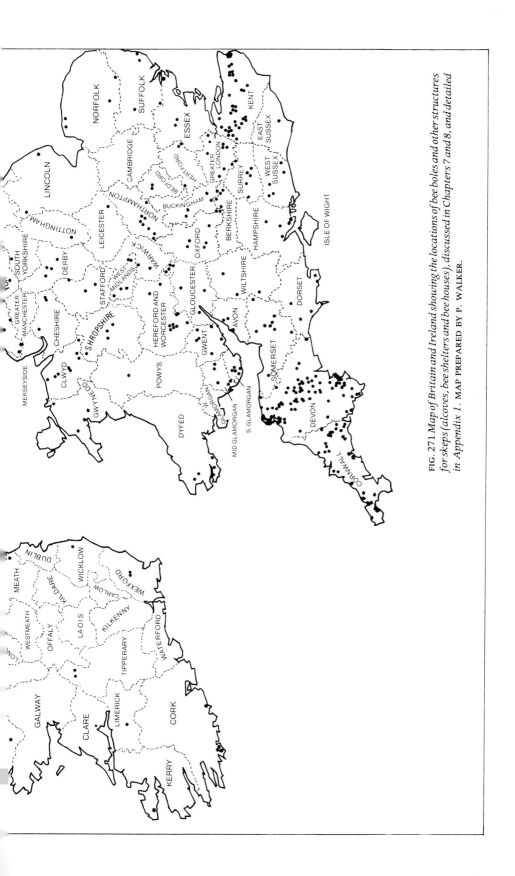

FIG. 271 *Map of Britain and Ireland showing the locations of bee boles and other structures for skeps (alcoves, bee shelters and bee houses), discussed in Chapters 7 and 8, and detailed in Appendix 1.* MAP PREPARED BY P. WALKER.

Appendixes

APPENDIX 1. A summary of records in the IBRA Register of bee boles, bee houses and other shelters for skeps, 1952 to June 1981

The records are summarised in the following tables:

Bee boles	p. 252
Alcoves	p. 310
Bee houses	p. 314
Winter storage for bees	p. 318
Shelters	p. 321
Other structures	p. 328

In each table, records for England, Wales, Scotland and Ireland are presented in that order, each in alphabetical order of county; the Channel Islands and the Isle of Man are listed with the English counties. Entries for bee boles (only) in France and Greece are on pp. 308–9.

Information in brackets ( ) is approximate and/or deduced; information with a question mark ? is tentative.

When the IBRA Register was started, metric units were not in common use in Britain. Most measurements are quoted in inches (1 in. = 2.54 cm); where necessary for clarity, inches are denoted by ″. Large measurements are in feet (1 ft = 30 cm), denoted by ′.

Column headings are designated by numbers, as follows:
1. IBRA Register Number.
2. Address of property.
3. Ordnance Survey National Grid reference, in England, Wales and Scotland to 6 figures if the exact location is known, otherwise to the 4-figure reference of the town or village. Irish National Grid reference used for Ireland. None for France, as no suitable system is available.
4. *Bee boles and winter storage:* number of recesses in set. If recesses in a set have different characteristics, these are given below the main entry.
   Alcoves: number at the site, then (in brackets) estimated total number of skeps accommodated.
   Bee houses and shelters: known or estimated number of skeps/hives accommodated.
5. *All recesses, including shelters:* dimensions in inches: height × width × depth.
   — = unknown.
   $x \times x \times x$ means size unknown, but height = width = depth.

Appendix 1

For a recess with a sloping roof, f and b are front and back height, respectively.
On line below, at left: height above ground of base and/or shelf.
at right: distance between adjacent recesses.
For *bee houses*, description of building given in column 5; ht = height.

6. Direction faced (aspect); N† means that the recess faces N, but a hole has been made in the wall to enable bees to fly out S. Column omitted for *bee houses, winter storage*.
7. Date of construction: most dates are only probable; those in italics are certain; many records indicate the century only; date known for dwelling only is given in column 13.
1700s = eighteenth century; 1700s-I = first half of eighteenth century; 1700s-II = second half of eighteenth century; *c.* = circa.
8. Situation of wall or structure within the property. Any bee boles in the wall of a building face outside; similarly for alcoves and shelters. For *winter storage* the entry gives either the type of structure or its situation.
9. Building material of wall or structure:

B = brick	Ga = granite	Ss = sandstone
C = cob	Gs = gritstone	Th = thatch (roof)
Ch = chalk	Ls = limestone	Ti = tiles (roof)
dS = dry stone	M = metal	W = wood
F = flint	S = stone	Ws = whinstone
	Sl = slate	

    If two materials are used, they are both given, e.g. BF.
10. *Bee boles* and *winter storage recesses*: shape of (vertical) opening.
    R = rectangular   T = with triangular arch at top
    A = arched (round arch springing from sides)
    D = domed (round arch springing from base)
    Any other shape described in column 13.
    Alcoves: column 10 omitted; all are arched unless stated in column 13.
    Bee houses and shelters: material of roof, symbols listed as in 9 above.
11. Condition at date of record:
    e = excellent, g = good, f = fair, p = poor, b = bad.
    If no longer extant (filled in or demolished), column 13 explains.
12. Photographs held by IBRA:
    B = black and white print, C = colour print, T = transparency, D = drawing.
13. Additional notes, including: any illustration in Chapter 7 or 8; local names; arrangement of tiers (e.g. 2 tiers of 6 & 5 means 6 in the lower row and 5 above). Any dimensions are given as in column 5: height (ht) × width × depth. NT = National Trust property.

BEE BOLES

1	2	3	4	5	6	7	8	9	10	11	12	13
ENGLAND												
Avon												
314	The Knoll Maternity Home, Chapel Hill, Clevedon	ST4071	1	24 × 20 × 36 17	S†	1739	garden	S	R			ht at back 12″; there is hole 3″ × 3″ at top back; wooden lintel and base; there may have been door
447	Olveston Court, Olveston, Bristol	ST598871	21	32–18 × 12–29 × 12 (48) $8\frac{1}{4}$–$19\frac{1}{4}$′	SW NW	1600s	garden & farmyard	Ls	T	g		described 1969[316]
750	Filton Rectory, Filton	ST6079	4	(20) × (20) × (20) (96) (10–12′)	2SSW 2ESE	1400	garden	dS	R			may have been 1 more (NNE); all demolished 1958
Bedfordshire—see below, 'Bee houses' and 'Winter storage for bees'												
Berkshire												
350	Mill Green House, Church Street, Wargrave	SU784786	3 1 2	as under 10 × 13 × 10 30 17 × 15 × 8 16 15	E	1500s	garden	BF	R T	g g		
406	Hitcham Old Garden, Burnham	SU9282	2	17 × 13 × 10 (84) (36)	S	1500s–II	garden	B	T	g		
449	Willow Farm, Oakley Green, Windsor	SU925763	4	11 × 11 × 11 38 27	S	1500s	garden	B	T	g	B	Fig. 136
523	Burnham Abbey, nr Maidenhead	SU9380	18			1500s–II		B	T	f	C	
Buckinghamshire												
333	The Old Rectory, Beaconsfield	SU944900	9	22 × 21–22 × 9–10 26–40	W	1500s	garden	B	A	f	BT	Fig. 141
448	Remnantz, Marlow	SU8586	6	24 × 18 × $8\frac{1}{4}$ 0.35 25	SE	1720s	garden	B	A	e		2 tiers of 3 & 3; two boles have been made into windows 24″ deep
550	Chilton House, Chilton, nr Aylesbury	SP6811	11 3	as under 15 × $15\frac{1}{4}$ × 10 (48)	E	1740	garden	B	A	g	C	some have mortar inside; distance apart varies between 20′ and 35′.

	Location	Grid ref	n	Dimensions	Dir	Date	Placement					Notes
			4	25½ × 16–19 × 9 40–48	E							
			2	26 × 18 × 9 40–42	S							
			2	26 × 18 × 9 19–26	W							

Cambridgeshire, Channel Islands—no records of any type

Cheshire

No.	Location	Grid ref	n	Dimensions	Dir	Date	Placement	T1	T2	T3	T4	Notes
543	Long Lane Farm, Bollington, Macclesfield	SJ936793	3	30 × 39 × 17 22 14	S	1700s	garden	dS	R	g	B	each for ?2 skeps
649	Pedley House, Pedley Hill, Rainow	SU950758	3	42 × 45 × 12 (24) (96)	W	1762	garden	Gs	R	f	B	each for 2 skeps

Cleveland—no records of any type

Cornwall

No.	Location	Grid ref	n	Dimensions	Dir	Date	Placement	T1	T2	T3	T4	Notes
317	North Wall Farm, Tideford Cross, Saltash	SX3461	6	18 × 20 × 21 6, 24, 42 (18)	S	pre-1880	field	Sl		f	B	3 tiers of 3 & 2 & 1 in 'beehive wall' adjoining field wall; shape = flat-topped arch
318	The Rectory, Sheviock, Torpoint	SX371551	8	30 × 21 × 18 (24)	S	1845	garden	S	T	f	B	slate alighting slab runs along all boles
335	Dannon Chapel Farm, Delabole	SX039824	5	16 × 19 × 19 (24)	E	pre-1850	garden	S		g		shape almost rectangular, rounded at back
356	Heligan House, St Ewe, St Austell	SW999466	14	24 × 24 × 26 30, 60 9	S	?1800s–II	next to garden	B	A	g		2 tiers of 7 & 7; each bole has pivoted wooden door with flight hole; concrete inside, rounded at back
399	Carthew Farm, Carthew, St Austell	SX003559	13	18 × 18 × 12 (18), ?24	S		farm building	Ga	D	g	B	2 tiers of 6 & 7
481	Tregreenwell, Michaelstow, St Teath	SX075804	2	18 × 18 × 18 21 32	SSE			S		f		probable shape = flat-topped arch; slate base
589	Fradd's Meadow, St Tudy, Bodmin	SX0676	4	18 × 18 × 16¼ 18 4¼	S		garden	C	R	g	C	brick piers, wooden lintel; slate base projects; house 1700s–I
627a	Godolphin House, Breage, Helston	SW601318	4	17–20 × 18–23 × 11½–15 27 11'–12'	S	1600s–I	stable	S	R	f		others may have been destroyed; recesses inside ruined building not for bees (627b)
628	Apple Orchard Farm, Gwithian	SW586414	3	14 × 11 × 17 >18 10½, 6	W	c. 1600	barn	Ga	R	f		

253

1	2	3	4	5	6	7	8	9	10	11	12	13
715	The Haven, Mount Hawke, St Agnes	SW715474	—				yard	C	R	g		C	one has wooden base; wall = cob + granite + copper slag; winter storage 718a
716	Ivey Farm, Altarnun	SX1276?	—										derelict farm
717	Garrow Farm, Altarnun	SX147780	—										
718b	Penpol House, St Erth, Hayle	SW5637	2	(11)×(13)×(13) 26 11	NE	c. 1750							
719	Girling Farm, Relubbus, Penzance	SW569325	—										
721	Vounder Vean, Jubilee Place, Pendeen, nr Penzance	SW3834	9	9–18×12–18×(12) 6–12 (72)	5S 4E		garden	Ga	R	g			
723	Upton Hall, Upton Cross, Liskeard	SX279724	3	22×16×14 (36) (36)	S	1500s	garden	S	R	f			slate base projects
724	Northcombe Farm, Northcombe, Linkinhorne, Callington	(SX2974)	—				under barn steps						
726	Pencroud Farm, Menheniot, Liskeard	SX2862	6	17×18×19 (36) 9	S	1825	garden	S		g		C	2 tiers of 3 & 3; shape quadrilateral, narrowing at top; rounded at back; slate base projects
727	Bathpool, North Hill, Launceston	SX2874	—										address incomplete
728	Bearah Farm, North Hill, Launceston	SX2874	—										
729	Uphill Farm, North Hill, Launceston	SX295740	7	10×15×18 (48) (72)	6S 1E		garden & farmyard	S	R	4g 3f		B	
731	Trevorian, Sancreed, Penzance	SW4228	—										none found in 1979
732	Chapel Euny Farm, Sancreed, Penzance	SW420293	—										
733	Church Row, Sheviock, Torpoint	SX370551	6	18×22×14 15	S	1500s	garden	S	R	f			
734	Polgreen Farm, St Mawgan-in-Pydar	SW863664	—	12×12×12				C					

No.	Location	Grid Ref	n	Dimensions	Orient.	Date	Situation	GaB	R	f	B	Notes
735	Killivose, Camborne	SW646388	5	16 × 16 × 16, 13	?S		farmyard			f	B	wall of slate stone; probable shape quadrilateral, narrowing at top, rounded at back; there was probably 1 more
737	Trenouth Farm, St Ervan, Wadebridge	SW904702	9	28½ × 23 × 23–24, 30 20	SW		garden	dS		8g 1b	B	plastered inside; slate base
774	Newton Farm, Launceston	SX345830	6		SSW				D			
776	Tempellow, Liskeard	SX2564	(5)	(20) × (20) × (18)			yard					

Cumbria

No.	Location	Grid Ref	n	Dimensions	Orient.	Date	Situation	GaB	R	f	B	Notes
34	Hill Top Farm, Near Sawrey, Ambleside	SD3796	3	as under: 12 × 19 × 23, 13½ × 21 × 22, 20 × 66 × 23 (36)	S	1500s–I	garden	dS	R	p	BT	next to Beatrix Potter's house—one bole pictured in *Jemima Puddle-Duck*;[238] largest held 3 skeps; NT but not for visiting
40	Greta Grove, Keswick	NY2624	14	18 × 15 × 14½, 0, 18 4½	N	pre-1850		SSl	R	p	BT	2 tiers of 7 & 7; brick piers; in 1979 wall collapsing, only 3 left
71	Seathwaite, Valley of Duddon	SD9623	1	21 × 39 × 24, 24	SE		garden	dS	R	g	B	wall of Lakeland stone; bole for 2 skeps has slate top and base with alighting strip
84	High Yewdale Farm, Coniston	SD3199	8	16 × 17 × 18, 32 ?21	E		?garden	dS	R	g	B	house 1500s
85	Holme Ground, Coniston	SD3098	6	20 × 20 × 20, 24 ?20	E		garden	S	R	4f 2p	B	house 1600s
86	High Water Head, Coniston	SD315984	2	18 × 20 × —, 30	E	1600s	garden	dS	R	f	B	boles filled in but outline clear; there was probably bee house (now ruined) in same wall
100	Bridge End Farm, Thirlspot	NY315194	3: as under; 1: 23 × 22 × 25; 2: 23 × 41 × 25 (24)		SE SW	1600s–II	?garden	dS	R	f	B	bees kept here till c. 1900; photo published 1953[101]; each for 2 skeps
106	Weir Cottage, Great Langdale	NY3106?	2	25 × 43, 60 × 23, 24 ?3	SSW	c. 1800	garden	dS	R	f	B	for 2, 3 skeps
107	Low Colwith Farm, Little Langdale	NY3103	2	22 × 24 × 18, 24 ?12	E	1700s	garden	dS	R	f	B	structure built out from wall; in use c. 1930

1	2	3	4	5	6	7	8	9	10	11	12	13
130	Church Style, Pennington	SD262774	3	(48) × (48) × 20 (12), ?36 742	SW	1782	building	S	R	p	B	whole structure projects; each bole originally had brick base, also shelf; held 4 skeps; 4th bole gone; bees at this address in 1920s; building now ruined
136	Albyfield Farm, Cumrew	NY548525	2	33 × 43 × 22 26 ?3	S		garden	Ss	R	f	B	whole structure projects; separate, shaped stone bases with alighting area were in boles; each for 2 skeps
137	Helme Farm, Cumrew	NY5550	3 2 1	23 × 21 × 18 23 × 18 × 18 29	N		house	Ss	R	g		one now blocked in to 11"
138	Meadow Park (was The Million), Hayton, nr Carlisle	NY5158	1	(18) × 22 × 19 36	S		garden	Ss	A	g	B	base projects 5"; described as 'not old'; known that bees were kept there; photo published 1930[45] shows skep in bole
150	The Croft, Bouth, Ulverston	SD3386	1	17 × 24 × 16 20	S		garden	dS	R	f	T	base projects
157	Rothvale, Kirkby Moor	SD255810	4	19 × 20 × 12 0	3S 1W	1850s	garden	S	R	p		unusual: wall specially built for bee boles
187	Stanegarth, Bampton	NY497177	2	27 × 36 × 27 15	E	1400s or 1600s	?garden	S	R	g	TD	roof stone projects 12", base 3½"; back corners rounded
190	The Old Manor House, Skelwith Fold, Ambleside	NY3502	1	38 × 44 × 17	S	?1600s	garden	dS	R	g		for 2 skeps
194	The Bield, Little Langdale	NY3103	2	as under (36) × (36) × 20 (54) × (54) × 20	S		garden	S		f		'shelf 20" above ground'
211	Ring House, Woodland, Broughton-in-Furness	SD8924	3	(24) × 42 × — (24)	SE		garden	S	R	g	B	whole structure projects from wall, with sloping roof; each for ?2 skeps; bees kept there c. 1920; now filled in
212	Kirkby Hall, Kirkby-in-Furness	SD2584	5	16½ × 20 × (24) 30, 48 ?4½	SE	1639	house	S	R	f	B	3 tiers of 2 & 2 & 1; slate base and top; 2 filled in
213	Dove Bank, Kirkby-in-Furness	SD235844	6	19 × 20 × ?16 40, 72, 8½ ?8	SW		barn	S		f	B	3 tiers of 3 & 2 & 1; shape = pointed arch; brick surround; slate base projects
214	Tenterbank, Grizebeck, Kirkby-in-Furness	SD2485	3	15 × 15 × 15 (72)	S		?barn	S	R	g	B	slate base and top

No.	Location	Grid ref		Dimensions	Orient.	Date	Setting					Notes
216	Ashlack Hall, Grizebeck, Kirkby-in-Furness	SD245855	?3	?24 × 46 × 20	SE		?garden	S	R	g	B	for ?2 skeps; triangular opening above, and may be another rectangular recess below (all adjoining); house 1300s
217a	The Hill, Heathwaite, Grizebeck, Kirkby-in-Furness	SD241870	15	20 × 20 × 20 (c. 36)	6S 6N 3E		garden	S	R	g	B	'bee stacks'; slate base and top; house 1641; in 1954 a 60-yr-old remembered bees in most boles; bee shelter 217d
217b	The Hill, Heathwaite, Grizebeck, Kirkby-in-Furness	SD241870	3		N		barn					
217c	The Hill, Heathwaite, Grizebeck, Kirkby-in-Furness	SD241870	3		?N		enclosure					
218	Anna's Croft, Chapel, Kirkby-in-Furness	SD239837	2	21 × 46 × 21 15 (14)	S		garden	S	R	g	B	slate top and projecting base; each for 2 skeps
219	Yeat House, Woodland, Broughton-in-Furness	SD248887	5	22 × 20 × 18 (24) (10)	SE		garden	S	R	g	BT	slate base and top
238	Drawwell, Howgill, Sedbergh	(SD6692)	2	17 × 18 × (24) (20) (60)	S	?1700s	?farmyard or garden	S	R	f		
241	Bank Ground, Coniston	SD318971	4	24 × 24 × 18 4	S	1750	garden	dS	R	g		
282	Low Bield, Knype Fold, Outgate, Ambleside	SD344997	2	13½, 16 × 16 × 24 24 28	SE	pre-1793	?paddock	dS	R	g	T	rawstone wall; flag base projects
285	Dovecote Farm, Sedbergh	SD6692	2	24 × 18 × 18 40	S	1840		S	?A	e		
313	nr Plough Inn, Kendal Road, 4 miles past Kirkby Lonsdale	(SD6278)	5	x × x × 0.7 x			orchard	dS	R	2f 3p	B	2 tiers of 2 & 3
326	High Bayles, Penrith Road, Alston	NY706449	3	(24) × (24) × 27 (24) ?9	E		garden	S	A	g	T	free-standing structure about 15' from house
454	Foulstone, Kirkby Lonsdale	SD562808	3	14 × 14 × 14 30	S	1655		S	R	g		
461	Fell Foot Farm, Little Langdale	NY299032	2	as under 18 × 18 × 14 (48) 20 × 27 × 14	E	?1700s		(dS)	?R	f	T	

257

1	2	3	4	5	6	7	8	9	10	11	12	13
482	Moor House, Warcop, Appleby	NY743170	4	33 × 26 × 10 18 16	SE	pre-1850	garden	Ss Ga	R	3g	C	flagstone base projects 2″–3″; 1 filled in; all may have been deeper
513a	Sykeside, Grasmere	NY341070	2	24 × 18 × 14 (12) (12)	S	pre-1800	yard	dGa	R	f	BT	slate base; bee shelter in same wall 513b; next to Dove Cottage where Wordsworth lived
517	Bank End, Millhouse, Hesket-Newmarket, Wigton	NY359381	3	26 × 24, 27 × 13 (24)	E	1784	house	S	A	e	T	wall is plastered; 2 boles rounded at back
519	Bradley Field House, Underbarrow Road, Kendal	SD495925	?3	(48) × (36) × — 0 3 — × 10 × —	E S	1670	yard garden	S ?A	A			possibly only 2; now filled in (1971) possibly rectangular; filled in
531	Curthwaite House, Curthwaite, nr Carlisle	(NY5049)	6	26 × 22 × 27 (12, 60) >12	SW	pre-1750	garden	BSs	R A	f		2 tiers of 3 & 3; top boles arched; base projects
532	Light Hall, nr Windy Hill, Rusland, via Ulverston	SD3489	5	as under 26 × 18 × 18 37 22 × 20 × 18 37 23 × 20 × 18 34 25 × 22 × 22 24 × 22 × 16 36	W		garden	dS	R			originally plaster inside; also 2 shelters 632a, b; not for visiting (1979)
533	Bowkerstead Farm, Satterthwaite, nr Ulverston	SD3491	6	smaller than 532			?field	dS			B	2 tiers of 2 & 4
536	Bee Boles (was Croft Cottage), Far Sawrey	SD380955	3	23 × 23 × 14 (24–36) 18	S		garden	S	R	g	T	
537	White Cottage, Cunsey, Ambleside	SD381941	6	27 × 21 × 21 (24–36) 12′	S		orchard	S	R	4f		2 blocked up
539	North View, Renwick	NY595435	3	30 × (36) × 18 (84)	SW		garden	Ss	T	g		?roofs project
540	Holly Bank, Renwick	NY5943	5	34 × 24 × 19 >60 >96	SW		orchard	Ss	T	g		?roofs project
541	Grassgarth, Crosthwaite, Lyth, Kendal	SD458869	2	21 × 42 × 21 (48) (24)	S		garden	dLs	R	g	BT	each for 2 skeps
588	Toddell, Brandlingill, Cockermouth	NY122264	3	23½ × 22 × 19 14 4½	SSW	1700s–I	garden	Ls	R	g	T	brick piers; shelf and ? top project

No.	Location		Dimensions		Orient.	Date	Site		R		C	Notes
593	Stonegarthside Hall, nr Penton, Carlisle		NY473804	3	S		garden	dS	R			
600	7 Tudor Square, Dalton-in-Furness		SD2374	4	SE		garden	dSs	R	g	BT	plaster-lined; base projects; roofing slates project over lintel
601	Orchard Rise, Cartmel		SD3879	4	S	c. 1800	garden	Sl	R	g	T	wall of Broughton stone
604	Lowick House, Lowick Green, nr Ulverston		SD298854	7	S		orchard	S	R			
				4						3f		
				1						1p		
				2						p		
										p		
605	Nettleslack Farm, nr Ulverston		SD284835	1	S	1809		dSl	R			there were 4 more
606	Little Urswick, Urswick		SD264738	2	SE		garden	Ls	R	g	B	projecting metal bar at one lower corner
607	Town Yeat, High Nibthwaite, nr Ulverston		SD294897	3	SW	c. 1700	garden	dSl	R	g	T	
609	Lake End, High Nibthwaite, nr Ulverston		SD295897	3	SE		garden	dSl	R	f	T	another recess of similar depth was probably bee bole; house 1700
610	in field opposite Abbey Farm, St Bees		NX967121	8	?S		?field	Ss	R	g	C	brick piers; base projects; believed used by Benedictine nuns
622	Cinder Hill, Finsthwaite, Ulverston		SD361871	8	SW		garden	dSl	R	f	T	'bee shells'; rounded at back; house 1600s–I
623	Appletree Farm, Blawith, Lowick, Ulverston		SD278887	4	?SE	1600s–II	garden	dSl	R	f	BT	2 tiers of 2 & 2; 'bee holes'

Row 600 dims: (24) × 21 × 21 / av. 24 (12)
Row 601 dims: 21 × 18 × 23 / 22 (12)
Row 604 dims: as under / 30 × 24 × 24 / 12 20 / 29 × 17 × 15 / 24 / 26 × 24 × 21 / 24 (24)
Row 605 dims: (24) × 19 × 19 / 15 (12)
Row 606 dims: 26 × 22 × — / 27
Row 607 dims: (24) × 18 × 21 / (24) 7½–9
Row 609 dims: as under / 18 × 22 × 19 / 9 / 22 × 22 × 18 / 11 / 16 × 27 × 19 / 12–18
Row 610 dims: (19) × (20) × ?16 / (20) ?4½
Row 622 dims: (24) × 22 × 18 / 21–24
Row 623 dims: 23 × 22 × 23 / 0, 22 17

259

1	2	3	4	5	6	7	8	9	10	11	12	13
624	Pithall, Blawith, nr Ulverston	SD286888	6 1 2 3	as under 17 × 34 × 17 −18 ?40 17 × 17 × 17 ?40 ?17 × ?17 × ?17 18 ? 8	SSE	1700s–II	field	dSl	R	f	T	2 tiers of ?3 & ?3, perhaps more (wall collapsing)
626a	Whetstone Croft, Woodland, nr Broughton-in-Furness	SD232899	1	22 × 23 × 23 20	SW	1753	garden	Sl	R	g	T	bee shelter 626b
629	Orchard House, Dean, Cockermouth	NY073253	2	22 × 21 × 13 42	1SE 1SW	c. 1860 or earlier	garden	S	R	g	BT	4 wider recesses in same wall
636	Thurston, Coniston Lake, Hawkshead	SD314965	5	20 × 20 × 16 25, (36)	S	1800s–I	garden	S	R	g	B	2 tiers of 3 & 2; ?slate base; 1 'similar' in same wall
640	18 Market Place, Dalton-in-Furness	SD225739	3	(24) × 19–20 × 19 (36) 8	S	?pre-1900	cottage	Ls	A	f	BT	wall has slate 'throughs'; brick arch
641	6A Market Place, Dalton-in-Furness	SD225739	7	19 × 18–21 × 15 42 12–16	SE	1800s–I	(garden)	Ls	R	g	BT	
642	Wellhouse or Wellside Farm, Netherhouses, nr Ulverston	SD274821	2 2 2	18 × 18 × 16 21 22 21 × 20 × 16 21 21 26 × (24) × 14 22 (24)	SE S ?SE	1800s–I	garden garden garden	dSl dSl dSl	R R R	lg f f	T T T	1 filled with rubble filled in to prevent wall collapse filled in
643	Holm Bank House, Great Urswick	SD271739	5	15–16 × 21–24 × 22 28	1SE ?4SW	1786	garden	Ls	R	f	T	slab bases; back now bricked in, but may have been open on both sides
658	Borrenthwaite Hall, Stainmore, Kirkby Stephen	NY855137	7	(54) × (30) × (30) (12)	5SW 2SE	1700s	garden	S	A	g		rounded at back; each had shaped sandstone base; small rectangular recess (10" × 8") between 2 pairs of boles
662	4 Orchard Close, Sedgewick, nr Kendal	SD515871	4	(24) × (24) × 19 (12) 54	SE		garden	Ls	R	g		course of green slate projected over lintel
667	Meaburn Hill Farm, ?Maulds Meaburn, Penrith	NY6217	—		E		orchard	Ss				
669	Craglands, Clawthorpe, Burton	SD535777	2	24 × 18 × 15 24 27	SW	1600s	garden	dLs	R	f	T	wall of granite and limestone
672	11 Rosemary Lane, Fellside, Kendal	SD513927	4	>24 × >24 × —				?A				

No.	Location	Grid ref	N	Dimensions	Orient	Date	Setting				Notes
674a	Great Hartbarrow Farm, Winster, Windermere	SD407906	4	16 × 17½–19½ × 18, 19 36 19	SE	1684 or earlier	house	S	R		B	2 pairs, each one bole above another in rendered wall; base of lower bole projects 1"; slate base of upper is set back 1" thatch projects over boles;
674b	Great Hartbarrow Farm, Winster, Windermere	SD407906	5	16 × 19 × 16			garden	dS	R	p	B	in 'monastery garden' (farm has links with Cartmel Priory); one bole is in an upper wall
675	Brinn's Farm, Shap	NY559171	3	x × x × — 0.7x			garden	dLs	R			
681	Whit Beck, Garsdale, nr Sedbergh	SD705908	1	(24) × (24) × 18 (24)	E	?1800s–I	garden	Ga	R	g	C	part of projecting stone base remains
683	Bridge House, Cartmel Fell	SD420884	2	as under 19 × 17 × 16 21 × 17 × 19 14 32	S	1600s	(orchard)		R	f	B	wall of slate and rock
686	Irvings Farm, Blencow, Penrith	NY4532	4	(36) × (24) × (12) (24)	?S			dLs				boles destroyed by 1979
687	Alma House, Wordsworth (?St), Penrith	NY5130	4	1.5x × x × 0.75–1x x 0.5x			garden	dS	R	f		? for bees; base and lintel project; demolished by 1979
688	Mansion House, Penrith	NY5130	3	>x x – x x			garden	Ss	A			
689	Brow House, Black Beck, Egremont	NY027069	3	20 × 16 × 16 14 (6', 12')	SW	1700s–II	road bank	S	R	g	T	bank faces garden; one base stone remains
690	Askham	?NY5123	1	?x × ?2x × —				dS				'has a fine lintel'
710	Town Head, Long Marton, nr Appleby	NY664247	4	30 × 23 × 14 23 (12)	SSE	pre-1820	orchard	Ss	T	p	T	slate base and gables projected 12"–15", now mostly gone
711	Sizergh Cottage, Sizergh, Kendal	SD499869	3				orchard		R			base projects, rounded at back; owner could not find in 1978
712	Ravensbarrow Lodge, Cartmel Fell	SD410880	2	18 × 24 × 11, 13 54 18 24 × 25 × 24 3, 27	E		garden	S	R	g		slate top; house 1700s–I
			2		E		?house	S	R	g		one above the other; house 1700s–I
713	derelict farm, Lupton, nr Kendal	SD554807	6				field	dS				2 tiers of 3 & 3
754a	High Longmire Cottage, Bouth, Ulverston	SD326870	3	as under 22 × 21 × 17 17 × 18 × 20	S		garden	S	R	f		base projects; not for visiting (1979)

1	2	3	4	5	6	7	8	9	10	11	12	13
754b	High Longmire Cottage, Bouth, Ulverston	SD326870	1	20 × 16 × 27 av. 22 av. 24 44 × 36–32 × 20 18	E		garden	dS	A	g		shape tapers up to spring of arch; arch projects slightly; described as winter storage bole
771	Croft Head, High Row, Haltcliffe, Hesket-Newmarket, Wigton	NY358355	3	as under 12 × 16 × 12 (36) 20 × 27 × 13 13 17 × 20 × 14 13	S		garden	S	R	f	B	sandstone base
772	Banks Foot Farm, Lanercost, Brampton	NY565643	4	(24–36) × (24–36) × 18 >36	2S 2?E		orchard	S	R	g	B	sandstone shelf; projection above
797	Beckside House, Cartmel	SD374805	4	(24) × 20 × 18 (24) 72, 10′	S	pre-1816	garden	Ga	R	g		2″ slate projection above
798	Grange Cottage, Grange-in-Borrowdale, Keswick	NY2517	1	(24) × (24) × 18 (24)	NW		farmyard	S	R	g	T	wall = granite + slate + sandstone; used recently as stand for dye water for sheep dip; use for bees not recalled
840	Mill House, Calder Bridge, nr Seascale	NY042061	2			1817		S	R			
841a	Bridge Field Farm, Spark Bridge, nr Ulverston	SD3085	5				field	Ga				Fig. 160 shows boles built step-wise up slope; bee shelter 841b
843	Gateside Barn, Coniston	SD301972	4 2 2	as under 51, 17 × 24, 28 × 13 0, 51 41, 18 × 33, 25 × 18 0, 41	E	1700s–I	ramp	S	R	g		built into side of ramp leading to barn; slate base projects slightly one above another similar; 51″ from first pair
847	Raw Head, Great Langdale	NY304067	2	19, 15 × 36 × 20 38, 61	S	1600s–II or 1700s–I	house	S	R	g		one almost over the other; 4″ slate base projects

Derbyshire

| 47 | Robin Hood, Baslow | SK281721 | 5 | 28 × 26 × 14 10 (6) | S | 1798 | garden | dS | R | p | B | |

No.	Name	Grid Ref		Dimensions		Orient.	Date	Location					Notes
48	Cuckoostone Grange Farm, Matlock	SK314624	6	36 × 42 × 14 18–30 (12)		S	pre-1870	orchard	S	R	p	B	'bee garths'; each for ?2 skeps
97	Millhouse, Millersdale	SK139733	1	44 × 36 × 18		S	*1800s*	garden	Ls	R	g	B	'bee niche' in rock buttress; recess converges to back
660a	Milford House Hotel, Bakewell	SK216687	5	39 × 20 × 16 18 7'–9'		S	*c. 1870*	house	Gs	A	f	C	wall early 1700s, refaced and boles built 1870; door fixtures
660b	Milford House Hotel, Bakewell	SK216687	5	30 × 18 × 12–14 (6) 9'		S		garden	Ls	R	f	C	

Devon

No.	Name	Grid Ref		Dimensions		Orient.	Date	Location					Notes
25	Zephyr Cottage, Lynton	SS7149	5	20 × 24 × 18 (12) (80–90)		S	*c. 1750*	garden	S	D	f	C	called ?'bee-holes'; projecting bases broken; probably used only pre-1900
37	by Brook Stores, Croyde, Barnstaple	SS450395	4	14 × 18 × 10 (24) 12		E	*c. 1800*	?garden	S	R	p	B	rounded at back, mortar inside; now damaged, but originally slate base projected
50	Pack of Cards Hotel, High St, Combe Martin	SS5946	12	18 × 18 × 15 18, (38) 4½		SE	?1644	garden	S	R	f	B	2 tiers of 6 & 6; brick piers; 4 boles broken
70	Horslake, Cheriton Bishop	SX773939	2	20 × (24) × 22 (84) ?18		SE	?1650s	house	C	D	e	B	photo published 1953[101]
122	Ford Farm, Manaton	SX7581	4	20 × 20 × 15 0		S	1700s or earlier	garden	S	R	p	BT	other wall niches on this farm
143	Brecan Cottage, Croyde, Barnstaple	SS4539	1	15 × 17 × 15 (48)		S		house	C	D	e	B	last used end of 1800s; now inside house
144	Gwynant, Chapel St, Georgeham, Barnstaple	SS4740	6	(18) × 18 × 15 (54) 8		WSW	1800s–I	garden	S		g	B	shape = flat-topped D — see Fig. 157; partly filled in at bottom.
152	The Old Post Office, Longdown, Exeter	SX8690	2	— >12'		SE		house	C	D	g	B	Fig. 161
153	Venn Farm, Morchard Bishop	SS780058	5			S		house	C	D	g		
165	Chawleigh Week Farm, Chulmleigh	SS683132	4	— (72)		E		stable	C	D	g	B	boles in use in 1930s
200	Sheepwash Farm, Molland, Bishop Monkton	SS791272	76	18 × 18 × 15 66		S		house	S	D	3f	B	?rounded at back, plastered inside; wood base; 2 filled in
204	North Heathercombe, Manaton	SX7181	6	(18) × (18) × (18) (24) ?7		E	pre-1500	garden	Ga	R	p	BT	

1	2	3	4	5	6	7	8	9	10	11	12	13
207a	Higher Westcott, Doccombe, Moretonhampstead	SX789870	7	14 × 17 × 16 48	W		garden	Ga	A		B	
207b	Higher Westcott, Doccombe, Moretonhampstead	SX789870	6	17 × 15 × 15 20	NE		garden	Ga	R		B	
208	Abbey Farm, Buckfast	SX7467	3	20 × 18 × 16 (30)	E	1100s–II	now garden	S	R	p	BT	'bee hives'; only one remained in 1982
231	Brook Farm, Dunsford, Moretonhampstead	SX815905	4	as under 16 × 18 × 17 35 15 × 18 × 19 36 31 15 × 18½ × 17 37 19 20 × 18 × 19 38 37	S	1600s–I	house	C	(D)	f		
304	Virginia Cottage, Trusham, Newton Abbot	SX8582	1	(21) × ?30 × — ?84		1700s	house	?C	D	g	B	above door
305	Penny's Cottage, Penny's Hill, Upton, Torquay	SX909654	2	— 12 (8')	S	1500s or earlier	house	?Ls	A	g		
306	Myrtle Cottage, Riverside Road West, Newton Ferrers, Plymouth	SX5448	5	18 × 18–20 × 16–18 9–24 10–14	E	1700s–II	garden	Ls	R	g		
308	Eastleigh Manor, Eastleigh, Bideford	SS4928	4 2	as under 18 × 22, 20 × 18 (60) 14 × 18 × 18 (60, 54)	S		barn	S	D 1D	g	T	also 5 alcoves (?for skeps) in yard walls (1860) one has flat-topped arch
309	Jasmine Cottage, School Rd, Stokeinteignhead	SX9170	2	18 × 14 × 12–14 (72–84) 16'	E	pre-1400	house	C	D	g		thatch projects over boles; one is now inside house
310	Grange Farm, Goodrington, Paignton	SX8958	4	30 × 30 × (24) (36) (12)	S		garden	Ls	?A			use for bees remembered; demolished 1958
311	Little Marland Farm, Petrockstow	SS5012	2	>18 × (48) × (18)			garden	S	A	f	B	each for 2 or 3 skeps; house 1400s–I, altered 1799
315	Aish House Farm, Stoke Gabriel, Totnes	SX8458	3	22 × 22 × 20 (36) 16	S		farmyard	Sl	D	g	B	?rounded at back; in wall nr orchard

No.	Location	Grid Ref	N	Dimensions	Orient.	Date	Setting					Notes
316	Aish House Farm, Stoke Gabriel, Totnes	SX8458	6	20 × 23 × 20 (24) 20	W	1651	garden	Sl Ls		g	B	shape quadrilateral, narrowing at top; wood and nails above each bole — ?for protective sacking
319	Gatcombe House, Littlehampton, Totnes	SX821625	5 1 4	as under 20 × 15 × 20 0 15 × 12 × 15 0 (15)	SSE	1687	garden	S	T	g	B	this bole in middle of row
320	Glebe House, Whitestone, Exeter	SX9486	—									not found in 1971
321	Lower Somel, Witheridge, Tiverton	SS8014	—									
322	corner of Clemnon Rd & Fore St, Barton, Torquay	SX9067	7	(24) × (24) × (24) 6–12 (12)	SE		garden	Ls	R	p	B	lintel projects; there may be another bole
332	Young's, Kenn, Exeter	SX9285	6	14 × 18½ × 17 52 (6)	S		garden	C	D	5f 1p	B	remains of wood plugs possibly for projecting lintel; top of wall is tiled
336	Westwood Farm, Longdown	SX874917	3	14 × 17 × 15 30 19	SE		garden	S	D	g	B	slate base projects 2"
338	nr Crediton, on Exeter Rd	SX841996?	—									on left before turn to Crediton Station
339	Tower House, Bideford	SS2745?	—									address insufficient
344	Farlacombe, Bickington, Newton Abbot	SX793712	6	17–18 × 21–22 × 17–19 26 15	ENE		garden	S	R	f	T	wall of basalt; wooden lintels; house staircase 1500s
347	Lake Cottage, Goodleigh, Barnstaple	SS6034	9	22 × 18 × 18 22 ?13	S	1600s or earlier	outhouse	SC	D	p	B	'skep rests'
413	Culleigh, Monkleigh, Bideford	SS4620	—	?20 × ?15 × —			?house	S				stone wall, covered with lime; probable shape = pointed arch; hole at back of at least one bole; wall demolished ?1952
442	Beare Farm, Uton, Crediton	SX825991	3	x × x × x × x 3x 0.5x	?E		outhouse or barn	?C	D	f	B	
446	Lower Well, Broadhempston, Totnes	SX8066	9	18 × 21 × 16–17 48 39	E	c. 1710	orchard	S	D	4g 5f	B	lime + sand inside, rounded at back; stone slab base projects; brick arch over each protrudes 4"
450	The Ring o' Bells, Cheriton Fitzpaine, Exeter	SX8706	3	18 × 20 × 10 (60–72) (72)	SW	?1300s	?stable or barn	?C	2D 1A	g	B	Fig. 150; wall may include wattle and daub

1	2	3	4	5	6	7	8	9	10	11	12	13
451	Lower Churchill Farm, Eastdown, Barnstaple	SS595409	4	16 × 19 × 21 100 (60)	S		barn	SC	D	g	B	plaster inside, rounded at back; slate base projects; barn overlooks walled garden
497	South Heale, High Bickington, Umberleigh	SS5920	5	(>24) × (24) × (18)	NE	pre-1700		SC	(D)	g		
503	Yeo Farm, North Tawton	SS653029	7	as under	SE			C	D	f		2 tiers of 4 & 3; photo on cover of *British Bee Journal* 16 Nov. 1968 larger bole in lower tier
			1	30 × 20 × 21 42								
			2	18 × 18 × 19½ 54 (12) 19 × 19 × 19 54								
			3	18 × 18 × 17 84 (12)								upper tier
555	Highfield, Lee, Ilfracombe	SS4946	3	as under 14 × 16½ × 22 15 × 19 × 21 16 × 16 × 21 0 (24)	SE		garden	dS	?R	g		
557	The Library, 21 Fore St, Cullompton	ST0207	4	14 × 17 × 14 34 18		1600s	garden	C	D	f	B	rounded at back; slate base projects; 8 others now destroyed
558	Great Wood, Merton, nr Great Torrington	SS548128	3	21 × 23 × 18 12	?S		barn	S	R	p		farm derelict, and wall may be ruined
559	Yarnley, nr Sandford, Crediton	SS807045	5	14–15 × 17½ × 15 34 (24)	S			S	R	p		
560	Newlands, Landkey, Barnstaple	SS60317 ?	1	18½ × 17½ × 18½ 40	SE		garden	C	D	g		trace of wax inside
561	Upcott Farm, Winkleigh	SS655092	4	17½ × 17–18 × 16½–18 (72)	S			S	R	f		
562	Upcott Farm, Winkleigh	SS655092	3	15–16 × 15–17 × 12–13 (72)	S			C	D	p		
564	Hurscott Farm, Landkey, Barnstaple	SS616318	2	as under 17½ × 18 × 18 (72) 14 × 19 × 16½ (24)	SE		garden	C	D	g		

565	site of electricity transformer, Braunton	SS4937	(3)	$14\frac{1}{2} \times 17-19 \times 15\frac{1}{2}-17$ 34 (24)	SE		C	D	f	now obscured by transformer
566	Martin Farm, Whidden Down, Okehampton	SX684929	2	as under $22 \times 20 \times 15$ (36) $21\frac{1}{2} \times 19 \times 15$ 42	SE	(garden)	S	A	g	owner (1972) remembers their being used
567	Nymett Bridge Cottage, Lapford	SS715093	2	as under $26 \times 23 \times 17$ $25 \times 22 \times 15$ $9\frac{1}{2}'$	SW	cottage	C	D	g	now covered by ?drain pipes
568	Coldharbour, Ashreigney, nr Winkleigh	SS6314	2	$14 \times 17 \times 15, 16$ (84) (30)	S	house	C	D	g	
569	Lake Farm, Cornborough, Abbotsham, nr Bideford	SS420281	5	$18 \times 18 \times 22$ (36)	S	(garden)	S	?D		
570	Victory Wood, Nymett Bridge, Lapford	SS7109	1	$20 \times 16 \times 20$ 9'			C	D	b	
571	Doch Cottage, Weare Gifford, Bideford	SS4822	1	$17 \times 17 \times 16$ 32	SE	house	S	A	g	
572	Higher Thorndean, Holsworthy	SS3403	—							
574	Blackness Farm, East Cornworthy, Totnes	SX8455	—							wall now destroyed
575	Landsend Barton, Colebrooke	SX744002?	6	$13 \times 14-17 \times 14-16$ 34 14	S	(garden)	C	D	e	rounded at back
576	Westacott, Barnstaple	SS5933	4	$17-22 \times 18-21 \times 19-21$ $24-34$?12			C	D	b	rounded at back; remnants of skep found
577	Mill Town Farm, Meshaw	SS755205	2	as under $16\frac{1}{2} \times 17 \times 15$ 39 (24) $17 \times 16 \times 17\frac{1}{2}$ 42			C	D	g	rounded at back
578	Taw Mill, Wembworthy, Winkleigh	SS669061	6 5 1	as under $14 \times 18\frac{1}{2} \times 17$ $13 \times 19 \times 16$ 23	?S	garden	C	D	f	
									B	rounded at back

267

1	2	3	4	5	6	7	8	9	10	11	12	13
579	29 Church Street, Braunton	SS4937	11	17–19 × 17–20 × 15–18	?S	1550	garden	Ls	D	f		2 tiers of 6 & 5; rounded at back
591	Upton Farm, Wright's Lane, Upton, Torquay	SX911654	2	16 × 17 × 17 40	S	1500s or earlier	farm building	S	A	g		'skip holes'
631	Elm Park, Broadhempston	SX8066	9	18 × 23 × 21 20–22 9	WSW	c. 1850	(garden)	S	D	f	BT	separate recesses in built-out structure with sloping slate roof; each had mortar inside, slate base
653	Oak House, Chard Rd, Axminster	SY2998	2	40 × (24) × (12) 14 34	SE	c. 1750	garden	B	D	g		rounded at back; recently cemented inside
656	Southdown Farm, Malborough, Kingsbridge	SX700385	6	18–20 × 21 × 18–20 30–34 21	SE		garden	S	R	g		wall of Devon schist and pink quartz; base projects
682	Lower Milton, Payhembury, Honiton	ST081006	1		S	1853	house					there were 2 others
761	The Gwythers, Lee, Ilfracombe	SS4847	7	24 × 18 × 20 0 18–20	S		garden	S	D	e	C	in stone/shale wall now just over 24″ high; shape = flat-topped arch
768	Warmington, Alverdiscott Road, Bideford	SS4826	2	14¾ × 27 × 12½ 54 66	S		yard	S	D			wooden base; some doubt whether bee boles; now made into windows
769	Rowes Farm, Colebrook, nr Crediton	SS7700	8	18 × 22 × 18 (24–48)	6S 2ESE	1500s	garden	C	D	f	C	rounded at back; wooden base; were possibly 2 more
770	Home Farm, Kingsnympton Park, Umberleigh	SS6720	6	15–19 × 15–19 × 15–19 67 15	S		farmyard	C	R	f		wooden base
777	Haye, Plymstock, Plymouth	SX5295452 ?	—	(12) × (12) × (10)			garden					
787a	Lower Soar, Malborough, Kingsbridge	SX706376	8	14 × 20 × 14 (30) (6)	S		was ?garden	S	R	f	C	wall of Devon schist
789a	Trehill Farm, Sampford Courtenay, nr Okehampton	(SX637973) or SS6301	2		SE	1500s	barn		D		C	used until late 1960s; winter storage 789b
791	Marlborough House, Church Street, South Brent	SX6960	6	15 × 16 × 12	SE	1500s	farmhouse	S		g	C	shape quadrilateral, narrowing at top, arched at back
800	Paradise House, South Street, Totnes	SX8060	2	23 × 30 × 27	SE	1500s–II	house	S	R	g	C	built under steps of Elizabethan cellar and garden walk; now at ground level; one has wood lintel; probably bee boles

Dorset

No.	Location	Grid Ref	n	Dimensions	Orient.	Date	Use					Notes
17	Little Westrow, Holwell, Sherborne	ST6810	1	32×33×29	E	1600	house	S	R	g	B	had oak lintel; ?originally arched
42	Lower Farm, Corscombe	ST514056	4	17×13×16 27 7,54	SSE	1600s	stable	S	A	g	B	base projects
265	Litton Cheney	SY5590	—									
266	Scowles Manor, Kingston Hill, Swanage	SY9579	8	as under	W	1200s–II or 1300s–I	hall (?chapel)	S	R	e	BT	2 tiers of 6 & 2 in Portland stone wall; Fig. 151
			6	16–19×16–18×21–24 (10) 8								this row surmounted by 3 load-spreading arches
			2	12×25, 17×17, 20 (60) 10								use not certain
479	Smedmore House, Kimmeridge, Wareham	SY924789	5	12–13×13–14×13 (48) 43–48	SW	pre-1800	?house	S	R	f	B	may date from 1600s
			1	24×28×13 48	SW	pre-1800	?house	S	R	f	B	in same wall, 41″ from above; perhaps enlarged

Durham—see below, 'Alcoves'

Essex

No.	Location	Grid Ref	n	Dimensions	Orient.	Date	Use					Notes
177	Pond Hall, Wix, Tendring	TM151297	9	15½×15½×25	SE	pre-1750	garden	B	A	g	B	3 tiers of 3 & 3 & 3; ?rounded at back; internal ht 20″, width 17″; structure protrudes at back of wall, with sloping slate roof
445	Tilty Hill Farm, Duton Hill, Dunmow	TL5926	6	17×24×22 0, 26 6	S			F	A	e	D	2 tiers of 3 & 3, renovated 1967 (Fig. 149); arches brick-lined; originally part of ?priory wall
457	Assington Hall, Assington, Colchester	TL936388	?3	15×15×11 45 10′	SW		?house	B	T	f	B	2″ bricks; revealed when house burnt down

Gloucestershire

No.	Location	Grid Ref	n	Dimensions	Orient.	Date	Use					Notes
94	Old Farm, Aston Magna, nr Moreton-in-the-Marsh	SP2035	4	16¼×16×16¼ 20 ?24	S	1600s	yard	S	A	f	B	each has alighting projection
95	The Manor, Weston-sub-Edge	SP126411	8	20×16×13¼ 43 ?5	E	*1600s–II*	garden	S	R	f	B	brick piers and back; whole row projects
96	The Folly, Gotherington, Winchcombe	SO9629	10	20×18×12 33 ?9	SE		back of garden	B	R	g	B	wooden ledge projects above whole row; wall is stone up to 30″

1	2	3	4	5	6	7	8	9	10	11	12	13
227	Upper Hall, Elton, Newnham-on-Severn, Westbury	SO693142	9	$26 \times 26 \times (12)$ 30	S	1700s	garden	S	R	f	BT	base projects; 'one gone' by 1978
			9	$26 \times 26 \times (12)$ 30	W	1700s	garden	S	R	f	BT	base projects; in wall adjoining above
341	rear of Old Forge, Westington, Chipping Campden	SP1438	?8	$x \times x \times -$ av.x 1.5x			?garden	dS	R	p	B	base projects; 'bee wall'
527	Manor House, Bourton-on-the-Hill, Moreton-in-the-Marsh	SP1732	3	as under		WNW 1500s–II or 1600s–I	garden	dS	R			wall of Cotswold stone; ?rounded at back; single ledge projects above
				$12 \times 17\frac{1}{2} \times 12$ $12 \times 13\frac{1}{2} \times 9\frac{1}{4}$ $14 \times 16 \times 12$ — 6, 10'								

Greater London, Greater Manchester—see London, Manchester

Hampshire

1	2	3	4	5	6	7	8	9	10	11	12	13
175	St Margaret's Priory Titchfield, Fareham	SU534061	4	$28 \times 26 \times -$ 33	SE	1500s	house chimney	B	A		T	2 in wall of each chimney: now bricked up; house, not a priory
594a	The Palace House, Bishop's Waltham	SU552174	3	$20 \times 20 \times -$ >27 30, 40	E	1600s–II	garden	B	A		T	bricked up: wall only c. 15" thick
594b	wall below Palace Meadow, Bishop's Waltham	SU552174	?2		E		garden	B				wall (below NE corner of Meadow) is now in builder's yard, and boles covered up, 1980
594c	Bailey of Bishop's Palace, Bishop's Waltham	SU552174	1	$26 \times 27 \times -$	NE				A			inside builder's shed, 1980
638	Little Dean, Bramdean	SU6128	3	$16\frac{1}{2} \times 17 \times 18$ 5, 25, 45	SE	1700s–I	garden	B	T	g	BT	Fig. 163 shows boles in purpose-built 'pillar' in angle of 2 walls; tiled base projects; rounded at back
684	Tudor House Museum, Bugle St, Southampton	SU420113	1	$20 \times 19 \times 11$ (26)	W	1600s–I	cottage	B	A	g	BT	Fig. 139; stone base projects; may be 1 more
709	Titchfield Abbey, Mill Lane, Titchfield	SU541067	4	$28 \times 20 \times -$ 14, 36 18	S	1500s	boundary	B	D		T	2 tiers of 2 & 2; wall thickened at back; filled in; Dept. Environment

Hereford and Worcester

No.	Location	Grid Ref		Dimensions		Orient.	Date					Notes
158	The Rock, Lugwardine	SO5541	2	?13 × (48) × 14 / 0		N		garden	S	A	g	each for ?2 skeps; one filled in
202	The Thorn, Glewston, Ross-on-Wye	SO550217	3	as under / 20½ × 22¾ × 17 / 19½ × 20 × 20 / 19 × 19½ × 22½ / (24) (14, 25)		SE		?garden or yard	S	R	f	B	whole bole projects
283	The Manor House, Porter's Mill, nr Claines, Droitwich	SO8660	13	20¼ × 20¼ × 15 / 16⅔ (11')		S		garden	B	A	e	B	Fig. 144; stone base projects; house 1300s–II, restored 1503; wall probably later on other side of same wall; now inside building
			13	probably as above		N		farmyard	B	A			

Hertfordshire

| 410 | next door to The Palace East, Much Hadham | TL4319 | — | | | | 1700s | | | | | | |
| 556 | The Old Rectory, Much Hadham | TL4319 | 7 | 15 × 12 × 9 / 39¼ | | S | c. 1630 | ?garden | B | T | f | | |

Humberside—see below, 'Bee houses'

Isle of Man

26	Ballachurry, Rushen	SC208698	11	21 × 18½ × 15 / 16½ (36)		S	c. 1550	?orchard	S	R	f	BD	wall of Manx stone; photo published 1953;[101] winter storage 141
43	Ronaldsway Farm, Malew	SC290682	11	x × x × 0.5x / 2x 1.5x		S	1506	garden	S	A		BD	may have been 12; demolished 1939
66	Ballakeighan, Arbory	SC256686	4	20 × 21 × 19 / 34		SE		garden	S	R	p	BD	wall of Manx stone; slate base projects; ?rounded at back; one bole still used in 1953
			6	18 × 18 × — / (15)		SW		garden	S	A		B	brick arch; filled in
103	The Vicarage, Laxey	SC4384	5	18 × 18 × 18 / 30 ?12		E	?1800s	garden	dS	R		BD	
119	Balladoole, Malew	SC2468	5	21 × 17 × — / 30		NE	?pre-1500	garden	S	R	f	D	wall of Manx stone; may be 1 more; all filled in; recesses on both sides of garden wall at nearby house
120	The Neary, Sky Hill	SC4393	7	15 × 15 × 15 / 24, 30		SE		farmyard	S	R	f		wall of Manx spar + slate + quartz; 2 tiers of ?4 & 3; rounded at back

1	2	3	4	5	6	7	8	9	10	11	12	13
142	The Cottage, New Road, Baldrine, Laxey	SC4281	2	as under 18–20×(72)×18 20–24 18×18×10 30	NE	?1850s	?garden	S	R	g	D	wall appears specially built for boles; top and base project; for 23 skeps; base projects; c. 9′ from larger bole
352	Baldromma, Maughold	SC4991	3 2 1	as under 21×18×15 16 17×14×12	NE		(garden)	dS	R		D	
353	Moorescroft, Ballafesson, Rushen	SC2070	1	18×15×12 ?12	W	1700 or earlier	garden	S	R		D	
354	Riverside, Colby, Arbory	SC2270	2	14×15×11 12	E	1800s or earlier	garden	?S	?R			

Isle of Wight—no records of any type

Kent

1	2	3	4	5	6	7	8	9	10	11	12	13
44	Roydon Hall, East Peckham	TQ665518	8	18×14×10 (48)	S	1530	garden	B	A	g	B	'bee garth'; described 1955[90]
49	Wrotham Water, Wrotham	TQ629598	2	21×16×17½, 20 7½, 40	S		house chimney	B	A	f	B	one above the other in chimney stack; described 1955;[90] NT—not open to public
78	Quebec House, Quebec Square, Westerham	TQ448540	3	17×14×10 30 12½	S	1500s–II		B	T	g	BT	described 1955;[90] NT
90	The Yews, Boxley, Maidstone	TQ7759	2	23×18×15 (36) ?24	S		garden	B	A	f	B	wooden base; each bole projects; house 1600s–I; described 1955[90]
91	Boroughs Oak Farm, East Peckham	TQ674499	7	19×16¼×15¾ 18 4	SE	1700s	brewery	B	A	g	B	Fig. 159; plastered inside; similar to 44; described 1955[90]
188	Nearly Corner, Heaverham, Kemsing, Sevenoaks	TQ5857	6	17–19×18×18 26 ?9	S	1650s	house	B	A	4g	B	3 tiers of 2 & 2& 2, but top 2 filled in; described 1955;[90] photo published 1930 (fig. 339)[145] shows skeps in boles
201	Scadbury Manor, Southfleet, Gravesend	TQ6171	10	17×10×9 30 ?5	E	1600s–I	garden	B	T	g	B	described 1955;[90] between recesses 9 & 10 is an unexplained arch (48″ × 58″ × 10¼″), perhaps a bee alcove
237	Eynsford Castle, Eynsford	TQ542658	4			c. 1100	castle curtain	F	A		B	?bee boles

No.	Location	Grid ref	Count	Dimensions	Aspect	Date	Setting					Notes
244	Pett Place, Charing	TQ960490	8	18 × 9½ × 10 43 (24)	SE	?1600s	garden	B	T	g	B	bricks 7″ × 2½″ × ?; described 1956[91]
245	Dane Court, Chilham	TR057531	4	12 × 9¼ × 9¼ 18 9	E	1580	house chimney	B	T	g	B	described 1956;[91] also 4 arched recesses 12″, 24″ from ground (?bee alcoves)—2 facing N (68″ × 54″, 72″ × 9″), 1W, 1E (both 100″ × 70″ × 4″)
246	Higham Hall, Higham, Rochester	TQ712727	2	17 × 28 × 9 44	E	1500s	garden	B	A	g	B	there were probably more; ?some filled in; described 1956[91]
247	Peckham Place, East Peckham, Paddock Wood	TQ654505	6	16 × 11 × 9 (48)	SE	1700s or earlier	garden	S	T	b	B	2 tiers of 3 & 3; there were 7 more; slots for separate bases visible, but no bases left; certainly in use in 1913, possibly till 1930s; described 1956[91]
248	The Tree House, Plaxtol, nr Sevenoaks	TQ6053	6	17 × 18 × 19 36, 54 ?13	SE	1400s	garden	B	D	g	B	described 1956[91]
250	Burton Farm House, Kennington, Ashford	TR0244	2	15½ × 14 × 9½ 38	?1600s		garden	B	A	f	B	brick surround; described 1956[91]
253	Austens, High Street, Sevenoaks	TQ5355	1	20 × 17 × 11 36	N	c. 1769 or earlier	garden	S	A	g	B	one is 80″ from ground
269	National Westminster Bank, 91 High Street, Maidstone	TQ7656	5	16½ × 14 × 14 63	E	1500s	(garden)	B	A	g	BT	2 groups of boles; may have been open on both sides
287	Cathedral Close, Canterbury	TR1558	6	12–16 × 9–11 × 9 42, 50 (78), (45′)	E	1600s–I	garden	B	T	g	B	3 (or 5) filled in; described 1958[92]
288	Memorial Gardens, Canterbury	TR1558	8	16 × 14 × 10 12 (84)	S	1500s	garden	SB	T	?5g	B	may have been more; house demolished, grounds overgrown; described 1958[92] with photo
289	Great Wenderton Manor, Wickhambreux, nr Wingham	TR2395922?	2	1.5× × × ×— 20.5×	W	1500s	(garden)		T			there may have been one other; described 1960[93]
290	Maidstone Museum, St Faith's Street, Maidstone	TQ7656	4	18 × 25 × 13 ?24	S	pre-1500	garden	S	R	f	BT	in enclosure known as 'Monastery garden'; 2 boles have ½ brick cut out at front; described 1960[93]
337	Cossington, Aylesford	TQ747597	3	22 × 18 × 11 (24) 13′	S	1500s or 1600s	garden	B	A	g	B	Fig. 152; wall of 2″ brick is scheduled as ancient monument; 1′ above boles, a row of bricks protrudes; monastery of Hales Place was near
375a	The Old Vicarage, St Stephens, Canterbury	TR1557	3	15 × 10½ × 10 36 42, 15′	S	1490	garden	B	T	g	B	on the back of wall 375a—described 1960;[93] between adjacent pairs is an arched niche down to the ground (?bee alcoves)
375b	Manor House, St Stephens, Canterbury	TR1557	20	18 × 10 × 10 12	N	1490	garden	B	T	g	B	

1	2	3	4	5	6	7	8	9	10	11	12	13
376	Simon Langton School, Canterbury	TR1557	2	18½ × 10 × 6	NNE			B	T			?bee boles; now demolished; described 1960[93] with photo
405	The Palace, Maidstone	TQ7656	1	15 × 15 × 10 (24)	S		(garden)	S	A			wall may include flint; boles filled in; (may be at no. 9 not 11)
471	Cramond House, 11 Harnet Street, Sandwich	TR3358	—				garden	B	T		B	brick or flint wall, probably garden wall originally
502	22 Church Street, St Mary's, Sandwich	TR3358	?1		S	1500s–l	house	B	T			one bricked up; may not be bee boles; described 1956[91] with photo
506	Richmond House, Charing	TQ9549	5	?18 × 10–12 × 5 40–45 20–40	S		garden	B	T			brick surround; at ground level
522	8 Cattle Market, Sandwich	TR3358	2	small			garden	F	T		B	brick surround
524	5 Strand Street, Sandwich	TR3358	1			1615	garden	?F	T			brick surround
525	7 Harnet Street, Sandwich	TR3358	1		N	1600s	house	?F	T			narrow brick; tiled base
526	Noud's Farm, Lynsted	TQ957612	1	21 × 15 × 18	W	c. 1578		B	A	g		all filled in
529	County Primary School, Sandwich	TR3358	7	21 × (24) × — 15	3S 4E		garden	BS	A			
534	1 Guildcount Lane, Sandwich	TR330583	1	12 × 13 × 9	S	1600s	garden	B	T	e		
739	Park Farm House, Rushmore Hill, Knockholt	TQ481605	—	(30) × (15) × — (30)	S	1800s–l	outbuilding	B	(T)			

Lancashire

1	2	3	4	5	6	7	8	9	10	11	12	13
88	Church House, Wray	SD604676	4	20 × 21 × 20 30 ?8½'	SSE		garden	S	R	g	B	house 1622
102	Lune Bank Cottage, Aughton, nr Lancaster	SD5563	3 1 2	as under 22 × 38 × 19 22 × 21 × 19 18 ?6	SSW ESE	1800s–l	house	S	R	e	B	projection above; bees kept there in early 1920s probably for 2 skeps

No.	Name	Grid ref	N	Dimensions	Orient	Date	Position					Notes
189	Ribby Hall, Kirkham	SD410319	6	36 × 29 × 29 / 26 ?9	S	1840	garden	B	R	f	B	roofs, if any, now missing; gardener says planks and cart sheets were used; recalls skeps there in early 1900s
			1	24 × (54) × ?29 / ?10'	S	1840	garden	B	R	f	B	high in same wall; for ?2 skeps; above large alcove (8½' × 42" × ?29", ?for skeps)
191	Entwistle Hall, Entwistle, [nr Bolton]	SD726175	7	26 × 43 × 14 / 26 34	S	1600s	garden	dGs	R	g	C	each for 2 skeps; house 1550 but wall may be later; postal address Greater Manchester
198	Heskin New Hall, Heskin, Chorley	SD5315	4	18 × 18 × 9 / 45	W	1653	house	B	A	e		
254	Blands, Wennington	SD625697	?12	42 × (36) × —		?1700s	garden	dS	R	f	B	?bee boles; curved lintel roof projects and has 1" hole; bole rounded at back
286	Hillside Farm, Wennington	SD6170	2	16 × 16 × 14 / 36	S	1750	garden	dLs	R	e		
581	Hall Green House, Upholland, Skelmersdale	SD5105	3	?36 × 43 × 23 / 18 4½	S		garden	B	R	f	B	'bee hives'; flagstone base, stone back; whole set projects 15"; each for 2 skeps; ? demolished 1973
635	Quaker House and Cemetery, Trawden, Colne	SD9138	3	30 × 17–19 × 8–11 / 16	2S 1N	1688	?cemetery	S	A		B	roughly filled in
661	Oak Cottage, Silverdale	SD461758	4	(18) × (30) × 28 / 18, (36)	E		garden	dLs	R	p		2 tiers of 2 & 2; whole set projects 14"
			(2)	14 × 18 × 12 / 30	E		garden	dLs		f		in same wall; rounded at back

Leicestershire

| 592 | The Bede House, Lyddington, nr Uppingham | SP8797 | 2 | 18 × 13 × 11 / 50 (18) | W | 1400s | garden | S | T | g | B | Dept. Environment; photograph on cover of *British Bee Journal* (1973) no. 4290 shows similar recess, but square, in back garden; large arched recess (?bee alcove) next to bee boles |

Lincolnshire

| 68 | Gainsborough Old Hall, Parnell St, Gainsborough | SK8189 | 4 | 15 × 12¼ × 12 / 23 | E | 1600 | house | B | A | e | BT | Tudor brick; photo published 1953[101] |
| 284 | Well Vale, Alford | SF4576 | — | | | | | | | | | not found in 1980 |

275

1	2	3	4	5	6	7	8	9	10	11	12	13
London, Greater												
251	Gatehouse, Eltham Place, Eltham	TQ4274	18	12×10–$15\frac{1}{2} \times 10$ 45 10'	W	*1500s*	garden	B	T	g	B	one set of 11, one of 7; 2 other recesses destroyed; may have been apiary of Eltham Palace; described 1956[91]
252	Well Hall, Pleasaunce, Eltham	TQ424751	15	$17\frac{1}{2} \times 9 \times 9$–10 22 (36–48)	E	*1500s*	garden	B	T	g	B	filled in; ?2 others destroyed; in public park; described 1956[91]
268	Church House Gardens, Bromley	TQ4069	1	$18\frac{1}{2} \times 11\frac{1}{2} \times 9$		*1500s*		B	T		BT	reconstructed bole, one of several in walled garden of Grete House, Bromley; described 1958[92]
401	? West Drayton	TQ0679	3	$12 \times 9 \times$ — 28 <12		?*1500s*	garden	B	T		B	Tudor brick; filled in
476	Northumberland Avenue, off Syon Lane, Isleworth	TQ1777	2	$44 \times 25 \times 13$ 15 $16\frac{1}{2}$	SE	*1500s*	garden	B	A	g	B	stone base projects; ?open on both sides; was in market garden/orchard, now in street; photograph in *Australian Beekeeper* 15 Nov. 1971, p. 147
478	Breton's Manor House, South Hornchurch	TQ517848	10	$17 \times 17 \times 10$ (36) 9', 22'	5S 5E	*1500s*	(garden)	B	A	3g	B	one destroyed; some repaired recently; 5 now inside lean-to
496	Eastbury House, Barking	TQ4584	6	av. 18×9–$12 \times (10)$ av. 66	4W 2N	1572	garden	B	T	f		were probably more; 1980, NT/Barking Corporation
510	30 Bark Hart Road, Orpington	TQ4665	12	$(16\frac{1}{2}) \times (10) \times (11)$ 54, 60		1650 or later	garden	B	T		B	filled in; some are in adjoining gardens—may have been more; described 1960[93]
Manchester, Greater—see 191 under Lancashire												
Merseyside—no records of any type												
Norfolk												
518	Middleton Hall, Mendham, Harleston	TM287837	4	$21\frac{1}{2} \times 13 \times 9$ (48)	S	*1500s*	garden	B	R		B	plastered inside; wooden top; possibly 1 more
585	Sea Peeps (was garden of 'Prince of Wales' inn), Burnham Norton, Kings Lynn	TF829439	6	$30 \times (24) \times 10$ (24) (12)	S	?1700s	garden	BF	A			
664	Lifeboat Public House, Thornham, nr Hunstanton	TF7343	6	17×19–20×10 18, (26) $4\frac{1}{2}$	W	?1700 or earlier	garden	B	A		C	(Fig. 164); 2 tiers of 3 & 3; skeps known to have been kept there
Northamptonshire												
228	The Stone House, Barnet's Hill, Eydon, Daventry	SP537513	18 9 9	as under 21–$23 \times 19 \times 20$ av. 6 14 17–$19 \times 19 \times 19$ av. 36 14	S	1700s–II	garden	B	A	f	B	2 tiers of 9 & 9

ID	Location	Grid ref	#	Dimensions	Orient	Date	Context					Notes
260	Fawsley Park, Daventry	SP5657	12	1.5× × × × —				B	A	p	BT	free-standing structure with ?tiled roof; 2 large compartments, each with 2 tiers of 3 & 3 boles
407	The Manor House, Lower Boddington, Daventry	SP4852	5	23 × 19 × 19 27 20–32	SSE	1300s	garden	dS	A	f	CT	wall now pointed; each recess brick-lined, and containing inverted 'saddle stone'
602	Dallington Grange Farm, New Duston, Northampton	SP730640	7	24 × 20 × 14 12–24 38	SW	1700s–I	garden	S	R	p	B	wall of Northampton sand ashlar; 3 filled in; was probably 1 more
611	Manor Farm, Brockhall	SP634627	3	22½ × 20½ × 12–18 ?24, ?48	SW	pre-1750	barn	S	A	g	B	wall of Northampton sand ashlar; 1 bole above 2
612	Manor Road, Kingsthorpe, Northampton	SP749633	—		SE							may now be overbuilt
613	garden wall at East Haddon, Northampton	SP6668	—									now demolished

Northumberland

ID	Location	Grid ref	#	Dimensions	Orient	Date	Context	B	A	p	BT	Notes
124	Lincoln Hill, Humshaugh, Hexham	NY9171	3	25 × 18–21 × 15 30 (25)	W	1700s	orchard	S	R	(g)	B	photo published 1953[101]
125	W. Woodburn Filling Station, West Woodburn, Hexham	NY8986	7	as under 19 × 30–34 × 15 0	SE	1800s–II	garden	S	R			skeps in 1890, sometimes 2 per bole
			1	17 × 16 × 13½ 0					A			rounded at back
206	4 Mill Cottages, Allendale	NY8355	2	as under 18 × 17 × 15 20 × 21 × 15 12	S	?1850s	garden	dS	R	p		
261	Castle View, Henshaw	NY7664	4	18 × 18½ × 18 16 3	S		garden	S	R	e		top projects 3″
262	Prestwick Hall, Prestwick	NZ1872	6	31, 26 × 17 × ?23 14, 43 ?4	N†		garden	B	R	g	B	bricks 8″ × 2⅜″ × 4¼″; door to each of the 3 vertical pairs; drip sill above; each bole has small hole through wall with sill on S side
264	Melkridge, nr Haltwhistle	NY7364	3	17 × 23 × 16 27, ?60 9	S		?garden	dS	R	g		1 is above; ?base projects

1	2	3	4	5	6	7	8	9	10	11	12	13
351	Longwitton Farm, Longwitton, Morpeth	NZ0788	3	22×20×17 10 (48)	S		garden	S	A	g	C	'skep holes'; base projects
408	Wester Hall, Humshaugh, Hexham	NY9171	6	17½×19½×18½ 15, (48)	3S 3E	1732–1840	garden	S	R	g	B	base projects 4"; had doors or flaps hinged at top; 1 filled in
464	Longhirst Hall, Morpeth	NZ224890	—									may have been a bee house or shelter; demolished by 1971
500	Holy Island	NU1343	1		S		house					'bowly hole'; several others seen on island
587	Close House, Kirkhaugh, Alston	NY693502	2	21×20×20 (24) 12	E	1700s	garden	S		f		postal address is Cumbria
620	Dyke Row Farm, Whitfield, Hexham	NY7758	5		S	1820–1860	garden	S	R	g		
803	Buteland Farm, Bellingham, Hexham	NY876816	7	as under	E	1850	garden	SB	A	g	C	in opposite (brick-faced) stone walls; ground slopes and some boles have wedge-shaped base-stone to give horizontal shelf
			2	30, 34×25, 22×15 (12–30) 21	E							
			2	33×24×15 (12–30) 21	E							
			3	27×25×15 (12–30) 21	W							a single bole + a pair
848	Plane Trees, Brunton Bank, nr Humshaugh	NY9272	2								B	base projects; each for 2 skeps
849	Ingoe	NZ0474	—									
850	nr Whitfield	NY7561	—									
851	in the garden of Mr Gutherson, Thropton, nr Rothbury	NZ0302	—									
852	North Barton, nr Whittingham, Alnwick	NU0612	—									
853	Catton, nr Allendale	NY8257	?5								B	shape quadrilateral, narrowing at top

Nottinghamshire—see below, 'Bee houses'

Oxfordshire

67	Southams Farm, Stonesfield, nr Charlbury	SP3917	2	$20 \times 16 \times 18$ (36)		S	pre-1800	garden	dS	R	g	BT	brick base, centre brick projects 3″; hole 4″ × 3″ at back of each bole
225	Netherton, Fyfield	SU420992	8	$15 \times 19\frac{1}{2} \times 17\frac{1}{2}$ 27 ?5		S	1855	garden	S	A	g	BT	house 1696
392	The Manor House, Warborough	SU5993	5	$13 \times 14 \times 9$ (48)	16′–18′	E		garden	S	T	g	BT	2 tiers of 3 & 3; threat of demolition, 1969
480	Bolney Court, Shiplake	SU768783	6	$?24 \times ?18 \times ?18$?15		W	c. 1600	garden	B	A	g	B	wall (now in garden) was in farmyard, possibly connected with Abingdon Abbey
821a	Champs Folly, Frilford, nr Abingdon	SU4497	3	$x \times x \times ?x$ (40)				farmyard	S	R	f		winter storage 821c
821b	Champs Folly, Frilford, nr Abingdon	SU4497	—					granary					

Shropshire

637	Preesgweene Hall, Weston Rhyn	SJ292359	6	$17 \times 18 \times 16$ (24)		S		garden	Ss		f		2 tiers of 3 & 3; shape = pointed arch; brick surround; slate base; structure built on to existing wall, has 3 'battlements' at top and 2 rectangular spaces below boles

Somerset

131	Charity Farm, Lovington	ST5930	11	$24 \times 20 \times 14$ 18 ?9		S	1600s	garden	S	R	e	BT	Fig. 137; rounded at back; lintel projects; boles shown in painting (c. 1700) in dwelling house (Fig. 138)
132	Laws Farm, Compton Dundon, Somerton	ST4833	(17)	$18 \times 20 \times 16$ 12 ?6		W		garden	S	R	b	B	lintel projects; all destroyed by 1976
133	The Cedars (was Rose Cottage), Compton Dundon, Somerton	ST4833	6	$16 \times 20 \times 18$ 20		S	1860	garden	S	R	g	B	base projects; arrow-slit hole over 1 bole to allow flight through high wall
199	Royal Oak Farm, Winsford	SS9034	3	$16 \times 18 \times 18$ (36) ?11		S		building	S	R	g	B	rounded at back; base projects; projecting stone a few inches above boles
275	Hillview, Ridgeway, Frome	ST745450	12	27×22–19×22 21 72		SSE	1600s	garden	S	R	f	B	4 boles have recesses above, 8″ × 8″ × 8″

1	2	3	4	5	6	7	8	9	10	11	12	13
312	Estate Farm, Pitney, Langport	ST4428	6	18 × 22 × 19 (48) 85	S	1694	garden	Ls	R	g	B	wall of blue lias limestone; base projects in 5 boles; traces of plaster inside
Staffordshire												
229	Pipe Ridware Hall, Hall Ridware, Rugeley	SK095175	5 4 1	as under 15–16 × 19 × 10 36 7–9½' 16 × 23½ × 10 43	S	c. 1680	garden	B	R	g	B	brick size irregular, from 2¼" to 2⅝"; wooden lintel above; filled in; described 1971/72[117]
763	Walbrook Cottage, Coton	SJ981319	8	17½ × 20 × 22–23 12			garden	SsB		f		shape round; small decorative carving above each
Suffolk												
41a	West Stow Hall, nr Bury St Edmunds	TL816708	3	20 × 30 × 11 (66) 25	NNE	1601	covered passage	B	R	g	B	one 24" wide; use for falcons suggested, but recesses seem too shallow
41b	West Stow Hall, nr Bury St Edmunds	TL816708	1	24 × 22 × 10	W or E	1485–1522	garden	B	A	b	B	?bee boles; 3 courses of brick possibly added later
Surrey												
162	Abbot Hospital of the Blessed Trinity, Guildford	TQ0049	6	15 × 22 × 10½ 36 15'	E	1621	garden	B	T	e	BT	
334	Batchelors Hall, 97 High Street, Thames Ditton	TQ162673	5	16 × 18 × 9¼ 40 (13'), 20'	2E 3S	1500s	garden	B	A	e	B	Tudor brick; 2 facing S are built over; there are 2 more facing E in adjoining gardens
346	Pitt House (ruins), Farley Green, Peaslake	TQ0544?	6	20 × 20 × 22 6–12 9	S		orchard	S	?D	p		ragstone wall; brick surround; ? rounded at back
359	Temple Elford House, Capel	TQ192398?	2	17 × 12 × 10 33 (72)	S			F	T			may have been more; name is Temple Elfande on map
633	The King John House, Guildford Rd. Gomshall, Guildford	TQ083480	3	21 × 18 × — 27 (72)	S		garden	F	A		B	brick surround; now filled in; house 1600s
784	Wrecclesham Farmhouse, Wrecclesham, Farnham	SU8245	1	19 × 19 × 10			house	B	A	g	C	

Sussex, East—see below, 'Bee houses'

Sussex, West

No.	Name	Grid ref		Dimensions		Aspect	Date	Site					Notes
409a	Denne Court, East Street, Petworth	SU9721	1	20 × 26 × 13 52		E	c. 1650	garden	S	T	g	CT	ragstone wall; brick surround; same wall as 409b; property has been divided
409b	Boles House, East Street, Petworth	SU9721	1	20 × 26 × 13 52		E	c. 1650	garden	S	T	g	T	ragstone wall; brick surround
			1	20 × 24 × 16 46		S	c. 1650	(garden)	S	D	g		ragstone wall; brick surround; hidden by building in 1979
507	New Hall, Small Dole, Henfield	TQ209133	5			SE	1700s–II	garden					were probably more
788	St Oswalds, Knock-hundred Row, Midhurst	SU8821	2	21 × 22 × 11½ (72) 42		1N 1W	1750	courtyard	S	R	g		brick-lined

Tyne and Wear

| 263 | West Farm, Black Callerton Lane, Westerhope, Newcastle-upon-Tyne | NZ1768? | 3 | 27½ × 24 × 16¼ ?8 33 | | SSW | | garden | B | A | e | B | Fig. 169; shaped stone base projects 6½″; bole rounded at back, cement lined; reports have used various addresses for this set |

Warwickshire

| 15 | Packwood House, nr Lapworth | SP174723 | 30 | 20 × 21 × 18 9 | (10) | SSE | c. 1660 | garden | B | A | e | B | Carolean brick; boles in pairs (Fig. 143); another wall has pipes for hot water; NT |

West Midlands

93b	The Rectory, Maxstoke	SP234869	4						S	R	f	B	
404	Manor House, Hall Green Road, West Bromwich	SP005943	4	22 × 24¼–26 × 10¼ (48) 15′–18′		ENE	1620	?garden	B	T	g	D	bricks 9″ × 4¼″ × 2⅜″; described 1963[115]
580	Wightwick Manor, Wolverhampton	SO870987	2	11 × 10 × 10 55 9		S	1500s	barn	B	D	f		grating at back; may not be bee boles; NT
648	Haden Hill House, Cradley Heath, Warley	SO959850	5	21½–24 × 21½–22½ × 21 21–24 9		SSW	1600s–I	dovecot	B	R	f		1 bole only 15″ deep; wooden top; fragments of wood bases and sides remain
795	nr Maxstoke	SP2386	6						B				2 tiers of 3 & 3; shape = pointed arch; top 3 taller; address not known but fig. 139 in Cowan[69] suggests this was free-standing structure; iron bars across each row and staples for padlocks (it was near road)

Wight, Isle of—no records of any type

Wiltshire

1	2	3	4	5	6	7	8	9	10	11	12	13
342	Farleigh Wick, Monkton Farleigh, Bradford-on-Avon	ST803642	6	14½ × 17 × — (18) 5				S		f		'holes'
355	Uffcott Farm, Broad Hinton	SU125775	20	15 × 16–18 × 14 (6) 4½	S	?1500s or later	garden	Ch	R	f	C	Fig. 170 shows boles in thatched wall; brick piers; 1980—some boles crumbling, but wall buttressed and recently rethatched

Worcester—see Hereford and Worcester

Yorkshire, North

1	2	3	4	5	6	7	8	9	10	11	12	13
1	Edgley, West Burton, nr Aysgarth	SE0287	3	20 × 20 × 16 (6) 54	SE	?1600s	garden	S	R	g	B	
2	Shaw Paddock, Hawes	SD870919	2	(24) × (36) × 27 (6) (24)	SW	?1700s	garden	S	R	g	B	whole bole projects; each for 2 skeps
3	nr Camshouse, Askrigg	SD912905	3	(30) × (24) × (16) (6) (24)	S	1700s–I	?orchard, garden	S	R	g	BT	in 'bee garth'
4	Warnford House, Thoralby	SE003087	8	27 × 20 × 18	S	1700s–II	garden	S	R	f	B	2 tiers of 4 & 4
5	Grange Farm (was Steepbanks), Linton, nr Grassington	SD9967	3	18 × 20 × 16 30 (6)	W	1820s	garden	S	R	g	B	
7	Nutwithcote Farm, Masham	SE231790	6	30 × 24 × 18 18 ?18	S	?1453 or later	orchard	S	A	e	BT	Fig. 147; rounded at back; house was monastic grange of Fountains Abbey
27	Parkside House, Follifoot, Harrogate	SE3452	8	24–30 × 21–24 × 18 15 ?6	S	1600	orchard	dSs	R	b	BCT	'bee garth'; base projects; good condition in 1973; mentioned 1930,[145] pp. 415–17
29	The Cottage, Ruddings, nr Follifoot, Harrogate	SE3353	4	29½ × 26 × 16 17 ?36	S		garden	S	R	g	B	brick-lined; stone base projects
35	Spen House, Askrigg, nr Leyburn	SD9491	5	36 × 24 × 24 20 ?12	S	1841	garden	Ls	A	e	B	'bee houses'; specially built structure at ground level, with shelves at 20″; probably last used 1860
39	West Scrafton, Coverdale	SE073836	12	31 × 21–25 × 17 21 15–31	S	pre-1880	rough ground	dS	R	b	T	
56	Bridge End, Arncliffe, nr Skipton	SD9372	2	33 × 24, 41 × 21 24 ?6	S		garden	dLs	R	g	B	built out from wall; Kingsley wrote *The water babies* here

282

No.	Name	Grid ref	Count	Dimensions	Orientation	Date	Location	Material	R/A	Code	B/BD	Notes
57	The Falcon Inn, Arncliffe, nr Skipton	SD9372	2 1	21×20×19 10 ?5 22×22×?22 18	S		garden garden	dLs dLs	R R	b p	BD B	'bee house'; whole structure projects; had sloping flagged roof—now collapsed; inside ht at back 33" in same wall as above; drip stone overhangs
58	Well House, Bankwell Rd, Giggleswick	SD8164	4	24×22×19 6 ?18	S	1700s	garden	S	R	e	B	base projects
59	Queens Rock, Bankwell Rd, Giggleswick	SD8164	4 2 2	as under 23×26×20 22 ?4 18×25×20 11 ?5			garden	S	R	e	B	in 2 groups ?4′ apart; house 1720 and 1776 base and ?top project
60	Tems House, Giggleswick	SD8164	2	as under 19×20×18 28 19×19×16 14	N E		garden	dS	R	p	B	boles in 2 garden walls; house (1400) belonged to monastery
61	Lane Top, Arncliffe, nr Skipton	SD9372	4 3 1	as under 17–19×21× — (36) 20×24× — (36)	E		yard	dLs	R	b	B	
62	Eldroth Old Hall, Giggleswick	SD764655	2	19×20×19 26	SSE	?1660s	orchard	dLs	R	e	B	
63	Eldroth Old Hall, Giggleswick	SD764655	8	22×20×(18) (24) (72)	SSE	?1660s	field	dLs	A	f	B	5 filled in
64	Taitlands, Stainforth, nr Settle	SD823669	14 8 1 1 1 1	as under 28×28×14 ?18 20×30×18 (18) 16×21×13 22×21×16 (18) 11×25×15	E		?field	S	R	f	B	slate base and lintel; some in 2 tiers; wall repaired with brick; 1 bole filled in; 1953—skeps found here; photo published[101]

1	2	3	4	5	6	7	8	9	10	11	12	13
75	Old Hall Farm, Feizor, Giggleswick	SD7968	1	24 × 25 × 15 (18) 1 21 × 18 × 15	S		garden	dS	R	f	B	widely spaced boles in boundary wall of old packhorse track; nearby house 1699; boles demolished by 1979
77	Stockdale House, Feizor, Giggleswick	SD791678	3	18 × 18 × 18 24	S		garden	Ls	R	e	B	nearby house (1700) belonged to Sawley Abbey, Clitheroe
83	Dale Head Farm, Westerdale, nr Whitby	NZ678045	3	27 × 25 × 25 32	S		garden	Ss	A	f	BCT	'the bee houses'; arch head, made from single stone, is shouldered; rounded at back; whole structure, built in front of wall, has sloping tiled roof; boles still used in summer in early 1950s; roof gone by 1979
			6	37½ × 25 × 22 15		1832						
99	(Wearingfold), Higher Wham Farm, Giggleswick Common, nr Settle	SD770622	3	17, 19, 20 × 18 × 17 30 33	S		outside cattle fold	dS	R	f	BC	top and sides project slightly; heather site only
116	Bilton Hall, Harrogate	SE334573	2	26 × 20 × 15 36 21	S	c. 1479	garden	B	A	e	B	Tudor brick
117	Field House, Darley	SE2059	3	as under 17 × 14 × 12 17 × 15 × 11 18 × 14 × 10 36	SE	1820	garden	S	R	g		wall 17″ thick; house built by Quakers
121	Cottage, Dacre, Harrogate	SE1960	4	'rather small'	SE		garden	dS	R	f		bees here c. 1920; now filled in
145	Sweet Briar Cottage, Starbotton, Skipton	SD9575	2	21 × 30 × 26 18	S		garden	S	R	f		house 1733
146	Bridge House, Starbotton, Skipton	SD9575	2	20 × 24, 20 × 18 16	NE		?garden	dS	R	p		house 1838
148	Wood House Farm, between Austwick and Feizor	SD778682	2		S							
176	East Camshouse, Askrigg	SD915905?	4	20 × 18 × 15 24 ?9	S		garth	dS	R	g	B	roof and base project; one is 28″ away from others; see also 477
178	Closed Garth, Conistone, Grassington	SD6798	5	30 × 23 × 20–24 18	SE		?garden	S	A	g		rounded at back; there were 2 others

No.	Name	Grid ref	n	Dimensions	Orient.	Date	Setting	Type			Cat.	Notes
179	Swinton Castle, Masham	SE2179	12	22 × 22 × 17 0, 30	S	?c. 1750	garden	SB	A		B	2 tiers of 6 & 6 in stone wall faced with brick; now filled in
180a	on Mukerside, Muker (above Appletreethwaite Farm)	SD9098	2	19 × 17 × 18 15 ?36	S	c. 1830	field	dS	R	e	B	base projects; altitude 1250′
180b	on Mukerside, Muker (above Appletreethwaite Farm)	SD9098	2	16 × 14 × 14 12 ?20	W	c. 1830	field	dS	A	e	B	Fig. 155; base projects; in same field as 180a
182b	West Ings, Muker	SD9198	—	x × x × — <x			garden	dS	R		B	filled in; probably in same wall as shelter 182a
186	Linley House, Barkston Ash, Tadcaster	SE491360	12 / 8 / 4	as under / 18 × 24 × 24 (36)(4) / 18 × 50 × 24 (54)(4)	S	pre-1800	garden	Ls	R	p	B	2 tiers of 8 & 4; base projects; rounded at back / each for 2 skeps (may have been subdivided originally)
209	Snayse Holme, Hawes	SD8387	—									
223	Overwood Farm, Wadsworth	SD9833?	—									
224	Hill Top Farm, Burton Leonard, Harrogate	SE325638	4	24 × 21 × 18 31 15	S	1700s–II	garden	S	A	g		2 filled in; near Fountains Abbey
240	Burton Hall (was Old Hall Farm), Burton Leonard, Harrogate	SE327639	8 / 4 / 4	as under / 24 × 20 × 18 30 12 / 18 × 20 × 18 66 12	S	1800s–I	garden	S	R	f		2 tiers of 4 & 4 (Fig. 148); there is preservation order on wall
255	Holly Cottage, Giggleswick School, Settle	SD8164	?1			?1850s or earlier	house					
340	East Witton, Masham	SE1485	1	12 × 14 × 18 16	E		garth	S	?R	e		
455	Holme House, Gargrave, Skipton	SD946544	8 / 2 / 3 / 3	as under / 24 × 22 × 12 13½ 9′ / 23½ × 25½ × 18 13½ 5 / 22½ × 22½ × 13¼ (12) 6¼	E / S / S / S	1800s–I	garden	dS	R	f	B	base projects 2″–3″; 1 filled in

1	2	3	4	5	6	7	8	9	10	11	12	13
469	Leyland House, Conistone, Skipton	SD982674	5	$30 \times 27 \times$ — 18 15	S	1500s–II or 1600s–I	garden	Gs	A	g	B	wall includes limestone; ?rounded at back
470	Milford Cottage, South Milford, Leeds	SE497318	10	$19, 17 \times 20\frac{1}{2} \times 21\frac{1}{2}$ 14, (35) $4\frac{1}{2}$	SE	1734	garden	B	R	f	B	2 tiers of 5 & 5 made from stone slabs and brick piers; base and top project; 1980—poor condition, some fallen in
473	Low Hall Cottage, Appletreewick, Skipton	SE0560 (SE051578?)		as under $18 \times 26 \times 20$ $18 \times 23\frac{1}{2} \times 18\frac{1}{2}$ 32	W		yard	Ss	R	g	T	'bogey-holes'
477	Lower Camshouse, Askrigg	SD914904	3					dS	R	g		base projects; was part of 176
498a	Countersett Hall, Askrigg, Wensleydale	SD919879	5	as under (78)	SE		outhouse (? was house)	S	R		D	all above door; 1 smaller than others; drip course runs above the row; now inside building; house 1650 filled in filled in
			2	$27 \times 37 \times$ —								
			1	$29 \times 52 \times 20$								
			1	$24\frac{1}{2} \times 48 \times$ —								
			1	$?24 \times - \times$ —								
498b	Countersett Hall, Askrigg, Wensleydale	SD919879	2	as under $10\frac{1}{2} \times 13\frac{1}{2} \times 12$ 50 $18 \times 13 \times 16$ 20	E		?garden	dS	R	f	B	base projects about 2"
504	Bent Hill Farm, Hazlewood, Bolton Bridge, nr Skipton	SE106547	1	$16\frac{1}{2} \times 34 \times 14\frac{1}{2}$ 24	S		field	dS	R	g	B	
639	6 Park Street, Ripon	SE309713	2	$10 \times 10 \times 10$ 74 19	NE	1700s–II	garden	B	R	f	B	in 15′ wall; boles unusually small for Yorkshire
644	Lythe House, Grassington, nr Skipton	SE0064	3	$(36) \times (24) \times 18$–$20$ 18 22′	S	1700s–II	garden terrace	dLs		f	D	shape triangular (Fig. 158); skep found in dwelling house 1956
652	behind 2 Keld Close, Scalby Road, Newby, Scarborough	TA0190	2	$25 \times 18 \times 20$ 43 12′–15′			?field					
657	Howe House (or Farm), Old Malton	SE806754	6	$33 \times 23 \times 15$ 16–24 $9\frac{1}{2}$	S	?1700s–I or earlier	farm cart-shed	S	A	g	C	brick arch; end bole repaired with brick; others rounded at back

No.	Location	Grid ref	n	Dimensions	Orient.	Date	Setting	Type	R/A		Code	Notes
663	Hogg Hall, Sawley, Ripon	SE256681	3	24 × 21 × 21 / 14 20, 24	S	1660	garden	dS	R		BT	base projects in irregular shape; boles filled in with loose stones
677	Bank End Farm, Lawkland, Settle	SD775664	2	32 × 27 × 13 / 18 5½	S		garden	dS	R	f	C	slate base remains in 1 bole
678	Lawkland Hall Farm, Lawkland, Settle	SD775658	4	16 × 16 × 17 / 31	SSW		farmyard	dS	R	g	C	
679	Lawkland Green, Lawkland, Settle	SD782656	6	27–29 × 20–23 × 17 / 21 8, 45	S		orchard	S	R	e	C	3 pairs; base projects 1″
740	derelict house, Preston-under-Scar, Leyburn	SE0791	6	24 × 19–28 × 22–25 / 22 6	S		garden	S	R		B	another recess above, ht 15″
756	The Café, Osmotherley, nr Northallerton	SE457973	6	29 × 24 × 23 / 16–25¼ (6)	S	pre-1780	garden	S	R	p	BC	whole structure projects
757	Loand House, Wrelton	SE759869	7	32 × 24 × 16 / 6 ?6	S		garden	S	A	g	BT	whole structure projects and has sloping tiled roof; possibly 1 more bole; house 1830
758	The Lund, Hutton-le-Hole	SE712904	8	30 × 24 × 15 / 37 4	SW	1600s-II	orchard	S	A	g	BT	Fig. 154; base slab has hole, ?for drainage
759	Tickhill Farm, Wetherby Road, Knaresborough	SE371551	6	23 × 19 × 14 / 26 5	SE		garden	S	R	g	BT	base projects 3½″; projection above
760	Steeton Hall Cottage, South Milford	SE483315	13	21½ × 22 × 20 / 23 (16)	S		garden	dS	R	f	BT	groups of 9, 4 in same wall; rounded at back; possibly medieval
762	Cropten Lane, Pickering	SE7984	6	(36) × (18) × — / (24)	S			Ls	R	f		rounded at back; sandstone base; in use up to 1939
804	Ings House, Hawes	SD874897	3	31 × 32 × 17 / 31 24	E	c. 1810	garden	dS	R	e	BT	base projects 6″; above row, slabs project 10″; ?cemented at back
809	Binks Cottage, Penn Lane, Hawes	SD874897	4	24 × 22 × 33½ / 29 10½	SE		garden	S	R	g	B	'bee house'; built against wall with large sloping slab for roof; house probably 1680
811	Rome Farm, Giggleswick	SD791628	2				house			e		near back door in small walled garden; may be 1 more

Yorkshire, South

No.	Location	Grid ref	n	Dimensions	Orient.	Date	Setting	Type	R/A		Code	Notes
499	Skyers House, Hemmingfield, nr Barnsley	SE387014	?36	28 × 20 × 20 / 24 ?4	?30N† 6W†	1600s-I	orchard	S	R		B	brick-faced; flight entrances at back, and possibly doors at front; restoration started 1976, but wall collapsed 1978

1	2	3	4	5	6	7	8	9	10	11	12	13
Yorkshire, West												
30	Reynard Ing, Ilkley	SD1147	4	28 × (24) × (24) 30 ?12	S	c. 1650	garden	S	R	e	BT	'bee holes'; last used 1880
31	Smallbanks (1), Addingham, Ilkley	SE078487	3	21 × (24) × 16 8 ?48	?SSE	c. 1800	garden	S	A	g	B	'bee holes'
32	Smallbanks (2), Addingham, Ilkley	SE078487	2	(24) × 21 × 21 32 ?24	1S 1SSW	1779	?paddock	S	R	g	B	?for bees
65	Wood Nook Farm, Horsforth, nr Leeds	SE257390	8	(24) × (24) × — (36) ?24	S	1800s–I	?garden	S	R	g	B	base projects; destroyed by 1953
104	Cowburn Farm, Silsden Moor, Silsden	SE0446	4	23 × 22 × 15 24 ?44	S	?1860	garden	S	R	g	B	'bee holes'
105	Prospect House, Brunthwaite, Silsden	SE0546	5	22 × 20 × 19 30 ?6	S	1700s or earlier	garden	dS	R	g	B	'bee holes'
127	Overwood Farm, Blake (or Black) Dene, Heptonstall	SD9531?	—									
139	Overwood, Walshaw, Hebden Bridge	SD9731?	—			1800s–II						may be same as 127
149	Kershaw House, Luddenden, Halifax	SE0426	—									
151	Swan Bank, Cragg Vale, Hebden Bridge	SD9923	—									
154	Thornton Hall, Thornton, Bradford	SE105327	4	37 × 25 × 18 35 24	S	c. 1640	garden	S	T	e	C	whole projects 3"
192	Moat House, East Keswick, Leeds	SE356442	3	38 × 84, 108 × 15 23	S	1700s–II	?garden	S	R			'bee or by-wall'; oak beam for lintel; two 108" wide, each for ?5 skeps; one 84" wide, for ?4 skeps; used until 1892
203	Catholes Jump Farm, Hudson Clough, Todmorden	SD9324	3	26½ × 33 × 19 0–7 ?8	S	1800s–I	yard	?S	R		B	may be bee bole
220	Turley Hole Farm, Cragg Vale, Hebden Bridge	SD9922	—					dS	R			
221	Higher House Farm, Cragg Vale, Hebden Bridge	SD9923	?15		E		field	dS	R	g	B	sizes irregular; ?specially built bee wall; house 1666

No.	Location	Grid ref		Dimensions		Aspect	Date					Setting	Notes
222	Lumb Bank Farm, Heptonstall	SE0321	(20)	—		SW						garden	
271	Owlet Cote, Ewood Estate, Mytholmroyd	SE0226	—										
272	281 Howe Croft Head, Greetland, Halifax	SE0821	1	23 × ?38 × 21 ?12–18		S	1835	Gs	R	?f	B	garden	probably had central dividing stone
358	Highgate House Farm, Clayton Heights, Bradford	SD1230	6	28 × 28 × 12 (48)		S	1700s–II	S	A	e	B	garden	base projects
456	Ponden Hall, Stanbury, Haworth	SD9836	8	38 × 26 × 26 24		S		dS	R	?p	B	(field)	built in depression in hillside
466	Rishworth Hall, Bingley	SD099409	4						R				
520	Lower Woodside Farm, Silsden	SE0446	7			?E							
552	Wilshaw Vicarage, Meltham	SE0911	4	(24) × (24) × (24) 30		N		S	R	g		garden terrace	
708a	in field nr Buckley House, Stanbury, nr Keighley	SD996368	—									field	?bee shelter 708b

WALES

Clwyd

No.	Location	Grid ref		Dimensions		Aspect	Date					Setting	Notes
92	Bronheulog, Llangollen	SJ2141	8	av. 18 × 22 × 18 24, 48 (12)		S	1700s	S	R	f	B	garden	2 tiers of 5 & 3; wall includes some brick
553	Garthgynan, Llanfair-Dyffryn, nr Ruthin	SJ143554	25	18½ × 10 × 11 42 80		17E 8W	1500s	B	T	g	B	garden	handmade bricks
582	Brynllithric Hall, Rhuallt, nr St Asaph	SJ072756	3	26 × (24) × 18 (36) 9		NW		S	A	g		garden	
706	Plas Ashpool, Llandyrnog	SJ107678	3	(20) × (23) × ?16 (24) (9)				B	A	g	B		stone base projects
707	Ty-Derwin, Llanfair-Dyffryn, nr Ruthin	SJ1355	1	$x \times 2.5x \times 0.5x$ x				S		g	B		base projects; for ?2 skeps
775	Aberwheeler Farmhouse, Bodfari, nr Denbigh	SJ1070	6	22 × 39 × 18 ?9				B	A	g	C	garden	2 tiers of 3 & 3, each probably for 2 skeps; house 1400s/1500s

1	2	3	4	5	6	7	8	9	10	11	12	13
820	Deiniols Ash, (was in Flintshire)	?	3	?15 × ?12 × — ?45–48 ?30			building	?B	T	f	B	?bee boles
Dyfed												
705	wall opposite Imperial Hotel, Tenby	SN1300	7	42 × 24 × 13				S		f	BT	shaped = pointed arch; central one 64″ high, 30′ wide; brick-lined; ?bee boles; (late information, not included in analysis
736	Amroth Castle Holiday Centre, Amroth, nr Saundersfoot	SN173073	9	26 × 20 × 16 (24)	E	1800s–I	?was orchard	S		g	CT	shape = pointed arch; rounded at back; 7 may have had drip sills; (late information, not included in analysis
793	Bee Hall, Lamphey, Pembroke	SN006019	3	17 × 15–18 × 12–13 60	NW	pre-1796	farmyard	Ls	R		T	may have been 2 more; John Keys lived here[171]
838	Southwood Hunting Lodge, Roche, nr Newgale	SM860217	4					S	R	g	B	slabs at base and top project
866	behind Tourist Information Centre, High St, Haverfordwest	SM9515	2	14 × 18 × 15 (54) 66	W	?1700s		S	R			boles probably made later than wall; sides and base lined with small yellow bricks; (late information, not included in analysis)
Glamorgan, Mid												
764	Lys Cwmllorwg, Lewistown, Bridgend	SS938892	4	18 × 19 × 24 (36) (19)	SE		farm garden	dS	R	f	BC	rounded at back; altitude 800′
815	Llancaiach Fawr, nr Treharris	ST114966	4	(20) × (20) × (20)	E		garden	S	R			
819	North Cornelly Hall, North Cornelly	SS821816	?2	x × x × —				S	A	g	B	2 or more; very bad condition 1974, presumably now gone; possibly a specially built structure, not very tall, with boles facing S and E or W
837	Blackhall Farm, Castle-upon-Alun, nr Bridgend	SS9174	12	21½ × 21½ × 21½ 12, (30, 60) 7	?S		garden	dS	R	g	B	Fig. 162; 3 tiers of 4 & 4 & 4
Glamorgan, South												
226	Ty Mawr, Llanbethery, Llan Carfan, Barry	ST0369	3	16–17 × 19–21 × 10–11 (24)	W		garden	S	D	f	B	rounded at back
325	Heol Las Farm, Llangan, Cowbridge	SS9577	3	20 × 24½ × 23 6 10	S	?c. 1900	garden	S	R	f	B	whole structure projects

No.	Name	Grid ref	n	Dimensions	Aspect	Date	Location	S	?A	g	B	Notes
753	Coach House, Llandough Castle, Cowbridge	SS995730	3	22 × 23 × — 42 13	E		coach house	S	?A	g	B	now converted to windows
813	Boverton Castle, Boverton Place, Llantwit Major	SS982683	9	30 × 15–28 × 14–21 ?0 ?7–9	S		garden or orchard	dS	A	p	B	may be more; some filled in
814	Caercady, St Mary Hill	SS949793	3		SE		house					
816a	Llantrithyd Place, Llantrithyd	ST044728	3	— × (40) × —	?S		garden					?for 2 skeps; stone-slabbed roof and 'gothic' arch; this set at E end of wall
816b	Llantrithyd Place, Llantrithyd	ST044728	3	— × (20) × —	?S		garden					at W end of wall; possibly later date

Glamorgan, West—no records of any type

Gwent

No.	Name	Grid ref	n	Dimensions	Aspect	Date	Location	S	?A	g	B	Notes
135	Troy House, Troy, Monmouth	SO509113	2	29 × 26 × 13 39, (55)	S	1500s	garden	S	R	e	B	one above another; 1' above recess, slabs project from wall
			2	29 × 26 × 13 39, (55)	W	1500s	garden	S	R	e	B	similar pair, now inside cowshed; on nearby outbuilding, ?bee bole (20" × 19" × —) facing SW, now filled in
163	Beaufort Arms Hotel, Tintern	SO5300	2	11 × 11½ × 12 8 (9)	S		garden	S	R	g	B	projection above; gone by 1976
166	Llanfair Court, Abergavenny	SO2914 (?SO3919)	4	30 × 34 × 27 0, 38	S	?1600s	?garden	S	A	e	BT	2 tiers of 2 & 2, each for ?2 skeps; was another pair
512	Whitebrook House, Llanvaches, Magor	SO421925	3	as under 13 × 15 × 12 12 × 14 × 16½ 12 × 15 × 12 48 (24)	S		courtyard	S	R	f		were 3 more; house 1500s
535	Abergavenny Castle, Abergavenny	SO300140	7	14 × 14 × 18 42 14	SSW	1200s or 1300s	castle courtyard	Ss	R	4g 3p	C	

Gwynedd

No.	Name	Grid ref	n	Dimensions	Aspect	Date	Location	S	?A	g	B	Notes
52	Pant Glas Uchaf, Upper Clynnog	SH4747	1	17 × 36 × 17 12	S		garden	Ga	R	f	B	'gardd'; slate base; house 1562

1	2	3	4	5	6	7	8	9	10	11	12	13
193	Tyddyn Adi, Morfa Bychan, Portmadoc	SH540381	2	20 × 30, 50 × 18 6 (6½)	SW	1600s	garden	S	R	f	B	probably for 2, 3 skeps
291	Bryn-y-Odyn, Trawsfynydd	SH7035 (?SH714294)	4	17–20 × 22 × 18 10 4	S		paddock	SSl	R	g	B	2 pairs, roughly constructed; house c. 1600

Powys—see below, 'Bee houses'

SCOTLAND

Borders Region

1	2	3	4	5	6	7	8	9	10	11	12	13
293	Ivy Cottage, Gavinton, Duns	NT7652	2	22 × 16 × 15 0 13	S	?1800s–I	garden	S	A	g	B	
435	St Andrews Churchyard, West Linton	NT1451	2	22 × 22 × 18 18 (3½)	S	pre-1780	garden	Ss	R	f	B	wall now in burial ground; bole projects slightly
468	Ivy (was Sorella) House, Gavinton, Duns	NT7652	?2									owners (same as 293) say no boles here in 1980
515	Dovecot, nr Romanno Bridge, West Linton	NT163480	3				garden	S				

Central Region—no records of any type

Dumfries and Galloway Region

1	2	3	4	5	6	7	8	9	10	11	12	13
616	Braewells, Lochmaben	NY0882	1	24 × 20 × (14) 24	W			Ss		g		rebuilt

Fife Region

1	2	3	4	5	6	7	8	9	10	11	12	13
12	The Auld Hoose, Friars Court, Crail	NO6107	3	24 × 26 × 17 6 9	SW	1686	garden	Ss	R	g	B	1 other recess
24	The Brae, Townhead Road (was 1 Kilcruik Rd), Kinghorn	NT2687	2	25½ × 22 × 24 25 66	SSE	pre-1800	garden	S	A	e	B	Fig. 166 shows rebates for front boards
101	Rosemount, Cupar	NO3714	1	20 × 20 × (13) 38	S	1817 or earlier	garden	Ss	R	g	B	base and top project; 2 others destroyed
118	Blue Stane, High Street, Crail	NO3714	4	24 × 22½ × 17 37 9	W	1700s–II	garden	Ss	R	g	B	

ID	Location	Grid Ref	No.	Dimensions	Orient.	Date	Setting	Stone	R/A	letter	B/C	Notes
235	14 Charles Street, Pittenweem	NO5402	1	36 × 38 × — 52	W	c. 1800	garden	Ss	R	f	B	c. 70′ away, in adjoining wall; for ?2 skeps; filled in
294	25 South Street, Elie	NO4900	2	28 × 15 × 12 21 69	SW	c. 1755	garden	Ss	R	f	B	smaller recess nearby probably charter bole (Fig. 177)
295	Kinglassie, Boarhills, St Andrews	NO559135	4	24 × 21 × 24 >12 6	S	1683	garden	Ss	R	g	B	fluting on lintel; one dividing slab gone
296	Old Post Office, Boarhills, St Andrews	NO5614	2	26 × 21 × 17 10 (13)	S		garden	Ss	R	g	BC	metal bars across
297	Edenshead, Gateside, nr Cupar	NO5614	3	21 × 24 × 24 (24) (33)	S		garden	dSs	R	f	BC	slab base; overgrown in 1980
		NO183092	6	19½ × 19½ × 19¼ 10, (36) (14)	SW	1500s	garden	Ss	R	e	B	2 tiers of 3 & 3; whole structure projects 15″; ?rounded at back: 4 iron hinges, probably for doors; may have been used by monks of Balmerino Abbey; bees there up to 1939
298	Seaview, Boarhills, St Andrews	NO5614	4 3 1	as under 26 × 22 × 26 6 3½ 19 × 20 × 15¼ 15	S	1749	garden	dSs	R	g	BC C	whole structure, made of large slabs, projects from wall in same wall, about 36′ away
362a	Castle Farm, Ceres, Cupar	NO4011	6	24 × 21 × 21 12, 36 4	S	after 1780	orchard	dSs	R	p	C	2 tiers of 3 & 3, built into bank; top 3 have collapsed
362b	Castle Farm, Ceres, Cupar	NO4011	2	28 × 42 × 8¼ 26	1S 1W	after 1780	paddock	Ss	R	g	C	top projects 10″, base 2″; reported use by Earl of Crawford for 'keeping bees under glass' (first such report in Scotland, date not known)
366	Scotscraig, Tayport	NO444282	4	36 × 29 × 34 12 9	S	1670	behind garden	Ss	?R	b		rebated edges; in one bole, stone slab with hole (?for bar); covered in ivy in 1979
380	The Priory, Nethergate, Crail	NO3714	1	33 × 23 × 17 0	SW	1600s–I	garden	Ss	A	f	B	may have had shelf ?18″ from ground; wall was in nunnery
386	16 Townhead Road, Kinghorn	NT2687	3 2 7 1	as under 20½ × 18 × 18 8¼ 21 × 19½ × 17½ 18	SSE	pre-1760	garden	S	R	g	B	54′ from the pair; had locking bar
425	Morton Farm, Tayport	NO466260	4	24 × 20 × 22 14 12	S	1640	garden	Ws	R	b	B	owner planning to restore (1979)

293

1	2	3	4	5	6	7	8	9	10	11	12	13
427	Lindores Abbey House, Newburgh	NO243185	1	$43 \times 50\frac{1}{2} \times 15\frac{1}{2}$ 30	N	1876	house	S	A	e	C	arch and projecting base have carved edges; for ?2 skeps; although facing N, well sheltered by trees; garden may once have been orchard; house adjoins ruined abbey
429	Old Logie Manse, Logie, Cupar	NO405205	2	$25 \times 22 \times 23$ 12	W			Ss	R	p	C	slab base; top of one bole gone
453	Newington House, by Cupar	NO347190	4	$15\frac{1}{4} \times 42\frac{1}{2} \times 19\frac{1}{2}$ 4 $11\frac{1}{2}$	S	1828	garden	Ws	R	e	B	Fig. 171; whole projects; top and base of Caithness flags; sides and back of small bricks; for ?2 skeps
544	Welltree, Kingskettle, Cupar	NO295064	2	$22 \times 21 \times 14$ (9) 4	?S		garden	S	R	g	B	base projects; remains of iron fitting at one side
548	Beech Walk (was Kirkmay Coach House), Crail	NO6107	5	$14 \times 13 \times 12$ 7 6	NE			Ss	R	g	B	unusually small for Scotland
549	Kirkmay House, Crail	NO6107	1	$47 \times (60) \times 12$ 15	NE	1700s	garden	Ss	A	g	B	plastered inside; for ?3 skeps
596	The Manse, Dunino, St Andrews	NO5311	6	$26 \times 19 \times 13$ 21, 13 6	S	pre-1720	garden	S	R	g	B	2 sets of 3 (30' apart); owner has wooden base and skep, Fig. 142
704	?141 South Street, St Andrews	NO5016	2	$21 \times 21 \times 17$ >16 8	S	pre-1580	garden	Ss	R	f	B	house number not certain
745	5 South Street, St Andrews	NO5016	1	$20\frac{1}{4} \times 41 \times 21$ $18\frac{1}{2}$	S	1600s–II or 1700s–I	garden	Ss	R	g	B	for ?2 skeps; projecting slab shelf with ground-level recess (ht 15″) below; decorative feature above (like aumbry) may also have been used for bees ($14″–23″ \times 24″ \times 24″$)
746	60 South Street, St Andrews	NO5016	3	$24\frac{1}{2} \times 22 \times 16$ $21\frac{1}{2}$ 5	S	c. 1600	garden	Ss	R	e	B	boles in part of a wall which originally ran from 2 to 62 South St and possibly contained many bee boles, but much of the wall now pulled down
747	St Salvator's College, St Andrews University, North Street, St Andrews	NO5016	1	$20–22 \times 20\frac{1}{4} \times 20$ 17	S	pre-1580	college garden	Ss	R	g	C	slab base projects
748	93 Market Street, St Andrews	NO5016	3 1 2	as under $17 \times 25 \times 22\frac{1}{2}$ $17 \times 20\frac{1}{4} \times 22\frac{1}{2}$ 22 $6\frac{1}{2}$	S	pre-1580	garden	Ss	R	2f	B	whole projects; sloping lintel
749	125 Market Street, St Andrews	NO5016	3	$20–21 \times 21–22 \times 16$ 37 5	S	pre-1580	?garden	Ss	R	f	B	slab base projects
751	149 South Street, St Andrews	NO5016	1	$19 \times 19\frac{1}{2} \times 14$ >20	S	pre-1580	garden	Ss	R	p	B	slab base

No.	Location	Grid Ref	n	Dimensions	Dir	Date	Setting	Mat			Type	Notes
752	151 South Street, St Andrews	NO5016	2	21 × 21 × 17 19 7	S	pre-1580	garden	Ss	R	g		slab base
755	133 South Street, St Andrews	NO5016	1	40 × 38 × 22 18	S	pre-1580	garden	Ss	R	g	B	shelf was sandstone slab, now wood; for ?2 skeps
766	Lover's Lane (new housing estate), Millgate, Cupar	NO3714	2	23 × 22 × 18 36 60′	S	?1700s		Ss	R	e	B	facings 6″ wide project 4¼″ all round
767	Barony House, The Barony, Cupar	NO3714	1	24 × 41 × 19¼ 23	S	c. 1800	garden	Ss	R	e	B	large slab facings project 6″ all round; for ?2 skeps
779	The Abbot's House, Maygate, Dunfermline	NT0987	1	42 × 36 × 19 26	S	1500s–II	house	Ss	A	g	B	ht at back 32″; for ?2 skeps; house was built for secular lordship of Dunfermline Abbey
785	Smithy House, Kingsbarns, St Andrews	NO5912	4	20½ × 22 × 23½ 20 4	S	pre-1700	field	Ss	R	g	C	whole set projects; 2 boles covered by 1 large slab
786	top of Abbey Street, St Andrews	NO5116	3	18¼ × 16½ × 14 28¼	W		garden	Ss	R	g	B	this old abbey wall is now by main road, but originally gardens adjoined it
796	Sligachen, Main Street, Strathkinness, St Andrews	NO4516	1	22½ × 38 × 14 20	S	1700s	garden	Ss	R	e	B	whole projects 9″; for ?2 skeps
801	Cathedral Cloister, St Andrews	NO514167	4	as under 15½ × 23 × 23 20 × 24 × 25 (66) (42) 2 25 × 33 × 24 (66) (72)	SE	1800s	garden/orchard	Ss			C	2 pairs of boles made in ruined 13thC cloister wall, probably by Victorian beekeeper who had garden (no longer there) in cloisters 1 slightly arched and rounded at back R f both have rebated surround and are rounded at back 1R f 1A
802	St Andrews Cathedral precincts adjoining end of North Street, St Andrews	NO514167	1	21½ × 16 × 9¼ (24)			cathedral grounds	Ss	R	g	C	?bee bole; rebated at sides; at left in sneb, at right 2 hinges; may have been deeper—now has top cemented in

Grampian Region

No.	Location	Grid Ref	n	Dimensions	Dir	Date	Setting	Mat			Type	Notes
109	Hatton Castle, Turriff	NJ758470	7	28 × 23 × 20 (15)	E		castle grounds	S	R	f	BT	
110	Tolquhon Castle, Tarves, by Ellon	NJ872286	12	25 × 25 × 20 30 (36)	W	c. 1620	castle grounds	S	R	e	BT	Fig. 140; Ancient Monument Dept., Scottish Development Dept.
112	Netherbow, Waterside St, Bishops Mill, Elgin	NJ216635	1	27 × 23 × 20 10	S		?garden	S	R		B	there was another

295

1	2	3	4	5	6	7	8	9	10	11	12	13
114	Pluscarden Abbey, Elgin	NJ142576	4	as under 2 30 × 24 × 24 42, 36 2 23 × 24 × 23 45, 46	S	?1200s	priory garden	Ss			BT	2 pairs in precincts boundary wall; rounded at back; date of 1 pair may be later; skeps still kept in 2 boles to attract stray swarms
360	Alma Park, Park House, Park	NO7898	3	19 × 22 × 20 0 (60)					D	f		slightly pointed arch; roughly tooled lintel and jambs probably added later
372	Chanonry Lodge, 13 The Chanonry, Old Aberdeen	NJ938086	2	as under 20 × 24 × 11 48 19 × 27 × 12 44	E		garden	dS	A	g	B	near 794
								S	R	g	C	
373	Glenmillan House, Lumphanan	NJ594053	6	34 × 31 × 31 0 20	?SE	1688 or 1872	garden	Ga	R	g		?bee boles, in rough red granite wall
374	Pitmedden House, Ellon	NJ883281	6	as under 2 25½ × 24 × 22 15 1 21 × 21 × 32 6 2 14 × 16 × 15 65–70 1 18 × 18 × 13 44	1N 1S E S S	1675	garden	Ga	R	e	BT	NT for Scotland
378	Balbegno Farm, Fettercairn, Laurencekirk	NO641730	6	30 × 23 × 23 6	SE	?1700s	garden	Ss	R	b		3 pairs c. 20′ apart in wall of boulders; 2 pairs probably had locking bars
422	Balbegno Castle, Fettercairn, Laurencekirk	NO639730	4	24 × 24 × — 22 ?8	SE	pre-1795		Ss	R		B	rebated for front boards; whole structure protrudes at back of wall; castle 1569; boles now filled in
501	Tarland	NJ4804	2	24, 32 × 24 × 18 20, 45	ENE	1600s or earlier	garden	Ss	R	g	C	'bee bowls'; one above the other; slate base projects
618	99 High St, Elgin	NJ2162	2	as under 23½ × 18½ × 11 40 (24) × 22 × 12 40	W	1700s–I	garden	S	R	f		20′ apart

No.	Location	Grid Ref		Dimensions	Orient	Date	Setting	Ga	A	f	B	Notes
630	Hillside House, Hillside, Portlethen	NO923977 (or 928977)	4	26–28 × 24 × 160 26′	SE	?1850	garden	Ss	A	f	B	brick arch projects; now rendered and painted, greenhouse built over boles
780	Fordyce House, Fordyce, nr Portsoy	NJ5563	2	(24) × (20) × (14) (6) (72)	S	1700s–I	garden	Ss	R	1f 1p	B	partly filled in; were probably 4 more: near 360
794	West Park Farm, Park	NO7898	4	36 × 20 × 180	S	1700s–II	garden	dS	R	p	B	large slab base projects 6″
805	The "Old School House, Fordyce, nr Portsoy	NJ5563	2	27 × 18½ × 12 20 57	S	1700s–I	garden	Ss	R	e	BC	if a bee bole, unusually high up, perhaps because in main village street
806	Fordyce Castle Fordyce, nr Portsoy	NJ5563	1	(15) × (30) × (12) (16″–17″)	W	1592	castle	Ss	R	g	BC	at least 2 boles; now almost at ground level, which has risen 18–24″; base is large slab
824	The Cottage, Drumoak, Banchory	(NO6995)	?2									

Highland Region

No.	Location	Grid Ref		Dimensions	Orient	Date	Setting	Ga	A	f	B	Notes
10	The Parks, Dingwall	NH5458	?6					SB	R		B	now demolished
108	Mackay's Sawmill, Cramond Brae, Tain	SH7782	2	33 × 55 × 19 27 8	SE		?castle garden	S	R	p		whole projects 7″; for ?2 skeps
270	Tulloch Farm, Bonar Bridge	NH6191	1	24 × 24 × 12 (12)	SW		garden	dS	R	f	B	overgrown; there was possibly 1 more
367	172 Easter Clynekirkton, Brora	NC9006	6	30 × 20–22 × 23 24					(R)			narrower at top than bottom; top projects 6″, and base 9″; plastered inside
368	Dunbeith (5 miles up glen)	ND1629?	—									Dunbeath on map
441	3 Wyvisview, Culliokis Conon Bridge	SH5455	6		N							on open moor
462	on roadside, Shebster, Reay	ND0164	7	42 × 26 × 24 9 22½	S	?1850s	garden	S	T	g	CTD	Caithness flag base and slate gables project 4″; rounded at back; near public road; bees kept there till 1920s
486	Croft of Ramscaig, The Doll, Brora	NC8803	5				dyke	dS				
487	croft at West Clyne, nr Brora	NC8805	2				dyke	dS				

1	2	3	4	5	6	7	8	9	10	11	12	13
488	Balranald, Brora	NC9003	3	28 × ?36 × 21 51 ?9	S		garden	Ss	R	g		'skips'; Caithness slate base
489	croft at East Clyne, nr Brora	NC9006	1									
490	Clynemilton Farm, nr Brora	NC9003	1				orchard					now demolished
491	National Bank of Scotland, Brora	NC9003	4				garden					bricked up
492	nr Dornoch	NH7797	10	(24) × (24) × 17 (12) 23	S	1850	garden	S	R	g	C	surround and base red sandstone; rounded at back; not for visiting
493	dyke on Rogart to Balnacoil road, 10 miles W of Brora	(NC7910)	1				dyke	dS				
494	farm at Golspie Tower, Golspie	NC836008	1				garden					
514	Ham Farm, Dunnet	ND2373	1	(15) × (15) × (24)	S		garden	S	R	f		not found in 1970
647	West Garty Lodge, Helmsdale	NC995125	4	23–24 × 22–25 × 12–14 18 6–8	SE	1780	garden	S	R	f	BT	top projects 3"; 1 filled in
781	Lieurary Mains, Bridge of Westfield, nr Thurso	ND073629	4	36 × 24 × 12 15½ 32	SW	1888	garden	S	T	e	BT	wall of local flagstone; base and gable slabs project

Lothian Region

1	2	3	4	5	6	7	8	9	10	11	12	13
20	cottage in Lothianburn, Edinburgh 10	NT250670	?6	x × x × — ?2x		?1700s	?building	S	R	f	B	base projects; 2 iron pegs ? for boarding up; known to be in use c. 1900
300	St Mungo's Manse, Penicuik	NT2359	2	25½ × 18½ × 18½ 19 (12)	S	1700s	garden	S	R	g	B	whole projects; surround rebated (? for boards)
301	Frostineb, Blackshiels	NT4360	4	18 × 18 × 18 6–12	S	1700s	garden	S	R	f	B	projection 9" above bole
302	Newmills House, Balerno	NT165671	6	18–20 × 23–24 × 17–19 <·6 (7)	S	?1700s	garden	S	A	f	B	rounded at back; ?base projects
303	Hillend, nr Lothianburn, Edinburgh 10	NT252665	8 6 2	as under 20 × 20 × 19½ 21 4 20 × 20 × 19 30 5½	S	?c. 1700	cottage	S	R	e	B	pair farther along same wall

No.	Name	NGR		Dimensions		Aspect	Date	Setting				Notes
343	Luffness, Aberlady	NT476804	6	24 × 24 × 19		S	1600s or 1700s–1	garden	Ss	R	f	C	2 tiers of 3 & 3 in structure projecting from 12′ wall; top and bases project a little farther; rebated (?for boards); 1 base gone
				10, ?36 ?3									
361	Glebe House, Cockpen, Bonnyrigg	NT327637	4	23 × 23 × 19		S	1796	garden	Ss	R	g	B	base projects; probably had locking bar; originally belonged to manse; fair condition in 1980
				21 (9)									
402	Ferrybank, 170 St John's Rd, Corstorphine, Edinburgh 12	NT197730	10	18 × 24 × 18			1700s–1	garden	S	R		B	had locking bars and ?doors; back was rounded and ?had flight holes; probably built by candlemaker in 1710; demolished 1963
				0 (26)									
431	Marchfield Lodge, Queensferry Road, Edinburgh	NT2075	1	34½ × 39 × —		SE		stable	S	R		C	fluted surround projects; was rounded at back; now filled in
				90									
432	Garvald Village Hall (was Free Church), Garvald, Haddington	NT5870	6	22 × 21–22 × 21–22		S		orchard	Ss	R	f	C	orchard wall with projecting boles was incorporated in church, built 1843; in use in 1929; tops had gone in 1980
				5–6 8									
433	Pilmuir House, nr Haddington	NT486694	4					garden					house 1624
434	Dirleton	NT5183	—										
463	The Manse (now Nursery Garden), Currie	NT1867	1	27 × 62 × 14		S	1837	garden	S	R	f	CT	'bee boll' for 73 skeps; slabs project ?5″ all round; two metal hooks below bole
				28									
505	Ford House, Ford, Crichton	NT389645	1	19 × 16 × 17		N	1680	garden	S	R	g	B	ruinous in 1971; gone by 1980
				33									
597	Bog's (or Boag's) Mill, Water of Leith, nr Slateford	NT217702	7	as under		S	1700s	water mill	S	R			2 tiers of 3 & 3: whole set projected
			6	x × x × —									
				?24									
			1	similar									near set of 6; there may have been another
598	Liberton House, Liberton, Edinburgh	NT267694	7	as under		?N		garden	Ss	R	p	T	?bee boles (unusual shape); 'face NW on NE wall'; filled in with brick and edged with decaying wood
			6	17 × 10½ × 18									
				7 13′–14′									
			1	< ?17 × 100 × ?18									
				> 7									
782	Fountainhall, Pencaitland	NT4569	2	18 × 20 × 14		SSE	1600s	garden	S	R	f	C	covered in creeper in 1980
				4 (60′)									

1	2	3	4	5	6	7	8	9	10	11	12	13
790	Cammo Estate, Cramond, Edinburgh	NT173746	6	$20 \times 24 \times 18$ (36) 28	S	?1600s	(garden)	S	R	p	T	on N side, wooden frame and door set in brick facing
807	Churchyard of Old Parish Church, The Vennel, South Queensferry	NT1278	4	$18 \times 20 \times 21$ (12) $6\frac{1}{2}$	S	1700s–I		S	R	f	C	top and base project 3"; may have had front boards; wall probably in manse garden originally

Orkney and Shetland Island Areas—no records of any type

Strathclyde Region

1	2	3	4	5	6	7	8	9	10	11	12	13
13	The Weaver's Cottage, The Cross, Kilbarchan	NS4063	1	$18 \times 57 \times 20$ (24)	SSE	1723 or earlier	garden	S	?R	f	D	for ?2 or 3 skeps; lintel has moulding; provision for boarding up; wall belonged to Priory (now NT for Scotland)
169	Bankhead Farm, Carmunnock	NS5957	5	$24 \times 22\frac{1}{2} \times 28$	S		?field	S	A	g	B	each arch made from 1 block of dressed freestone; farm 1756
170	Sweethope House, Bothwell	NO7058	4	$22 \times 22 \times 25$ 20	S			S	R	f	B	rounded at back; probably had doors
171	Craig Park Farm, Bellshill	NS7350	5	$21 \times 22 \times 20\frac{1}{2}$ 18	SW		garden	S	R	p	B	whole projects 7"; had ?doors or locking bars; built into curved garden wall
173	Castlehill Farm, Kitlochside, E. Kilbride	NS610562	2	$18\frac{3}{4} \times 19\frac{1}{2} \times 31$ 27	?S			B	?A			freestone facings; bee house 172
299	Eastfield of Wiston, Symington, Biggar	NS968333	3	$(42) \times (24) \times 22$ (24)	SE	pre-1800	garden	S	1R 2T	2b 1f	B	whole projects; was probably 1 more
328	Cathcart Castle, Lymnpark, Glasgow	NS5876601?	—	$18 \times 18 \times 18$			castle extension					castle 1000s, wall later
329	Pollok Castle, Newton-Mearns, Glasgow	NS5355										castle demolished and no sign of boles, 1971
331	Manse of Inchinnan, nr Renfrew	NS4868										boles destroyed by 1958
349	Ladyland, Beith	NS3257	4	as under $26 \times 21\frac{1}{2} \times 18$ 2 $20 \times 21 \times 20$ $5\frac{1}{2}, 8\frac{1}{2}$	SW	pre-1530	garden	Ss	R	f		top projects 3", base $4\frac{1}{2}$"; walls were part of monastery
436	Puckylands Farm, Biggar	NT0437	2	$29 \times 19 \times 12, 17$ 7 (48)	S			Ss	A	g		

No.	Location	Grid Ref	No.	Dimensions	Aspect	Date	Setting	Ws	R			Notes
438	15 Main Street, Ochiltree	NS5021	2	24 × 18 × 24 72 54	S		house			b		sandstone surround; access from inside 'like cupboards'; may now be demolished
440	Churchyard, Ochiltree	NS5021	—									
516	St Columba Hotel, Iona	NM2823	—		?E		courtyard					'several' reported in 1957, not found in 1970
590a	Merryton Cottage, Colzium Gardens, Kilsyth	NS729788	10	13, 17 × 27 × 13 20 0	N†	pre-1820s	garden	S	R	p	C	2 sets of 5 in wall 30″ high on N side; at lower corners of some there is metal bracket with bar to back of bole—possibly for holding front boards; boles may have been open on S side of wall (11′ high, brick and plaster); adjoining bee house 590b
Tayside Region												
8	Shian, Amulree, Dunkeld	NN8439	3	26 × 19 × 19 9	SW	1705	enclosure	dS	R	f	B	
9	Ethie Castle, nr Arbroath	NO688469	3	24 × 24 × 15 18 9	S	1784	garden	Ss	R	e	BT	had an anti-theft device; ground level now raised—boles close to ground
11	Mains of Letham (Old Letham House), St Vigeans, Arbroath	NO6343	4	18¼ × 18¼ × 18½ 28 9	S	1756 or earlier	orchard	Ss	R	e	B	whole projects 4″
19	March of Lunan, Inverkeilor, Arbroath	NO646482	3	18 × 17½ × 21 1–10 (9)	S	1780	garden	dSs	R	g	BT	rounded at back; had locking bar; whole bole projects; demolished by 1979
21	Hollybank, Dickmontlaw, Arbroath	NO653434	3	26 × 23 × 19 29 4¼	S	?1700s–II	garden	Ss	R	g	BT	whole projects and base projects further 2″; shed in front in 1980
22	Boysack Mill, Inverkeilor, Arbroath	NO625492	3	25 × 19 × 20 24 9	S	c. 1750	garden	S	R	g	B	securing bar
23	East Church Manse, Carmyllie, Arbroath	NO5542	3	24 × 23 × 20 18	S	1820	garden	Ss	R	e	B	base projects; ornamental locking bar—Fig. 168
28	The Neuk, Marywell, Arbroath	NO651440	2	18 × 16 × 16 7 23	S	?1750s	garden	dSs	R	f	B	whole projects 7″; had securing bar, now gone
80	West Newton, Arbroath	NO656463	4	as under	SE	1750	garden	Ss	R	f	BT	2 pairs c. 60′ apart; whole projects 8½″; in winter, wire or rope across front secured straw packing; used within living memory (1954)
			2	30 × 20 × 20 12 3								
			2	30 × 20 × 20 8								central division now gone

1	2	3	4	5	6	7	8	9	10	11	12	13
134	Brax Toll, Old Forfar Road, nr Arbroath	NO600432	4	$22\frac{1}{2} \times 17\frac{3}{4} \times 18\frac{1}{2}$ 15 ?6	S	c. 1825	garden	Ss	R	e	B	'bee houses', built against wall; probably had 2 cross-bars; in use c. 1890
147	originally at Cairnie Farm, Arbroath now at Arbroath Museum, Signal Tower, Ladyloan, Arbroath	NO6341 NO6440	5	$23 \times 26 \times 19$ 42 5	was ?S now W	?1650s	farmyard	Ss	R	p f		whole set projected 8", with moulded surround; rebated jambs and metal rings probably for boarding up; in 1969 farm demolished and boles moved to park, and in 1975 to museum grounds; ht from ground now 24"
232	Waulk Mill, Inverkeilor, Arbroath	NO636490	3	$28\frac{1}{2} \times 21\frac{3}{4} \times 22$ 31 ?3	S	1780	?garden	Ss	R	p	BT	structure against wall; ornamental locking bars; ?had doors; one vertical division gone
233	Bandoch, Inverkeilor, Arbroath	NO6649	4	$27\frac{1}{2} \times 21-24 \times 20\frac{1}{2}$ 0 (6)	S	1780	field	Ss	R	e	BT	
234	The Welton, Carnoustie Road, Forfar	NO472499	1	$21 \times 19\frac{1}{2} \times 16\frac{1}{2}$ 25	SE	1801	(garden)	Ss	R	f	B	'bee hive'; whole projects
236	Lawton Mill, Inverkeilor, Arbroath	NO6448	2	$23\frac{1}{2} \times 45\frac{1}{4} \times 18\frac{1}{4}$ 13 (4)	S	1787	garden	Ss	R	e	BT	whole projects 12"; ornamental locking bars; for 2 skeps
239	Balglassie, Aberlemno, Forfar	NO538577	6	$26\frac{1}{4} \times 19\frac{1}{2} \times 20$ $24\frac{1}{2}$?9	S		?garden	S	R	g	B	3 pairs projecting from wall, 14' apart; locking bars (Fig. 167); house 1636, wall later
256	Cairndrum Farm, Edzell, Brechin	NO589665	3	$26\frac{1}{2} \times 21 \times -$	SE	1750	garden	Ss	R			whole projects 6"; had locking bar; filled in
257	Witton Farm, Edzell, Brechin	NO563700	2	$27\frac{1}{2} \times 40\frac{1}{2} \times 19\frac{1}{2}$ $16\frac{1}{2}$	SE	c. 1800	garden	Ss	R	g	B	walls of whin boulders; each for 2 skeps; sloping lintel to shed rain
258	Hawthorn Cottage, Castle Street, Brechin	NO5959	2	$28 \times 24 \times 19$ 33 60'	SSE	c. 1800	garden	Ss	R	p	B	whole projects 10"; plastered at back
259	Cairnton, Arbroath	NO6442	2	$25 \times 19\frac{1}{2} \times 21\frac{1}{2}$ 12 (4)	SE	1764	garden	dSs	R	f	BT	whole projects $15\frac{1}{2}$"; rounded at back; had locking bar
267	Mill of Colliston, Gowan Bank. Arbroath	NO599498	2	$32 \times 23\frac{1}{2} \times -$ 6	SSW	1825	garden	Ss	R			?was plastered inside, later filled in; no trace in 1980
276	Nether Dysart, Craig, Montrose	NO698534	4	$28 \times 21\frac{3}{4} \times 18$ 18 (6)	SE	1827	garden	S	R		B	?2 pairs 75' apart in wall of igneous boulders; sandstone lintel and jambs; whole projects $5\frac{1}{2}$"
277	101 Montrose Street, Brechin	NO5960	2	$26 \times 24 \times 20$ $4\frac{1}{2}$	S	1750	garden	Ss	R	p	B	whole projects 9"; had locking bar, and ?board; roughly filled in by 1980

No.	Name	NGR	n	Dimensions	Aspect	Date	Setting					Notes
278	Fairneyknowe, Arbirlot, Arbroath	NO612413	1	24¼ × 81 × 20½ 12	SE	1700s–I	garden	S		e	B	whole bole projects slightly; had bar across; for 4 skeps
279	Old School House, Rescobie, Forfar	NO508522	1	22½ × 54¼ × 18 (14)	S	pre-1700	garden	Ss	R	f	B	whole bole projects 11"; mortar at back, slab base; for ?2 skeps
280a	Pinehill, Auchterhouse, Dundee	NO346385	6	22½ × 18 × 17 > 5 (10–18)	SSW	pre-1750	garden	Ss	R	f	C	contained earth and plants 1980; bee shelter 280b
292	Blackhill, Dunkeld	NO0242	1	15 × 17 × 13 22	N		garden	dS	R	g		'boley holes'; house 1700s–I; possibly 2 others, now destroyed
			1	16 × 21 × 18 10	S		garden	dS		g		possibly another, now destroyed
363	Chapelton	NO6248	—									demolished
364	Cotterhouse, Seafield Farm, nr Arbroath	NO6340	—									1980—wall demolished
365	Birkhill	NO3534?	—									
370	Inchbrayoch, Montrose	NO718561	—									demolished
371	Willanyards Farm, Farnell, Brechin	NO625533	1	39 × 66 × 21 15	S		garden	S	R	f		flagstone wall; for ?3 skeps
377	Kirkton Cottage, Auchterhouse, Dundee	NO343381	1	22 × 43½ × 26 0	S	pre-1800	garden	Ss	R	g		built against wall; slab base; for ?2 skeps; filled in by 1979
379	Denfield Farm, Arbroath	NO617421	2	20½ × 20½ × 18 av. 13 (4½)	SSE	1732	garden	Ss	R	e	B	rebated for front boards; rounded at back
382	9 Union Street, Brechin	NO5960	2	as under 13 × 12 × 9 44	SSE	pre-1775	garden	dSs	R	g		about 130' apart; wall gone by 1980
				41 × 53 × 14½ 21								for 2 skeps
383	Hedderwick House, Montrose	NO701604	4	as under	S	1680s or 1740	garden/orchard	S	R			2 pairs 15' apart; rain-shedding cope over lintel; had locking bars; rebated for front boards; originally connected with Montrose Abbey
			2	29½ × 35½ × 19 24 6¼							b	whole projects; now broken
			2	28 × 35 × — 20								whole projects 6"; now filled in

1	2	3	4	5	6	7	8	9	10	11	12	13
384	Cliffburn Hotel, Cliffburn Road, Arbroath	NO6440	3	28 × 24 × 17½ (33) 3½	SW	1840	garden	Ss	R	e	B	whole projects 11½″; pedimented lintel; shelf consoles; slots for front boards
387	Kinnell Manse, Friockheim, Arbroath	NO609502	4	as under 2 23 × 19¼ × 21½ 6 12 2 20¾ × 19, 20¾ × 21½ 5½ 12	S	pre-1800	paddock	S	R	e	B	2 pairs 50′ apart; sandstone surround projects 6″, rebated for front boards; locking bar
388	Kirkden Manse, Letham, Forfar	NO531485	5	11½ × 17 × 13 18	ESE	1749		Ss	R	g		whole projects 5″; inside outhouse (probably since 1783); another separate recess
389	Dubton House, Montrose	NO704666	1	25½ × 66 × 18 31	SSE	?1700 or later	garden	S	R	e		whole projects 8″; sandstone surround; had locking bar; for ?3 skeps
390	Parkhead, Auchterhouse, Dundee	NO348385	6	(21) × (20) × (18) 24	SSW	pre-1700	garden	dSs	R	b		roughly built, sizes vary
391	Bryanton Farm, Inverkeilor, Arbroath	NO656483	2	22 × 18 × 18 22 (4)	S	pre-1750	garden	dSs	R	b		whole projects 8″; had 2 locking bars
393	Tullybaccart Farm, Coupar Angus	NO264362	2	30, 26 × 64 × 25 0, 33½	SW	pre-1750	garden	dSs	R	b		one above another; had locking bars; each ?for 3 skeps
395	Hermonhill Terrace, Perth Road, Dundee	NO3829?	2	as under 13½ × 12½ × 13½ 27 12½ × 12½ × — 36	SW	pre-1800	estate boundary	S	R	g		120′ apart
396	11 Denhead of Gray, Invergowrie, Dundee	NO347317	3	20 × 19 × 18 16 3	SSW	pre-1770	garden	Ss	R	f		filled in
397	49 High Street, Brechin	NO5960	2	30½ × 18 × 13½ 12 1½	SSE	pre-1800	garden	Ss	R	f		rounded at back
398	50 East High Street, Forfar	NO4550	1	20 × 16 × 11 (36)	S	pre-1800	garden	Ss	R	g	B	whole projects 3″
412	Glentarkie Farm, Strathmiglo	NO190120	8 6 2	as under 20 × 20 × 20 (36) ?6 20 × 20 × 20 (36) ?60	S		garden	S	R	f g	B BC	bricked in, ivy-covered, by 1980 rebated for front boards; postal address is Fife locking bars

No.	Location	NGR	n	Dimensions	Setting	Date	S	Mat	Cond	Gr	Notes
414	East Haven, Arbroath	NO6440	—								no walls left by 1980
415	Whitewell, Arbroath	NO650445	—								now called Robin Hill; altered c. 1973 and boles gone
416	East Mains Farm, Auchterhouse, Dundee	NO3237	3	18 × 22 × 25 10 (4)			dS	R	2g	C	structure built out from wall, with base built up to 10"; 1 bole overgrown
418	Baldovie, Broughty Ferry	NO4533	—								boles, if they existed, demolished
419	Greensland, Drumgeith, ?nr Dundee	NO4433	—								demolished
421	Pitkerro House, Broughty Ferry	NO453336	—								none found in 1971; walls now demolished
423	Struan, nr Blair Atholl	NN8165	—								
424	Craigentaggart, nr Caputh, Dunkeld	NO0840	—								
542	Turfhills House, Kinross	NO1102	3	23 × 17 × 23 18 6, (36)	garden	pre-1890	S	?R	g		
545	Balconnel Farm, Menmuir	NO521640	5	as under	?outhouse	1700s–I	S	R	f	B	2 sets of 3 and 2 in same wall 8' apart; top and base project
			3	22 × 18½ × 14½ 12 3							
			2	22 × 18½ × 14½ 7 5½							probably had locking bar or door
546	Mill of Cruick, nr Brechin	NO567627	4	28 × 23 × 24 6	garden		Ss	R	g	B	2 pairs 3' apart; 8 metal rings in wall probably for 2 bars across each pair
547	Balkello Smithy, nr Auchterhouse	NO366383	1	30 × 60 × 18 20	garden		S	R	g	C	flagstone wall; top of bole projects 6"; for ?3 skeps; repaired in 1963; Fig. 146
614	101 Montrose Street, Brechin	NO6060	2	25 × 23 × 20 4 6			Ss		p		
617	Duntrune House, nr Broughty Ferry	NO366383	4	26 × 25 × 21 31 (3)	(garden)	1800s–II	S	R	g	C	boles covered in ivy in 1979
619	Gagie Den, Wellbank, nr Broughty Ferry	NO4537	2	22 × 42 × 24 27 3	(garden)		S	R	f	B	remains of fittings for iron bars in front; for 2 skeps; cottage deserted by 1980, boles in poor condition

1	2	3	4	5	6	7	8	9	10	11	12	13
825	Annacroich House, nr Cleish, Kinross	NO116988	6	18 × 20 × 22 20	SSW	1749	garden	B	R	e	BC	2 sets of 3 c. 18′ apart; whole structure projects (Fig. 165); each bole rounded at back; between the sets is a large recess to top of c. 7′ wall (?bee alcove)

Western Isles Island Area—no records of any type

IRELAND

Co. Antrim

1	2	3	4	5	6	7	8	9	10	11	12	13
528	nr Red Bay	D22	4	(30) × (21–24) × (15) 0	SW		field	S	A	b	D	Fig. 156; 'keep holes'
812	Moat House, Donegore Hill, Antrim	(J1387)	6	20 × 17 × 13 5–8 20	S	1700	orchard	S	R	f	BC	bee wall with preservation order; 1 bole only 17″ high; ground level risen, and orchard now garden

Co. Armagh

1	2	3	4	5	6	7	8	9	10	11	12	13
183	Ashfort, Middletown	H7638	?8	as under 18 × (36) × 21 ?3 ?5 22 × 18 × 21 24	S	1800s–I	garden	B	A	p	BT	2 tiers of 3 & ?5 in wall (ht 54″) built against inside back wall of rectangular brick building, overall 9¼′ × 10¼′ × 5¼′, with tiled roof; 2 large arches make it open-fronted; one upper bole has had door fitted
702	Killeary (or Killeavy), Slieve Gullian (or Gallion)	J0323?	?1				house					1 or more niches in house gable; nearby was 'bee house' (?shelter) now fallen down but containing many skeps in 1800s; 120′ from Killeary Old Church which was convent or monastery

Co. Carlow, Co. Cavan—no records of any type

Co. Clare

1	2	3	4	5	6	7	8	9	10	11	12	13
697	Dromoland Castle, Newmarket-on-Fergus	R3967	3 4	20 × 16–18 × 14 40 9 as above	E	1500s–II	garden	B	A	f	BT	slightly pointed at top; cement or plaster inside
					SE	1500s–II	orchard	B	A	f	BT	rounded at back; cement or plaster inside

Co. Cork

1	2	3	4	5	6	7	8	9	10	11	12	13
583	The White House, Castletownshend, Skibbereen	W1831	6	22 × 21 × 21 6 (24)	SE	*1700s–II*	(garden)	dS	R	g	BT	base projects

No.	Name	Grid ref	Count	Dimensions		Orientation	Date	Location			Type	Description
685	The Castle, Castletownshend, Skibbereen	W1831	1	27 × 20 × 27 18		S	1700s–II	castle boundary	S	R	f	rock forms back of bole; may have been 5 more
695	Ardbrack House, Kinsale	W6451	4	18 × 18 × 20 8–12	9	SW		paddock	S	R	B	cement base

Co. **Donegal**—no records of any type

Co. **Down**

38	Harmony Hill, Drumaness, Ballynahinch	J3653	5	19 × 17 × 18 <6	(18)	ESE	pre-1800	garden	S	A	B	wall of blue stone; known to have been used for bees
443	Timpany, Ballynahinch	J3653	1	16 × 14 × 14 27		S	1780	house	S	R		'the bee's bole'; had wood front, with slit, and alighting board; were 3 more, 2 now made into windows

also bee tower (Chapter 8, p. 185)

Co. **Dublin**, Co. **Fermanagh**, Co. **Galway**—no records of any type

Co. **Kerry**

799	deserted house on Commons of Dingle, Dingle	Q4403	2	as under 29 × 20 × 27 16 28 × 21 × 30 11		SW S		?orchard	dSs S	g g	BC	shape = arch with curved sides, inside rounded at back and sides; shelf of large slabs; 9' apart in angle between 2 walls

Co. **Kildare**, Co. **Kilkenny**, Co. **Laois**, Co. **Leitrim**—no records of any type

Co. **Limerick**

698	Ballyneale House, Ballingarry	R4235	20	22–24 × 18 × 15–17 19–24	(24)	S		?garden	S		BT	Fig. 145; gothic-shaped arch, brick surround; rounded at back; base projects 3″; at top of wall, slabs overhang 5″; house 1820

Co. **Londonderry**, Co. **Longford**—no records of any type

Co. **Louth**—2 bee towers, see Chapter 8, pp. 184–5

Co. **Mayo**, Co. **Meath**—see below, 'Winter storage for bees'

Co. **Monaghan**, Co. **Offaly**, Co. **Roscommon**, Co. **Sligo**—no records of any type

1	2	3	4	5	6	7	8	9	10	11	12	13
Co. Tipperary												
167a	Ballingarry House, Ballingarry	S3248	18	21 × 18½ × 18 18, (48, 68)	SE		garden	Ls	A	g	BCT	structure (9′ × 17′ × 18″) built (?1800s) in front of wall dated ?1000; contains 3 tiers of 6 & 6 & 6 boles with brick arch; evidence of plaster inside; along each row, base stones project 2″; winter storage 167b
699	Milford House, Borrisokane	R9293	10	16 × 14–16 × 31 0 20–24	NNE	c. 1650	?garden	S	R	f	T	wall specially built for boles, has sloping roof of slabs; inside plan is circular
Co. Tyrone												
159	Lettry, Ballygawley	H6358	5	(24) × (24) × (24) (12, 60)	W		farmyard	S	R			2 tiers of 3 & 2; brick inside; destroyed by 1953
160	Martray, Ballygawley	H6358	8 or 10	same as 159	E		farmyard	S	R			2 tiers of ?4 & ?4; brick inside; destroyed by 1953
Co. Waterford, Co. Westmeath—no records of any type												
Co. Wexford												
184	Ballyhyland, Killanne, Enniscorthy	S8542	6	24 × 20 × 15 30 (9)	E	1830	garden	SB	A	p	B	wall faced with brick, boles brick-lined; 1 projecting base remains
185	The Daphne, Enniscorthy	S9740	3	22 × 24 × 20 36 (9)	W	1800	garden	S	A	e	B	brick surround; arch projects slightly; projecting base runs along row; cement or plaster inside; used until 1934
Co. Wicklow												
696	Warble Bank, Newtown Mounkennedy	O2606	5	30 × (24) × (18) 30 (24)	W	pre-1798	garden	S	T	f		
FRANCE												
F1	'Le Toron', Lurs-en-Provence, Alpes de Haute Provence		6	(40) × (40) × 32 0	S		?garden	S	A	g		were probably 2 more
F2	'La Chartreuse de Bonpas', Caumont-sur-Durance, Vaucluse		30	30 × 19–20 × 19–26 32, ?105 30	S	1600s	orchard/ vineyard	S	R	e	BC	2 tiers of 15 & 15 (Fig. 174); last used before 1920; described 1975[304]

ID	Location	No.	Dimensions	Orient.	Date	Setting				Notes
F3	Autoroute Aix-en-Provence to Salon, Halte de Ventabren, Bouches-du-Rhône	12	28 × 16 × 16 6 84–100	S		meadow	dS	R	g	C	Fig. 172: 350 m E of rest area at Ventabren; described 1975[304]
F4	Cuises-Chatillon, 51.700 Dormans, Marne	15	32–40 × 32–40 × 24–32 20–32 24–40	S	c. 1800	garden	S	R	g	C	described 1975[304]
F5	'Bois de Boulaise', Vicq-Exemplet, nr La Châtre, Indre	13	as under 9 22 × 18 × 21 16 8 4 13 × 14 × 20 16 24	S E		garden	S	A	p	B	surround is recessed; ?rounded at back; described 1975[304]
F6	Lycée de Valognes, Valognes, Manche	8	24 × 13 × 18 52 8	SSE	1600s or earlier		S	A	e	C	Fig. 173; rounded at back; in wall of college, previously manor house; described 1975[304]
F7	Château de Lacoste, Bonnieux, Vaucluse	16	32 × 20 × 20 40 60	SE	1785	castle grounds	dS	A	g	C	rounded at back; castle belonged to Marquis de Sade; described 1975[304]
F8	Entrecasteaux, nr Salernes, Var	6	x × x × —				dS	R	f	BT	6 or more boles near the ground in low wall
F9	Cotignac, nr Salernes, Var	3	>x × x × —			retaining wall	dS	R		T	
F12	Beaumont de Pertuis, nr Pertuis, Vaucluse	—									'several units' in the hills, below the ruins of Mouhalier le Vieux, nr 'Le Moulin de Giono' at le Contadour
F14	Forcalquier, Alpes de Haute Provence	—									'several in the town'

GREECE

| | near Mycenae, Peloponnesos | 98 | | | | garden | dS | R | | | Fig. 175; top and base project; destroyed by late 1960s |

ALCOVES (See notes on pp. 250–1)

1	2	3	4	5	6	7	8	9	11	12	13
ENGLAND											
Cheshire											
16	Daresbury Hall, Daresbury	SJ584824	1 (?4)	(120) × (84) × (18) (48)	S	1760–1775	garden	B	g	B	whole structure projects from wall; 1 wooden shelf; Fig. 181 shows pillars in front and ornamentation above; known use for bees; mentioned 1930,[145] pp. 418–19
Cumbria											
129	Neals Row, Urswick, Barrow-in-Furness	SD2774	1 (4)	54 × 44 × 18 39, 265	NE		garden	Ls	f	B	brick arch; base projects; there was shelf (now gone) 26" above base
670	Booth's Car Park, Highgate, Kendal	SD515925	3 (?6)	44 × 47 × 16 0 (40)	S			Ls		B	probably had shelf
Devon											
308	Eastleigh Manor, Eastleigh, Bideford	SS4928									see above, 'Bee boles'
460	Allaleigh House, Blackawton, Totnes	SX809537	1 (3)	66 × (72) × 24	N		garden	S			brick arch; domed shape inside; no evidence of shelf
645	Kilworthy House, Tavistock	SW482770	3 (?6)	(48) × 30 × 18 (24)			garden	S	g		domed shape inside; recesses mentioned in *Ghosts in the South West* by J. Turner (1973); also a larger alcove called the Judge's Seat; house 1598
Co. Durham											
475	High Green, Mickleton, nr Barnard Castle	NY9623	1 (?3)	(48) × (78) × 18–24	S	1752	garden	S			skeps stood on mushroom-shaped stone stools; floor flagged; records of owner's ancestors paying tithes in bee swarms (1792–1804); demolished 1944
Kent											
196	Underberg (was The Ferns), High St, Eynsford	TQ5465	2 (10)	60 × 42 × 23 9, 230, ?50 9	S	c. 1770 or earlier	garden	B	f	BC	structure (overall 10′ × 9′ × 28″) in front of brick and flint wall; contains 2 adjacent alcoves, each with base and 2 shelves; above is a dome-shaped recess for 1 skep, partly flint-lined; described 1955;[90] in poor condition 1980

310

201 Scadbury Manor, Southfleet, Gravesend	TQ6171								see above, 'Bee boles'
246 Higham Hall, Higham, Rochester	TQ712727								see above, 'Bee boles'
249 Farningham (near by-pass)	TQ5466	1 (?2)	37 × 26 × 13 (24)	B	g	garden	SW	1700s	brick arch and ?base; possibly had shelf halfway up; described 1956[91]
375b Manor House, St Stephens, Canterbury	TR1557								see above, 'Bee boles'
823 Farningham	TQ5466	1 (4)	x × 0.6x × — 0.5x, x	SF					structure with gabled top in front of wall; opening rectangular; has shelf; photo published 1930,[145] fig. 392
Lancashire									
189 Ribby Hall, Kirkham	SD410319								see above, 'Bee boles'
Leicestershire									
592 The Bede House, Lyddington, nr Uppingham	SP8797								see above, 'Bee boles'
Northumberland									
123 Otterburn Towers, Otterburn	NY8893	1 (?4)	(72) × 80 × 30 ?30	S	g	garden	S	1700s	domed shape inside; shelf gone and replaced by seat; bees here in 1890s
Somerset									
509 Sand Hall, Wedmore	ST429463	1 (?5)	80 × 42 × 19 38, 57	S	g	garden		?1700s or earlier	brick arch projects slightly; 2 wooden shelves, and possibly another 19″ above base
Suffolk									
551 175 High St, Aldeburgh	TM4656	4 (?16)	56 × 42 × 10 0 18	B	2g 2f	garden	S	1700s or earlier	possibly had shelf; 2 now faced with cement; were 3 more, now demolished

1	2	3	4	5	6	7	8	9	11	12	13
Wiltshire											
654	St Denys Convent, Vicarage St, Warminster	ST8745	1 (?2)	40 × 30 × 15 20	NE	?1700s	garden	B	g		built against wall; rounded at back; no evidence of shelf; repaired 1960—now wider and shallower than formerly
Yorkshire, North											
6	Hutton Mount, Thirsk Road, Ripon	SE3474	2 (4)	(54) × (48) × (36) (12) ?12	SE	?1600	garden	S	f	B	'bee farm'; 2 adjacent recesses; wall extended at back with sloping tiled roof; shelves originally stone slabs; mentioned 1930,[145] pp. 416–17
54	Paradise Field, Topham House Farm, Cracoe, Skipton	SD9760	2 (?8)	65 × 39 × 7–12 (24), (48)	SE		garden	dS	f	B	brick arch; top shelf projects, lower one worn; skeps here c. 1910 when this was a garden (now field); between the alcoves is a large ruinous structure (another recess?)
181	Slack Hills Farm, Hurst	NZ0402	8	48 × 42–(60) × 48 0 (?20)	S		field	dS	5f	B	a row built into hillside; outer 4 each 5′ wide; no evidence of shelves; bees here c. 1900; 3 now gone
205	Holly Hill, Well, Bedale	SE262818	1 (6)	72f, 66b × 72 × 54 18, 48	E		?garden	S	g	B	'bee house'; not arched; were 2 shelves, now gone
WALES											
Clywd											
51	Cristionydd, Pen-y-cae	SJ281449	3	2x × 3x × x x, 2x		1800s		S		B	specially built arched structure on raised base in front of wall; Fig. 182 shows the 'gothic' arches, each divided in two by shelf
SCOTLAND											
Grampian Region											
115b	Midmar Castle, Midmar, Inverurie	NJ704052	2 (4)	73 × 38 × 24–27 20, 50 30′	S		castle garden	S	f	BT	2 identical structures with rectangular opening; each projects 14″ from wall, and is divided horizontally to give 2 recesses each 23″ high; outer edges rebated for door, and hinges remain; the whole is surmounted by a classical pediment; winter storage 115a

Lothian Region

595	Beech Hill, Haddington	NT542705	1 (2)	70 × 55 × 20	NNW	1700s–II	garden	s g C	domed shape inside; wall thickened at back; steel shelf 28″ from ground is probably modern

Tayside Region

825	Annacroch House, nr Cleish, Kinross	NO116988							see above, 'Bee boles'

IRELAND—no records of alcoves

BEE HOUSES (See notes on pp. 250-1)

1	2	3	4	5	7	8	9	10	11	12	13
ENGLAND											
Bedfordshire											
45	27 Rothsay Road, Bedford (land now owned by Girls' High School)	TL0449	?6	hexagonal in plan, each side $3\frac{1}{2}'$ wide	c. 1900	garden	B	?Ti	g	B	1 flight entrance on each side, alighting shelf 10" from ground, and projection over it; inside shelf at same level; alighting shelves gone
Berkshire											
621	Berkshire College of Agriculture, Hall Place, Burchetts Green, Maidenhead	SU825818	9	10 sides, each 5' wide, $6\frac{1}{2}'$ high	1880 or 1890	garden	W	M	e	BC	sides of zinc and wooden lattice, boarded at top; curved wood and zinc roof: each side has window with flight hole below: hives raised on stands; Fig. 189 taken after recent restoration
Hereford and Worcester											
742	'The Home of Industry' (location unknown)		?4	tall, rectangular			WB	?Th		B	known from photo WC 1652;* wooden structure on brick base; were probably 2 shelves with cupboard space over, somewhat similar to German bee houses
743	Caerswall, Much Marcle, Ledbury	SO641336	10	rectangular shed			?W	?M		B	known from photo WC 1651;* one side has 2 rows of flight entrances c. 30" and 60" from ground; each has small alighting board; 2 shelves inside, each for 5 skeps/hives
783	plan of bee house by T. Baylis, Ledbury			hexagonal							Fig. 187 shows this design, c. 1840: each of 5 walls has a vertical row of 6 flight holes indicating 6 shelves, each for a collateral hive
808	Bretforton, Evesham	SP093439	6x	rectangular $7\frac{1}{2}' \times 32\frac{3}{4}' \times 5\frac{1}{2}'$			W	Ti	g	T	hipped stone tile roof; each of the 6 openings along one side is like a window but with a metal grille; semi-circular wooden alighting board below each (2 still remain); would hold a multiple of 6 skeps/hives; used within living memory

*WC = Watkins Collection (of photographs taken up to and in 1920s by Alfred Watkins; now in Hereford City Library).

827	Mr Anthony's bee house, Mansion House, Widemarsh St, Hereford	SO5139						'fine bee house built for Mr Anthony, founder of the *Hereford Times*'

Humberside

72	Appleby Hall, Appleby, Scunthorpe	TE950151	32	octagonal in plan, each side 6' wide outside, 4½' inside; ht 14'	1830–1854	grounds	W Ti g	B	flight entrances in 7 sides, with alighting boards (some missing) at 2¼', 6¾', 11' from ground; inside are wooden shelves (each for 2 skeps) at 3 levels; 3 old Marriott skeps found here in 1953

Kent

126	Arnold Hill, Leeds, nr Maidstone	TQ813533	12 +8	rectangular	1870		W Th b	B	flight entrances on E and W sides; each has alighting board and small projecting gables above; described 1955;[90] mentioned 1930,[145] pp. 418–21

London, Greater

839	Tegetmeier's Experimental Bee House, 101 St James's Lane, Muswell Hill, Hornsey	TQ2889	10	rectangular	c. 1860	garden	W ?Th	D	no longer exists, but known from the (probably flattering) lithograph, reproduced in Fig. 186

Northamptonshire

161a	in market garden of Castle Ashby House, Castle Ashby	SP863588	12	rectangular 6' × 14' × 5¼'		grounds	Ls M f	B	Fig. 184; internal ht 62"; door at back gives access to shelf 27" deep, 33½" from ground; lower skeps 3" from ground; at front are 2 rows of 6 & 6 arched openings facing SE, each 24" × 18" with wooden shutter; lead roof (not known if original) now gone; garden attributed to Godwin 1865
161b	in market garden of Castle Ashby House, Castle Ashby	SP863588	12						identical with 161a, and facing it, i.e. openings face NW; Marquess of Northampton tried, unsuccessfully, to keep bees in these houses in 1920s or 1930s

Nottinghamshire

55	Watnall Hall, Watnall, Chaworth, nr Nottingham	SK499460	5 or more	rectangular	pre-1870	?garden	W Th g	B	black and white timbered building; photograph shows 5 flight entrances, each with alighting board, in one long side; may have been more in other side(s); in 1930, original Nutt hives, Fig. 234, still in use; by 1953, converted to greenhouse; mentioned 1930,[145] pp. 420–1

315

1	2	3	4	5	7	8	9	10	11	12	13
Oxfordshire											
444	Grey's Piece, Rotherfield Greys, Henley-on-Thames	SU718829	12	rectangular $10\frac{1}{4} \times 15\frac{1}{4} \times 6\frac{1}{4}'$	pre-1890	garden	W	Ti	g		on S side are 2 rows of 6 & 6 flight entrances 24", 48" from ground; also 2 small shutters above; doors and windows on N side; used for bees until 1914 or 1939
Shropshire											
46	Attingham Park, Atcham	SJ550100	12	rectangular $>7 \times 18 \times 8\frac{3}{4}'$		orchard	W	Sl	e	BT	Fig. 190 was taken after restoration in 1960s; inside on S side are 2 rows of shelves, each for 6 skeps, at 18", 44" from ground; slate floor; skeps there 1981; NT
Somerset											
140	Butleigh Court, Street	ST520340	?12	rectangular		garden	W	Ti		B	on S side are 2 rows of 6 & 6 flight entrances, each with protection above; demolished by 1976
Staffordshire											
98	Ecton Lees, Manifold Valley	SK0958	21	rectangular $- \times 11' \times 6'$	pre-1850	?garden	S	Ti	p	B	'honey house'; on S side are 2 rows of 10 & 10 flight entrances, with alighting board along each row at 12", 45" from ground; skeps were said to stand over bowls 'used for tipping the skeps over when they emptied them'; another flight entrance with alighting board on E wall, and shelf inside
Sussex, East											
634	Cherry Croft Farm, Herstmonceux	TQ639108	12	rectangular $10' \times 10\frac{1}{2}' \times 6'$			B	Ti		C	in E wall, at c. 6' from ground, 6 flight entrances, and 6 in W wall; each has ?wooden surround, small alighting board; inside, wide wooden shelves held by iron supports to old roof timbers; there are doors at ends; may be contemporary with farmhouse built 1600s

316

Sussex, West

No.	Name	Grid ref	Count	Shape	Date	Location	S/W/B	Mat	Cat	Description
495	The Bee Master's Cottage, Burpham	TQ0409	10 +?4	generally rectangular	c. 1900		W	Th	B	The Rev. Tickner Edwardes' bee house, known from the photo he published in 1907 (Fig. 188);[105] now demolished, and cottage replaced by Burgh Cottage

WALES

No.	Name	Grid ref	Count	Shape	Date	Location	S/W/B	Mat	Cat	Description
830	Berthddu, Llandinam, Powys	SO015858	6				S		B	all we know of this replica castle is the photograph Fig. 185; perhaps $8\frac{1}{2}'$ high overall

SCOTLAND

Strathclyde Region

No.	Name	Grid ref	Count	Shape	Date	Location	S/W/B	Mat	Cat	Description
168	Gadgarth House, Aunbank	NS408223	12	octagonal in plan	?1800s-II	garden	B	Ti g	B	built of black and white bricks; in each of 6 sides are 2 flight entrances at $7\frac{1}{2}''$, $7\frac{4}{8}''$ from ground—each is carved, skep-shaped stone, with a semi-circular wooden alighting board; inside is wooden staging on cast iron frame; ventilation holes near top of each wall
172	Castlehill Farm, Kitlochside, East Kilbride	NS610562	8	circular in plan		garden	S	g	B	stone building, rendered on outside; curved door (N) and window (S); each flight entrance has semi-circular alighting shelf; inside each are 2 stone pillars to support hive (on shelf?); garden wall dated 1767 contains bee boles (173)
590b	Merryton Cottage, Colzium Gardens, Kilsyth	NS729788	—	rectangular	?pre-1820s	garden	S	Ti f	C	building adjoins N side of wall with bee boles (590a); steps down to bee house with loft above; 2 flight entrances on N side

IRELAND—no records of bee houses

WINTER STORAGE FOR BEES (See notes on pp. 250–1)

1	2	3	4	5	7	8	9	10	11	12	13
ENGLAND											
Bedfordshire											
472	Old School House, Studham	TL0215	3 2 1	as under $13 \times 18 \times 13$ (36) 90 $19 \times 22\frac{1}{2} \times 13\frac{1}{2}$ (36)	c. 1700	cellar	?B	e A		B	shape = pointed arch
Cornwall											
307	Radland Mill, St Dominick, nr Saltash	SX399684	12	$15–17 \times 17–20 \times 16$ 45, 72, 9, 26 95, 119	pre-1800	yard	S			BC	Fig. 194 shows the 4 tiers of 5 & 4 & 2 & 1 recesses (shape = flat-topped dome) on inside of SE wall of building (now roofless); each has a flight hole, probably added later
718a	Penpol House, St Erth, Hayle	SW5637	9	$(12) \times (14) \times (14)$ 30 7–10		yard		R	g	C	recesses in 3 walls of small building made of ?copper slag blocks; bee boles 718b
Cumbria											
79	Appleby Castle Conservation Centre, Appleby-in-Westmorland	NY684200	—		1650–1676		?S		f	BC	traditional name 'Lady Clifford's bee house', but no evidence of recesses; probably first built as guard house; building rather like Isle of Man 141, below (Fig. 191)
666	Hill House, Maulds Meaburn, Penrith	NY6216	12	$12–38 \times 22–31 \times 10–18$ > 30	1700s	dairy	Ss	R		C	recesses in 2 walls of building; most are in 2 tiers
Devon											
650	The Old Rectory, Clannaborough Bow, Crediton	SS7402	6	$15 \times 15 \times 15$ 36	1870 or earlier	yard	C	D			recesses on 3 inside walls of outbuilding with sloping thatched roof; recesses rounded at back; above each are 3 small holes (?for fixing cover) and there was (horizontal) batten at 4' on S wall; 2 recesses now filled in
789b	Trehill Farm, Sampford Courtenay, nr Okehampton	(SX637973) or SS6301	1		1500s	barn		R			'winter home'; recess is inside threshing barn, 6' from ground; has door with hole in; bee boles 789a
792	Court Barton, Aveton Gifford, Kingsbridge	SX695412	1	$72 \times 36 \times 18$?18	1650	farm kitchen	S	A	g		rounded at back; has 3 wooden shelves 18" apart; may not have been for bees

Isle of Man

141 Ballachurry, Rushen	SC208698	31	20 × 13¼ × 12		S	R	B	windowless building, area 15' × 10' externally; tile or slate roof; Fig. 191 shows doors to upper and lower rooms—in each are recesses in 2 tiers in all walls (Fig. 192); bee boles 26

Oxfordshire

821c Champs Folly Frilford, nr Abingdon	SU4497	2		outbuilding			building mainly for other purposes; bee boles 821a, b

Yorkshire, North

82 New Wath Apiaries, Goathland	SE815001	—		moor	dS	1700s–II	B	built in hollow of hill; walls collapsed, only 3' remains; no recesses—may have had wooden shelves or stone stands for skeps in winter; stands found nearby
651 Stonegrave House, Stonegrave, York	SE656778	6	34 × 20¼ × 10¼ 42	garden	S	1600s–II	A e	B	building (internally 8¼' × 7¼') attached to garden wall near dwelling house; door at SW; window SE (?was another recess); recesses on 3 walls; wooden bases project slightly; old wooden hive and scraps of comb found here in 1980

WALES

Dyfed

817a Butterhill Farm, St Ishmaels	SM830087	21	14 × 14 × 14 0, 30	farmyard	Ss	?1750	R p	B	in a row of farm buildings; recesses in 3 walls of building 8 × 12¼' internally; one has projecting semi-circular base; roof gone, now overgrown 1980; Fig. 193

SCOTLAND

Grampian Region

115a Manor House, Midmar, Inverurie	NJ704052	40	12–24 × 17 × 15	wooded land	Ga	?1700s–II or earlier	R p	BT	free-standing building 11' high at front, 3½' at back; had steeply sloping slate roof (now gone); in front (S) wall were 2 tiers of 4 & 4 rectangular openings c. 2 × 1–2" which once had doors/boards; inside N wall are 2 tiers of ?10 & 10 recesses, E and W walls each have 4 tiers of 3 & 3 & 3 & 1; recesses made from large horizontal granite slabs and brick piers; not for visiting; alcoves 115b

319

1	2	3	4	5	7	8	9	10	11	12	13
Strathclyde region											
703	Inverlussa House, Crinan	NR7785 or 7894	—	12–15 × 12–15 × —		outhouse		?R			one wall of outhouse 'full of' symmetrically placed recesses
IRELAND											
Co. Mayo											
584	ruins next to Castlecarra, nr Ballinrobe	M1575	46	15–21 × 13–20 × 14–19 0, ?, ?		outbuilding	S	R		CTD	recesses are in E & W walls; other 2 walls are gabled, 1 has small opening high up; on E wall are 2 groups of recesses—3 tiers of 4 & 5 & 6 and 3 tiers of 4 & 5 & 5; on W wall, S of door are 2 tiers of 3 & 10, and N of door 2 tiers of 2 & 2; dimensions of most of the recesses lie within ranges quoted; roof gone and interior very overgrown 1978
Co. Meath											
701a	Ninch House, Ninch West, via Laytown (nr Drogheda, Co. Louth)	O1672	?18	15 × 15 × 15 0, ?22, ?45	1750s	near orchard		A			building 10′ × 15′, outside walled orchard; 3 tiers of red brick recesses in W & N walls; roof gone and (1979) W wall accidentally demolished
701b	Ninch House, Ninch West, via Laytown (nr Drogheda, Co. Louth)	O1672	4	15 × 15 × 15	1750s	farmyard	S	A			2 recesses above 2 others in one wall of intact outhouse; each has wooden base 1″ thick
Co. Tipperary											
167b	Ballingarry House, Ballingarry	S3248	8	34 × 23 × 32 0, 42, ?84		outbuilding		A		g	3 tiers of 2 & 2 & 4 on wall now inside dwelling house; originally there may have been 4 in each row; all 8 similar—only 1 measured; bee boles 167a

SHELTERS (See notes on pp. 250–1)

1	2	3	4	5	6	7	8	9	10	11	12	13
ENGLAND												
Buckinghamshire												
403	The Limes, Lower Winchendon	SP741139	3	66f, 82b × 58 × 34	S		garden	BLs	Ti	g	C	built in limestone wall which also has tiled roof; shelter protrudes 2½′–3′ at back of wall; skeps there c. 1900; repaired recently but shelf gone; (late information, not included in analysis)
Cumbria												
73	Fairfield, Lorton, nr Cockermouth	NY154268	5	52 × (120) × 29 ?6	SE	1860s	garden	B	Sl	g	B	free-standing structure (Fig. 201); wooden base; 5 compartments, each 28″ × 20″ × 29″ with hinged wooden door at front (with flight entrance) and back; colony records written inside
74	Oak Hill, Lorton, nr Cockermouth	NY159262	5	(40) × 120 × 18¼ 10	SW	1850s	garden	B	Sl	g	BC	similar to 73; each compartment 21″ × 18″ × 21″
76	Combe Cottage, Borrowdale	NY2614	—		S	c. 1850	garden	S		b	B	free-standing structure, in use in 1890s; in 1953 half gone and the rest crumbling, overgrown; by 1979 garage built on site
87	Whitestock Hall, Rusland	SD330890	?4	66 × 54 × 24	S	pre-1847	garden	S		e	B	free-standing structure with sloping roof; probably had 2 shelves; front covered with trellis by 1953
155	Nab Cottage, White Moss, Grasmere	NY355064	8	54 × 96 × 24 20, 50	S	1702	garden	S	?Sl	f	B	solid base 20″ high makes lower shelf; upper shelf gone; De Quincey, and later Hartley Coleridge, lived here
156	Well Know, Cartmel	SD373793	20	10 compartments (20) × 36 × 24 18, 40	S	1700s	garden	S	Sl	p	B	2 tiers of 5 & 5 with roof sloping down to front; top compartments 36″ high at back; each for 2 skeps
195	Oaks Farm, Langdale	NY3105	?4	66 × 45 × (24) 18	SE	1670	garden	S		e	B	there were probably 2 shelves
210	The Hill, Grizebeck, Kirkby-in-Furness	SD2485	?3	38 × 76 × 24 14	S		garden	S	Sl	g	B	Fig. 197; slate shelf
215	The Cottage, Grizebeck, Kirkby-in-Furness	SD2485	6	?36 × 58 × 21 24, 42	S		garden	S		p	B	on sloping ground; solid base 24″–16″ high; slate shelf; top has collapsed; cottage 1704

1	2	3	4	5	6	7	8	9	10	11	12	13
217d	The Hill, Heathwaite, Grizebeck, Kirkby-in-Furness	SD241870	?4		SE		garden	s	?Sl	?f	B	built in angle of 2 walls of 'bee garden' which also has 15 bee boles (217a); house 1641
281	Manor House, Town Head, Newby (?nr Penrith)	NY5821?		?10 ?96 × ?102 × 21 (48, 72)	S	1720		s	Sl	f	B	solid base 24″ high, and 2 large slab shelves 24″ and 48″ above this; roof gone
513b	Sykeside, Grasmere	NY341070	3	48f × 63 × 25	S	pre-1800	yard	Ga	Sl	f	BT	structure projects at back and front of wall; ridged roof (?repaired recently); plastered at back; shelf gone; Wordsworth Trust; bee boles 513a
530	Low Nest Farm, Keswick	NY227291	4	48f, 54b × 96 × 36	E	1700s–II	garden	Ga	Sl	b	T	only the ends (18″ thick) remain
608	Laurel Bank, High Nibthwaite, nr Ulverston	SD294897	3	54 × 78 × 22 26	SE		garden	Sl	Sl	f	T	'bee house'; structural beams are oak; shelf was probably slate slab, now wooden; used until 1906
626b	Whetstone Croft, Woodland, nr Broughton-in-Furness	SD232899	3	50 × 72 × 48 26	SE	1753	garden	Sl	Sl	?g	T	solid base 26″ high; sloping roof; bee bole 626a
632a	Light Hall, nr Windy Hill, Rusland, nr Ulverston	SD3488	4	44f, 62b × 42, 38 × 33 27	S	c. 1860	garden	Sl	Sl	g	T	'bee butts'; 2 adjacent compartments with 4″ dividing slab; sloping roof; shelves gone, but iron pegs for holding them remain; used for winter storage by packing with bracken and fitting doors; not for visiting; bee boles 532
632b	Light Hall, nr Windy Hill, Rusland, nr Ulverston	SD3488	?5	54 × 52, 50 × 32 ?21	E		field	?Sl	Sl		T	2 adjacent compartments with vertical slabs for sides and 18″ dividing wall which has holes (for pivots?) at bottom; shelves gone, but iron pegs remain; known to have been used for bees, and probably for winter storage (cf 632a)
646	Goody Bridge House, Grasmere	NY323079	4	75 × 46, 52 × 36 ?24	SW			Sl	Sl	g	B	'bee house'; 2 adjacent compartments under one sloping roof; part of a shelf remains
665	Boot Cottage, Troutbeck, Windermere	SD411013	6	54 × 36, 36, 36 × 30 18	S	?1700s–I	orchard	Sl Ga	Sl	p	T	3 adjacent compartments with 16″ dividing walls; shelves and part of roof gone
668	Fern Bank, Main Street, Burton	SD529764	3	53 × 71 × 22 30	S	1800s–I	garden	Ls	Sl	p		ends 12″ thick; roof gone, but recess 8″ high in wall would have held slab(s); shelf gone but slot extends 17″ frontwards; garden is next to orchard
671	4 Prospect Terrace, Fellside, Kendal	SD5193	?3	60f, 66b × 58 × 36 18	S	1700s–I	garden	S	Sl	f	T	recessed into wall, and projecting 20″ from it; single roof slate projects 4″ further; solid base 18″ high; ?slate shelf

322

No.	Location	Grid ref		Dimensions	Aspect	Date	Situation				Notes
673	Sidey Bank, Troutbeck, nr Windermere	SD415037	6	84f, 96b, × 60 × 24, 54	SSW	1600s–l	cottage	Sl Ga	e	T	solid base 24" high, and slate shelf 30" above this; next-door cottage 1623
691	Moss Dyke, Mungrisdale	NY369315	3	30 × 72 × 36 >12	S		garden	Sl	f	T	oak lintel
810	Tower Head (or Bank) House, Near Sawrey, Ambleside	SD370956	6	57f, 87b × 75 × 39 15, 41	S	c. 1780	orchard	Sl	g		slate shelves 2" thick (now gone) were 18" deep, with gap behind, possibly for air circulation; ends 18" thick
833	Clubs (near Roman wall) nr Carlisle		3			1774		Sl			slate roof slopes down to front; 1 shelf; photo published 1930, fig. 391, shows 3 skeps;[45] we cannot trace Clubs
841b	Bridge Field Farm, Spark Bridge, nr Ulverston	SD3085	—								faces back door of farmhouse; bee boles 841a
844	Townfoot, Troutbeck, Windermere	NY406020	3	(36) × (60) × (27)	E	?1800s	?garden	Sl	f		roof slopes down slightly to rear; solid base 12" high, 6" deep and wooden beam in front of it; probably used for wintering skeps
845	Low Fold, Troutbeck, Windermere	NY407026	3	43f, 50b × 40 × 25 15	SE	?1700s	house	Sl	g		ends 12" thick; slate shelf 1" thick
846	Town End Farm, Troutbeck, Windermere	NY407022	4	30f, 75b × 46 × 46 29	S	?1700s	field	?Sl			ends 14", 30" thick; slate shelf 2" thick, set back 9" from front; demolished in 1980
Devon											
230	Brownberry Farm, Dunnabridge Pound	SX643743	2	21 × 52 × 30				Ga			flat slab as roof; there was wooden door, hinged at bottom; 'used within living memory' (1953)
Essex											
18	Chatley House, Great Leighs	TL7371	9	67f, 88b × 67 × ?20	S	1736 or earlier	house	W	T f		'bee house' of wood and plaster, built against house; had 3 shelves resting on iron pegs (1 shelf now gone); used until late 1920s
Gloucestershire											
242	was at Hive House, Church St, Nailsworth now at Gloucestershire College of Agriculture, Hartpury	ST8499 SO7924	28 + 25 or 10	(300) × 120 × – ?27, ?48		pre-1500	?garden	S S	p g	B BT	'hive shelter'; elaborately carved structure of Caen stone with 2 tiers of 14 & 14; each compartment is 19", 16" × 19" × 24"; stone roof slopes on both sides; has carvings on back, suggesting it was built free-standing; at bottom are 5 wide recesses (24" high)—each had shelf; Fig. 199 taken after removal and careful restoration in 1968

1	2	3	4	5	6	7	8	9	10	11	12	13
Hereford and Worcester												
741	Bee shelter (location unknown)		?6				house	W			B	WC 1653;* high up on house wall under projecting roof tiles; 2 shelves
744	Bee shelter (location unknown)		?8	54f × (48) × ?15 (18, 36)			house	S			BT	Fig. 198; WC 3111;* 2 shelves; roof slopes down to front
Lancashire												
14	The Willows, Allowdesley, Ormskirk	SD4108	>12					W	Th	p	B	tall free-standing shelter, roughly constructed with rustic poles supporting a thatched roof; probably open on all sides; shelf (?24″ from ground) held many hives and skeps; some also in gable ends
89	Church House, Warton, nr Carnforth	SD504725	4	(66) × 40 × 24 20, 46	SE	1600s	garden	S	Sl	g	B	no shelves, but 2 oak beams 20″, 46″ from ground; house 1674; bees kept here till 1946
Norfolk												
828	North View, Larling, Norwich	TL9889	8	3x × 4x × ?x x, 2x				W	M		B	probably portable (see Fig. 200); 2 shelves, and a hinged shutter to shelter each row of skeps
Shropshire												
274	Home Farm, Cruckton	SJ430104	4	56 × 24 × — 20, 37	SW		house	B	Ti	g		wooden shelf
Somerset												
829	Westport, Ilminster	ST3614	?8				?orchard	W	M		B	probably similar to 828, Norfolk (above); access at back by 2 pairs of hinged doors
Surrey												
832	Holmwood	TQ1745	9			?1600s–I		W	Th			3 shelves, each for 3 skeps; decorative wooden carvings include depiction of farmstead; mentioned 1930[45] with photo, fig. 389

*WC = Watkins Collection (of photographs taken up to and in 1920s by Alfred Watkins; now in Hereford City Library).

Warwickshire

485	Great Alne, Alcester	SP1159	6	65 × ?32 × — ?12, 27, 45		cottage			p		shelter in angle of 2 walls has sloping roof; 1 wooden, 1 stone shelves remain; may have been more; cottage several centuries old, may have been inn; photo in *Birmingham Post* 24.2.37

Yorkshire, North

53	Barlow Hall, Barlow, Selby	SE644284	8	22f, 33b × 177 × 24 14	S	garden	B	Ti	f	B	decorative wooden pelmet; brick support in middle; solid base; shelter not as old as wall; skeps there in 1880s
128	Hollins Farm, High Farndale, (nr Church Houses)	SE6899 ?	10	?60f × ?200 × — 215	?S	?field	S	Th	p	B	against dry stone wall; 3 stone posts at front support cross poles for tiled roof sloping down to wall; in 1953 skeps on wide slabs on shelf, and thatched roof being replaced by tiles
182a	West Ings, Muker	SD9198	4	54 × (84) × (24) (24)	SSE	?1800s-II	garden	S	S	b	B	'bee holes'; each side is a vertical slab, and similar support in centre; slab roof slopes down to wall; shelf rests on two supports (not attached to sides); in use c. 1900; only upright slabs in position 1954; bee boles (182b) may be in same wall
197	Great Fryup [?Ellers House], Lealholm, ?Danby	NZ731053	3	(72) × (72) × (36)	SSE	1800s-II	farmyard	?S	?Ti	p		?partly recessed into wall with roof (now gone) sloping down to wall; there was no shelf, skeps stood on ?stone stands
831	Muker	SD9198	?9	?90 × ?70 × —		?garden	S	?S	g	B	one side adjoins house; solid base 12″ or 15″ high; were probably 2 shelves (lower one missing); whitewashed inside; full address not known, but see shelter (182a), bee boles (182b)

Yorkshire, West

708b	in a field next to Buckley House, Stanbury, nr Keighley	SD996368									'little holes in a building of their own' — probably a shelter; bee boles (708a) in next field

WALES

Dyfed

818	Brownslade Home Farm, Castlemartin	SR907976	?8	24 × (156) × — ?12			S		p	B	free-standing structure near walled garden; row of 4 compartments each 36″ wide with ?6″ dividing

325

1	2	3	4	5	6	7	8	9	10	11	12	13
Glamorgan, Mid												
411	Cefn Hengoed Farm, ?Hengoed	ST1495?	?6					S		p		photo in *British Bee Journal* (1961) 89:270 shows supports for 2 shelves and ?slate roof sloping down to front; now demolished
Gwent												
323	Penycacau Farm, Crumlin	ST232983	6	40*f*, 80*b* × 132 × ?56 (12)	SW		field	dS	Ti	f	B	oak support pillar at centre front with crossbeam for roof, sloping down to front; originally a similar adjacent shelter; altitude 1100′
Gwynedd												
164	Plas Isaf, Llansantffraid, Glan Conwy	SH815757	10	?60*f* × 102 × 36 ?24, ?48	SW		?garden	dS		p	B	shelves probably supported on 'rails'; in use at end of 1800s; roof and shelves now gone
SCOTLAND												
Grampian Region												
111a	Kinnermit, Turriff	NJ7249	2	55 × 55 × 24	S		orchard	S		p	BT	mentioned in book describing garden in 1880s; roof and shelf now gone
111b	Kinnermit, Turriff	NJ7249	2	55 × 55 × 24	S		orchard	S		p		identical with 111a, and 50′ away from it
113a	Gordonstoun House, Duffus	NJ184690	10	100*f* × 100 × 30 24, 60	?SSE		wood	S		g	B	free-standing structure with sloping roof; bottom shelf has 3 sockets, possibly for door, shutter or upright rods; said to have been used for bees
113b	Gordonstoun House, Duffus	NJ184690	10	100*f* × 100 × 30 24, 60	?SSE		wood	S		p		identical with 113a, possibly 180′ away; roof gone
Highland Region												
834	Strathglass, Beauly Bridge, nr Inverness	NH3834 approx.	?4					W	?M	g	T	decorative pelmet; roof slopes down to back and is hinged at front to allow access to hives
835	Strathglass, Beauly Bridge,	NH3834 approx.	?4					W	M	g	T	decorative pelmet; corrugated iron roof slopes down to back

836	Strathconan, Muir of Ord	NH4055 approx.	3			W	T	Fig. 202 shows 3 skeps; 'submerged' by 1960
Tayside Region								
280b	Pinehill, Auchterhouse, Dundee	NO346385	3	35 × 68 × 26 10	garden	SSW	?1800s	f locking bar; iron hinges at side probably for door; bee boles 280a
381	Bridge Street (derelict site), Brechin	NO599599	5	36 × 108 × — 24		SE	pre-1775	b recessed 6" into wall, but projection destroyed; wall gone by 1980

IRELAND—see above, 'Bee boles', Co. Armagh. 702

OTHER STRUCTURES (See notes on pp. 250–1)

ENGLAND

1	36	Beehive Cottage, Bladon, Oxon.	SP4414	5	2 pairs of wooden supports jutting out from cottage wall, level with first floor windows; each pair for 1 skep on stand; Fig. 203	12 B
273		Nag's Head, Avening, Glos.	ST893982		above ground-floor window, sculptured facade incorporates 4 entrance holes for skeps that are under hinged lid of window-seat in bedroom above	BT
826		Benthall Church, Benthall, Broseley, Shropshire	SJ6602		curiously similar to 273 above, but no known connection; 2 hives in window-seat above door; bees enter through pipes leading to mouth of carved lion on outside wall; NT	T

WALES

324		The Cottage, Llandough, Cowbridge, S. Glamorgan	SS9972		Fig. 204; housed a Nutt hive now in IBRA Collection, B74/23	B

SCOTLAND

369		69 Belmellie Street, Turriff, Grampian Region	NJ7249		free-standing cylindrical structure with arched recess for 1 skep	BT

328

APPENDIX 2. Beekeeping museums in Europe and elsewhere

Readers can see examples of many of the hives and other beekeeping material described in this book in beekeeping museums. Europe is the continent with the richest heritage of beekeeping history, and most of the museums are to be found there, especially in the German Federal Republic and Poland. The museums are not widely publicised, and a complete list of those known is given here, in alphabetical order of country. Europe is covered first, then the Americas and Asia. The language spoken (if not English) is indicated for each country, or for individual museums.

Treasures of interest to the beekeeping archaeologist can also be found in many national museums, some of which are mentioned in various chapters of the book. No guide has, however, been written to them, and any reader who has the interest and time to search for such material may well make interesting finds. A record of bee-related material in even one museum would be welcome at IBRA, and could help others to locate it.

The list here is updated from the *Directory of the world's beekeeping museums* by Eva Crane (IBRA Reprint M977[77a]), which also includes information on holdings, catalogues, published articles and material for sale.

EUROPE

AU AUSTRIA (German)

AU1 Oesterreichisches Bienenzuchtmuseum, 2304 Orth an der Donau (Bienenzuchtverein Orth an der Donau). Founded 1972; write to Othmar Happel, 2304 Orth an der Donau 345, tel. 02212-441, or Walter Lack, 1040 Wien, Margaretenstr. 5/18, tel. 0222-5719445.
Open daily (except Monday) 9–12, 13–16 h.

AU2 Schloss Trautenfels, 8952 Irdning, Innstal (Landschaftsmuseum). Founded 1955; tel. 03682-2233.
Open daily in summer 9–17 h.

AU3 Dom-Museum, Salzburg. This museum recently staged an exhibition 'Köstlich Wachsgebild', and is likely to have a collection of folk art based on beeswax.

AU4 Heimatmuseum, Spittal, Kärnten.
Open daily.

AU5 Schloss Tellet bei Grieskirchen. No detailed information.

Appendix 2

BE BELGIUM

BE1 'Apicultura' Bijenteeltmuseum, Antwerpsesteenweg 92, 2800 Mechelen. Founded 1973; language Dutch (also French, English, German).
July–August, open every afternoon; March–October, Sat./Sun. only; other days by appointment.

BE2 Musée de l'Abeille, rue du Bihet 9, 4040 Tilff. Founded 1974; language French; write to Secretariat, tel. (041) 882263.
Open 10–12, 14–18 h daily in June–August; Sunday only April, May, September; otherwise by previous appointment.

BE3 Musée Apicole des Ardennes Brabançonnes. Write to S. Meulemans, Rue de la Procession 45, 1310 La Hulpe; language French.
Museum closed at present for lack of accommodation.

CZ CZECHOSLOVAKIA (Czech/Slovak)

CZ1 Zemedelske Muzeum (Agricultural Museum), Kacina, nr Kutna Hora. Well displayed exhibits of early beekeeping material.

CZ2 Agricultural Museum, Nitra. No recent information except article by J. Vontorcick in *Vcelar* 46(6): 132–133 (1972).

CZ3 J. G. Mendel Museum, Material in Old Brno (in Moravian Museum, and bee house in garden of Queen's Monastery), and in Hynčice (birthplace).

CZ4 Slovak National Museum, Martin.

DE DENMARK

DE1 Herning Museum, Museumsgade 1, 7400 Herning (95 km W. of Aarhus). Language Danish; write to Mag art. Ulla Thyrring, Director, tel. (07)-123266.
Visits by previous application.

FI FINLAND

FI1 Beekeeping Museum of Finnish Beekeepers' Central Organization (SUME), Rahkoila, 13880 Hattula. Founded 1973; language Finnish (English, German, Swedish, Russian also used); write to P. Maki-Kihnia.
Open 6 June – 30 September daily, 10–18 h; otherwise open on application.

FR FRANCE (French)

FR1 Gazette Apicole, 84140 Montfavet, Vaucluse. This museum in the Alphandéry home 4 km from Avignon was founded in 1899; write to G. Alphandéry, Montfavet, tel. 31-00-29 or R. Alphandéry, 20 rue de Montevideo, 75116 Paris, tel. 504-05-89.

FR2 Musée National d'Histoire Naturelle, 57 Rue Cuvier, 75231 Paris. Write to Serge Bahuchet.

FR3 Parc Naturel Régional des Landes Gascogne, Route de Mont-de-Marsan,

Appendix 2

40630 Sabres (Ecomusée de Marqueze). Write to J. Albisetti, Laboratoire de pathologie apicole du Sud-Ouest, 40630 Sabres.

FR4 Musée des Arts et Traditions Populaires, Route du Mahatma Gandhi, 75016 Paris. Believed to have some bee material.

FR5 Musée de l'Abeille, St-Agnan, 58230 Montsauche (Nièvre), 400 m from Lac de Saint-Agnan. Write to A. Mellini, tel. 17.
Open daily.

FR6 Musée Maison de Miel, Volvic, 15 km north of Clermont-Ferrand.
Open daily 10–17 h.

Musée de l'Homme, Palais de Chaillot, Paris 16, may well have historical beekeeping material on display.

GD   GERMAN DEMOCRATIC REPUBLIC

GD1 Deutsches Bienenmuseum Weimar had an important collection, but in 1977 the museum was reported to be closed, the material stored, and the premises used by another museum.

GF   GERMAN FEDERAL REPUBLIC (German)

GF1 Niedersächsisches Landesinstitut für Bienenforschung, Wehlestr. 4A, 3100 Celle. Founded 1927; write to H. Geffcken, tel. 05141-22456.
Visits, with guide provided, by previous appointment (2–3 weeks), or on summer Wednesdays 9–12 h, and on open days, usually Ascension Day (May), 2nd or 3rd weekend in August, 1st in September.

GF2 Bayerische Landesanstalt für Bienenzucht, Burgbergstr. 70, 8520 Erlangen. Founded 1925; write to Dr K. Weiss, tel. 09131-21913.
Visits by previous application.

GF3 Bienenkunde-Museum, Altes Rathaus, 7816 Münstertal/Schwarzwald. Opened April 1978 by the Bienenzüchter-Verein, Münstertal Schw.; write to E. Pfefferle, Kurverwaltung, 7816 Münstertal, tel. 07636-213 or -235.
Likely to be open to the public Wed., Sat., Sun., 14–17 h; for other times, and for visits by groups, contact Mr Pfefferle.

GF4 Oberpfälzisches Bauernmuseum, 8470 Nabburg-Perschen (Regierungsbezirk Oberpfalz). Founded 1964; write to Alfons Haseneder.
Open (days not known).

GF5 Westfälisches Freilichtmuseum, 4930 Detmold 1, Krummes Haus (Landschaftverband Westfalen-Lippe). Founded 1962; write to Stefan Baumeier, tel. (05231) 23964.
Open in summer.

GF6 Münchner Stadtmuseum, St-Jakobs-Platz, 8000 München 1. Write to Mrs. Ch. Angeletti, tel. 233-4615. This museum has an important reserve collection of beeswax votives and related folk-art material. It is not on display, but can be seen by previous arrangement.

GF7 Museum für Naturkunde, 7800 Freiburg/Breisg., Gerberau 32. Write to Dr P. Lögler. No other information.

GF8 Heimatmuseum, Rosenheim, 60 km SE of München. No information except

Appendix 2

for the book *Viel köstlich Wachsgebild*[141] which shows many votive offerings and other beeswax models in this museum.

GF9 Museumsdorf Cloppenburg, Niedersächsisches Freilichtmuseum, 4590 Cloppenburg. In summer 1978 an exhibition was staged on beekeeping in north-west Lower Saxony.

GF10 Volkskundliche Abteilung, Naturwissenschaftliches Museum Osnabrück, Hegertor Wall 28, 4500 Osnabrück. An illustrated catalogue was published in 1977, *Imkerei des Osnabrücker Landes*.[272]

GF11 Stormarnsche Museumsdorf, Sprenger Weg 1, 2071 Hoisdorf, 25 km ENE of Hamburg. Write to Adolf Christen, Hoisdorf, tel. 04107 4881.
Open 9–12 h, Sat. 15–18 h.

The following folk museums, which are open to the public, have small beekeeping collections:

GF12 Hamaland-Museum (Kreismuseum Borken), Butenwall 4, 4426 Vreden/Westfalen. The Museum mentions also the Heimathaus in 4423 Gescher (15 km SE) and the Openluchtmuseum Erve Kots, Lievelde über Lichtenvoorde (20 km W in the Netherlands).

GF13 Freilichtmuseum am Kiekeberg (Helms-Museum, Hamburg), Ehestorf Hamburg-Harburg.

GF14 Landesmuseum Koblenz, Festung Ehrenbreitstein, 54 Koblenz.

GF15 Bomann-Museum, Schlossplatz, 7, 3100 Celle.

GF16 Verdener Heimatsmuseum, Grosse Fischerstrasse 10, 2810 Verden.

GF17 Nienburg-Museum, Leinstr. 4., 3070 Nienburg (not yet opened).

GF18 Heimatmuseum Biberach, Biberach, Schwarzwald.

GF19 Freilichtmuseum Sobernheim, Postfach 180, 6553 Sobernheim, Rheinland-Pfalz.

GF20 Heidemuseum 'Dat Ole Huus', Wilsede, Lüneburger Heide.
Open daily in summer.

GF21 Landwirtschaftsmuseum Lüneburger Heide, 3113 Suderburg 2, Hösseringen, 15 km SW of Velzen. Founded 1975; write to H. W. Löbert.

GF22 Naturwissenschaftliches Museum, Berlin; proposed 1978; write to Dr I. Jung-Hoffman, Zoologisches Institut der Freien Universität Berlin, Berlin-Dahlem, Königen-Luise-Strasse 1–3. A collection of beekeeping material from the Institut für Bienenkunde, Berlin-Dahlem, is now on display in the Institut für Allgemeine Zoologie.

HU   HUNGARY (Hungarian)

HU1 Kisállattenyésztési Kutatóintézet, Division of Beekeeping, Méhészet, 2101 Gödöllö. Founded 1902; write to Dr Z. Örösi-Pal (in German or English). Open to visitors, by appointment if possible.

HU2 Neprajzi Museum (Ethnographical Museum) Budapest VIII, Konyves Kalmankorut 40. Visited 1963; collection of traditional hives then on display; no later information.

Appendix 2

IT   ITALY (Italian)

IT1   Osservatorio di Apicoltura 'Don Giacomo Angeleri', Strada del Cresto 2, 10132 Reaglie-Torino (Istituto di Entomologia agraria e Apicoltura, Università di Torino), 5 km NE of city centre along Corso Chieri. Founded 1969; write to Prof. Carlo Vidano, tel. (011)-659150.
Visits by previous appointment only.

IT2   Museo Apistico Didattico, Via Memegardo 12, Bregnano, Como, Founded 1950; write to Cav. Angelo Capelletti, tel. 771-661.
Open last Sunday of each month from 9.30 h.

IT3   Museo Zoologia 'La Specola', Università degli Studi, Via Romana 17, 50 125 Firenze. This museum has an important collection of medical preparations in wax, made in the seventeenth and nineteenth centuries to show the anatomical construction of the body.

IT4   Museo de Istituto di Zoologia Sistematica dellà Università di Torino, Via Giovanni Giolitti 34, 10123 Torino. Founded about 1811; write to Dr Passerin d'Entrèves, tel. 011-879211.
Visits by appointment only.

NE   NETHERLANDS

NE1   Dutch Open Air Museum (Het Nederlands Openluchtmuseum), Schelmseweg 89, Arnhem. Founded 1912; language Dutch; Director J. H. Jager-Gerlings, tel. (085)-452064-68.
Open daily 9–17 h, Sundays and holidays 10–17 h, April–October.

NE2   Beekeeping Antiquities (Stichting Drents Bijenteeltmuseum), Hoeve Bekhof, de Hoek 5, Vledder. Founded 1974; language Dutch; write to K. P. de Bruijn, tel. (05212)-1641.
Open for individuals 14–17 h Thursday, Saturday, 15 June–31 July; by appointment for groups May–September.

NE3   Collection Atze Dykstra, Kampingerhof 6, Oosterwolde, Friesland. Language Friesian (also Dutch, German, English); write to Atze Dykstra, tel. 05160-2439.
Open by arrangement.

NE4   Bijenschans Corversbos, Schuttersweg, Hilversum. Language Dutch.
Open 10–15 h Saturday only, May–August.

See also GF12.

NO   NORWAY

NO1   Norges Birøkterlags Museum, Bergerveien 15, 1362 Billingstad. Founded 1890; language Norwegian (also English).
Open (days not known).

PO   POLAND (Polish)

PO1   Skansen Pszczelarski (Beekeeping Skansen), ul. Poznanska 35, 62-020

333

Appendix 2

Swarzedz, near Poznan (Veterinary Institute). Founded 1965, language Polish (also German, English, Russian); write to M. Jeliński, tel. Swarzedz 341.
Open daily.

PO2 Skansen Pszczelarski, w Pszczelej Woli, 23–109 Pszczela Wola (Panstwowe Technikum Pszczelarskie). Founded 1973.
Open (days not known). No previous application necessary.

PO3 Muzeum im. Jana Dzierzona, ul. 15 Grudnia 12, 46-200 Kluczbork (under the Ministry of Culture and Art). Founded 1957; write to Mgr. R. Pastwiński, tel. Kluczbork 707.
Open to the public; previous application needed to see material not on display.

PO4 Państwowe Muzeum Entograficzne, 1 Kredytowa St., 00-056 Warsaw. Founded 1946.
Open to the public.

PO5 Pasieka Uli Klodowych, Muzeum Wsi Opelskiej, Opole-Bierkowice. Founded 1970.
Open to the public.

PO6 Muzeum Narodowe Rolnictwa w Szreniawie (National Museum of Agriculture in Szreniawa), 65-052 Komorniki. Founded 1964.
Normally open 9–16 h except Mondays.

PO7 Muzeum and Skansenmuzeum, Locowicz.
Open daily.

PT PORTUGAL

PT1 O Museu De Ovar, Ovar, started a beekeeping section in 1978.

RO ROMANIA

RO1 Expozitia Permanentă de Apicultură a Asociatiei Crescătorilor de Albine din România, Bd. Fiscusului 42, sect. 1, Bucureşti. Founded 1965; language Romanian (also English, French, German, Russian, Spanish, Hungarian by prior arrangement); write to the Association at Str. Iulius Fucik 17 sect. 2, tel. 122010 or 123750.
Open daily 7–15.30 h.

SW SWEDEN

SW1 Biodlingsmuseet, Hembygdsparken, 29700 Degeberga. Founded 1975; languages Swedish and English.
Open afternoons except Monday, mid-May–September (for other times tel. 045050657 or write to Kerstin Sinha).

SZ SWITZERLAND

SZ1 Naturhistorisches Museum Bern, 3005 Bern, Bernastrasse 15 (Burgergemeinde Bern). Founded 1954; language German (also French, English); write to H. D. Volkart, tel. 4131-431839.

Appendix 2

Open daily 9–17 h (Sundays 10–17 h). Most of the beekeeping material (Sektion Die Biene) is on display; the rest may be viewed by previous appointment only.

SZ2 Sammlung K. A. Forster 'Die Biene', 8700 Küsnach-ZH, Himmeli-Str. 20. Private collection founded 1930; language German; write to Dr K. A. Forster, tel. 910 76 61. It is possible for those interested to see the collection in Dr Forster's house; it has been on exhibition in a number of German and Swiss towns.

SZ3 Société Romande d'Apiculture. The material in the Museum of this Association is now in the Bern Museum (SZ1).

SZ4 Apistisches Museum des VDSB, Gasthaus Rosenberg, 6300 Zug. Language German; write to Peter Theiler.

SZ5 Musée Paysan et Artisanal, La Chaux-de-Fonds, Neuchâtel.

SZ6 Museo Bleniese, Lottigna, Ticino. Language Italian.

SZ7 Castle Burgdorf, Burgdorf. Language German.

UK UNITED KINGDOM

UK1 IBRA Collection of Historical and Contemporary Beekeeping Material, International Bee Research Association, Hill House, Gerrards Cross, Buckinghamshire SL9 0NR. Founded 1952, main language English (others understood); write to Dr Eva Crane; tel. (0753)–885011.
Viewing by appointment only; office hours Monday–Friday, 9–17.30 h, except Bank Holidays. Much of the material is at UK2.

UK2 Museum of English Rural Life, University of Reading, Whiteknights, Reading, Berks. RG6 2AG. Founded 1951; language English; write to the Keeper; tel. (0734)-85123 (ext. 475). See end of entry UK1.
Open daily (except Sundays and Mondays) 10–13, 14–16.30 h.

UK3 National Museum of Antiquities of Scotland, Queen Street, Edinburgh. Founded late 1940s; languages English, Scots dialect. Write to Gavin Sprott; tel. (031)-556-8921.
Museum open daily, but the Country Life Section, with the bee material, can be seen only by previous appointment; it may be moved soon to the new Agricultural Museum at Ingliston near Edinburgh.

UK4 Welsh Folk Museum (Amgueddfa Werin Cymru), St Fagans, Cardiff CF5 6XB. Founded 1916; languages Welsh and English; tel. (0222) 569311. Open daily 10–17 h, except Sunday 14.30–17 h. Beekeeping material in Department of Farming and Rural Life.

A note on 'The British Museum beeswax treasures' was published in *Bee World* 1978.[32]

UR USSR

UR1 Central Beekeeping Research Station, Rybnoe, Ryazan Province, has an important collection of historical beekeeping material, visited 1962, 1971. Founded before 1900; language Russian. Write to G. Bilash, Director. Viewing by appointment only.

Appendix 2

UR2   Exhibition of Economic Achievement, Moscow, has a display in the Beekeeping Pavilion.
UR3   Georgian Beekeeping Research Station, Okrokama, Tbilisi, Georgia. Languages Georgian, Russian; write to Z. A. Makashvili.
UR4   Lithuanian National Park, Ignalina, 100 km north of Vilnuis.

YU   YUGOSLAVIA

YU1   Apicultural Museum (Muzeji radovlijiške občine), 64240 Radovljica, Linhartov trg 1, Slovenia (Cultural Community of Radovljica & Medex (Ljubljana)). Founded in 1959; language Slovene; write to Miss Marusa Avgustin, tel. (664) 75–188. Radovljica is 7 km SE of Bled.
Open daily 1 April–31 October, 10–12, 16–18 h.
YU2   Pokrajinski Muzej Celje, Muzejskitrg 1, 63000 Celje. Founded 1952; language Slovene.
Open daily in summer; bee collections by appointment.
YU3   Muzej Grad, Skofja Loka, Slovenia.
Open daily in summer.

NORTH AND SOUTH AMERICA

AR   ARGENTINA (Spanish)

AR1   Museo Apicola Nacional e Internacional, SADA, Rivadavia 717, Piso 8, 1392 Buenos Aires.
No information.

CA   CANADA

CA1   Ontario Agricultural Museum, Box 38, Milton, Ont. L9T 2YE (Ontario Ministry of Agriculture and Food). Founded 1973; languages English, French; write to R. W. Carbert, tel. (416)-878-8151.
Open daily 8–16.30 h Monday–Friday (weekends soon).

US   USA

US1   Dadant Library, Hamilton, IL 62341 (Dadant & Sons, Inc.). Founded 1863 by Charles Dadant; write to Howard Veatch, tel. (217)-847-3324. Used mainly for research and study, but beekeepers and other adults may view it by previous application; no school groups.
US2   Michigan State University Museum, East Lansing, MI 48824. Write to Dr R. Hoopingarner.
Periodical exhibitions are open to the public (otherwise apply in advance).
US3   P. J. Hewitt Jr Apicultural Collection, University of Connecticut, Waterbury Campus, 32 Hillside Ave., Waterbury, CT. Founded 1953; write to Prof. A. Avitabile, tel. 203-757-1231 (ext. 38).
Visits with or without previous appointment.
US4   University of Massachusetts, Amherst, Mass.

Appendix 2

	Visited 1953; no recent information.
US5	Phillips Collection, Cornell University, Ithaca, NY.
	Visited 1953 and later; no recent information.
US6	Henry Ford Museum, Dearborn, MI 48121 (Edison Institute). Founded 1929.
	Open 7 days a week.
US7	Shelburne Museum, Shelburne, VT 05482. Rt. 7, 7 miles S of Burlington. Founded 1956; tel. 8029853344.
	Open daily 9–17 h, mid-May to mid-October.

ASIA

IN INDIA

IN1 Central Bee Research Institute, Khadi and Village Industries Commission, 839/1 Deccan Gymkhana, Pune 411 004, Maharashtra State (3 km from Pune Railway Station). Founded 1962; language English; write to Shri C. V. Thakar or Shri R. P. Phadke, tel. 54019.
Open in working hours, normally 10.30–17.30 h Mon.–Fri. and 1st and 3rd Saturdays.

IS ISRAEL

IS1 Ministry of Agriculture, Extension Service, Beekeeping Division, Hakirya, Tel Aviv; languages Hebrew, English; write to N. Yekutieli, tel. 259411.
There is now a project for creating a Beekeeping Museum at the Beekeeping Division of the Ministry, near Tzrifin, 20 km E of Tel Aviv.

JA JAPAN (Japanese)

JA1 Bee House, Livestock Center Park, Tsubakibora-Nakano, Gifu-shi, 502. Founded 1981; tel. 0582 942002.
Open 9–16.30 h except holidays.

Bibliography

1. Allchin, B. (1966) *The stone-tipped arrow. Late Stone-age hunters of the tropical Old World.* London: Phoenix House
2. Allchin, F. R. and Allchin, B. (1962) A neolithic pot from Andhra Pradesh. *Antiquity*: 302–3 + 1 pl.
   Almeida Correia (1981) See Correia (64)
3. Anpilogov, G. N. (1964) [Bee tree marks as a historical source (based on manuscripts from Putivl' and Ryl' of the late 1500s and early 1600s).] *Sov. Arkheol.* 4: 151–69 *In Russian*
4. Armbruster, L. (1921) Bienenzucht vor 5000 Jahren. *Arch. Bienenk.* 3(1/2): 68–80
5. ———(1926) Der Bienenstand als völkerkundliches Denkmal. *Bücherei Bienenk.* 8: 1–147
6. ——— (1928) Die alte Bienenzucht der Alpen. *Bücherei Bienenk.* 9: 1–184
7. ——— (1929) Die alte Bienenzucht Italiens. *Arch. Bienenk.* 10(6): 185–207
8. ——— (1931) Klassische Bienenzuchtgebiete im Lichte der historischen Betriebslehre und Völkerkunde. *Arch. Bienenk.* 12(1): 1–36
9. ——— (1933) Süddeutsche Bienenstände auf alten Holzschnitten. *Arch. Bienenk.* 14(5): 224–31
10. ——— (1934) Imkerei-Betriebsformen. IV. *Arch. Bienenk.* 15(2/3): 117–32 (see 25. Erdbienenzucht, 126–32)
11. ——— (1935) Litergewicht der einzelnen Bienenwohnungen. *Arch. Bienenk.* 16(1): 8–10
12. ———(1938*a*) Alte Graphik und Imkerei. *Arch. Bienenk.* 19(6/7): 185–248
13. ———(1938*b*) Die Zeideln und die Baiwaren. Imkerei, Bienenrecht, Siedlung und Forstnutzung in Altbayern. *Arch. Bienenk.* 19(8): 256–304
14. ——— (1952) Bienenbilder und Verwandtes auf antiken Gemmen. *Arch. Bienenk.* 29: 68–73
15. ——— (1954) How old are English bee boles? *Bee Wld* 35(3): 50–2
16. ——— (1955) Die Biene auf neueren Münzen. *Arch. Bienenk.* 32(1): 17–30
17. ——— (1956*a*) ¿Existen en España, muros de abejas? *Apicultura, Madrid* (48): 8
18. ——— (1956*b*) Veröffentlichungen von L. Armbruster. *Arch. Bienenk.* 33(1): 47–53
19. ——— (1957) Imkerkünste in Bienen-Röhren, -Urnen, -Steintunneln, -Mauern. *Arch. Bienenk.* 34: 1–8, 13–22
20. ——— (1970) Erdbienenzucht. *Imkerfreund* 25(12): 353–6

21. Avery, M. (1936) *The Exultet rolls of S. Italy.* Vol. 2. Princeton, USA: Princeton University Press
22. Avgustin, M. (1979) Das Bienenzuchtmuseum in Radovljica und bemalte Bienenstockstirnbretter. Pp. 145–54 from *Bienenmuseum und Geschichte der Bienenzucht.* Bucharest: Apimondia
23. Bagster, S., Jr (1834) *The management of bees* ... London: Bagster & Pickering
24. Bailey, P. (1971) *Orkney.* Newton Abbot, Devon: David & Charles
25. Balassa, I. (1957) A vadméh befogása Abaujan. [The taking of wild bees in the comitatus of Abauj.] *Népr. Közlem* 2(1/2): 148
26. ——— (1975) Waldbienenzucht im Karpatenbecken. *Acta ethnogr. Acad. Sci. hung.* 24 (1/2): 117–28
27. Baltrušaitis, J. (1967) *La Quête d'Isis.* Paris: Olivier Perrin
28. Bănăteau, T. (1959) La chasse aux abeilles en Roumanie. *Schweizer Arch. Volksk.* 55(4): 279–84
29. Banchereau, J. (1913) Travaux d'apiculture sur un chapiteau de Vézelay. *Bull. monum.*: 403–15
30. Batsylev, E. G. (1972) [Bees on arms of towns in Tambov guberniya.] *Pchelovodstvo* (10): 28–31   *In Russian*
31. Beck, B. F. (1938) *Honey and health.* New York: Robert M. McBride & Co.
32. Bee World (1978) The British Museum's beeswax treasures. *Bee Wld* 59(1): 39–40
33. Beetsma, J. (1977) Bee markets and exhibitions in the Netherlands. *Bee Wld* 58(2): 92–3
34. Beltran, A. (1961/62) Peintures rupestres du levant de 'El abrigo de los recolectores' dans le ravin de 'El Mortero' (Alacon, Teruel, España). *Bull. Soc. préhist. Ariège* 16/17: 15–50
35. Berlepsch, A., Baron von (1860) *Die Biene und die Bienenzucht.* Mühlhausen in Thüringen: Heinrichshofen
36. Bessler, J. G. (1886) *Geschichte der Bienenzucht.* Stuttgart: W. Kohlhammer
37. Best, H. (1857) *Rural economy in Yorkshire in 1641, being the farming and account books of H.B. of Elmswell in the E. Riding* ... Ed. C. Best Robinson, Durham: G. Andrews, for the Surtees Society
38. Bevan, E. (1827) *The honey-bee; its natural history, physiology, and management.* London: Baldwin, Cradock & Joy. 2nd ed. 1838
39. ——— (1870) *The honey bee: its natural history, physiology and management.* Revised, enlarged and illustrated by W. A. Munn. London: John Van Voorst
40. Bewick, T. (1820) *Select fables* ... with woodcuts by T. & J. Bewick. Newcastle: S. Hodgson
41. Bielby, W. (1977) *Home honey production.* E. Ardsley, Yorks.: EP Publishing Ltd
42. Birch, C. A. (1959) Mazers. *Bee Craft* 41: 3, 12–13
43. Blank-Weissberg, S. (1937) *Barcie i kłody w Polsce.* [Bee tree-cavities and log hives in Poland.] Warsaw: Z Zasiłku Ministerstwa Rolnictwa i Reform Rolnych
44. Böcker, M. (1930) Das Imkerbeil, ein merkwürdiges Gerät des alten Heideimkers. *Arch. Bienenk.* 11(3): 97–126

Bibliography

45. Bonner, J. (1795) *A new plan for speedily increasing the number of bee-hives in Scotland* ... Edinburgh: J. Moir and others
46. Bray, W. (1979) *The gold of Eldorado*. London: Times Newspapers Ltd
47. Brekelmans, T. (1979) *Korfvlechten*. [Making baskets [with coiled straw].] De Bilt, Netherlands: Uitgeverij Cantercleer b.v.
48. Brethenoux, D. (1980) Des abeilles dévotes ... *Abeilles et Fleurs* (305): 7
49. British Bee-Keepers' Association (undated, c. 1890) *Modern bee-keeping series No. 1—Skeps*. London: British Bee-Keepers' Association
50. Broneer, O. (1958) Excavations at Isthmia, Third Campaign, 1955–1956. *Hesperia* 27: 1–37
51. Brunskill, R. W. (1970) *Illustrated handbook of vernacular architecture*. London: Faber & Faber
52. Brüweln, J. (1719) *Brandenburgische bewährte Bienen-Kunst ... nach dem 4. Buch Georgicorum* ... Berlin: Johann Andreas Rüdigern
53. Buchanan, R. A. (1972) *Industrial archaeology in Britain*. Harmondsworth, Middx: Pelican Books
54. Büll, R. (1959–70) *Vom Wachs. Hoechster Beiträge zur Kenntnis der Wachse*. Frankfurt am Main: Hoechst AG

Vol	Part	Year	Pages	Title
I	1	1959	1–60	Einführung: von der Fülle und Vielfalt des Wachses; Was ist Wachs und wie Verhält sich Wachs? Antworten in alten und neuen Zeiten; Zur Wachsklassifizierung
I	2	1959	61–90	Textilornamentik nach dem Wachsreserveverfahren
I	3	1959	91–142	Bronze- und Feinguss nach dem Wachsausschmelzverfahren
I	4	1960	143–190	Zur Geschichte des Wachshandels
I	5[a]	1961	191–236	Wachs als Erzeugnis der Natur und Technik
I	6[a]	1961	237–318	Wachs als Gegenstand der naturwissenschaftlichen Forschung und der Anwendungstechnik
I	7/1	1963	319–416	Wachsmalerei: Enkaustik und Temperatechnik unter besonderer Berücksichtigung antiker Wachsmalverfahren
I	7/2	1963	417–526	Keroplastik: ein Einblick in ihre Erscheinungsformen, ihre Technik und Ästhetik
I	8/1	1965	527–678	Zur Phänomenologie und Technologie der Kerze unter besonderer Berücksichtigung der Wachskerze
I	8/2	1967	679–784	Zur Phänomenologie und Technologie der Kerze: Die Kerze heute
I	9[b]	1968	785–894	Wachs als Beschreib- und Siegelstoff; Wachsschreibtafeln und ihre Verwendung
I	10/11	1970	895–1012	Wachs und Kerzen im Brauch, Recht und Kult; Zur Typologie der Kerzen

[a] with G. von Rosenberg [b] with E. Moses and H. Kühn

I   12ᶜ   1977   1013–1097 Verzeichnisse, Register, Impressum, Gesamtinhaltsverzeichnis
ᶜwith E. Moser and with two supplementary papers by F. Klemm

55. Butler, C. (1609) *The feminine monarchie, or A treatise concerning bees.* Oxford: printed by Joseph Barnes
56. Cannon, H. (ed.) (1980) *Utah folk art.* Provo, Utah: Brigham Young University Press
57. Carreras, R. (1944) Los hornales. *Valencia avic.* 1(7): 22–4
58. Chakrabarti, K. and Chaudhuri, A. B. (1972) Wild life biology of the Sundarbans forests: honey production and behaviour pattern of the honey bee. *Sci. Cult.* 38(6): 269–76
59. Christ, J. L. (1802) *Anweisung zur nützlichsten und angenehmsten Bienenzucht für alle Gegenden* . . . Frankfurt and Leipzig: publisher not stated
60. Church, L. (1980) Beekeeping and honey production in Cyprus. *Proc. II int. Conf. Apic. trop. Climates*
61. *Congresso Internazionale, Ceroplastica nella Scienze et nell'Arte, I* (1977) [Ceroplasty in science and the arts. Proceedings of the 1st International Congress, Florence, 3–7 June 1975.] Florence: Leo S. Olschki 2 vols *In Italian*
62. Cooke, S. [c. 1780] *The complete English gardener: . . . to which is added The complete bee-master, or, Best method for managing bees, both for profit and pleasure.* London: Cooke
63. Cooper, J. Fenimore (1848) *The bee-hunter; or, The oak openings.* London: Bentley
64. Correia, M. de L. M. de A. (1981) Contribution à l'étude de la biologie d'Heriades truncorum L. (Hymenoptera, Apoidea, Megachilidae). II. Aspect écologique. *Apidologie* 12(1): 3–30
65. Cotlow, L. (1957) *Zanzabuku.* London: Robert Hale
66. Cotton, W. C. (1842) *My bee book.* London: Rivington
67. Country Curate, pseud. Filleul, P.V.M. (1851) *The English bee-keeper* . . . London: Rivington
68. Cowan, T. W. (1882) *British bee-keeper's guide book* . . . 2nd ed. London: Houlston
69. ——— [1907] *British bee-keeper's guide book* . . . 19th ed. (rev.) London: Madgwick, Houlston and British Bee Journal
69a ——— (1908) *Wax craft.* London: Sampson Low, Marston & Co. and British Bee Journal
70. Cox, J. C. (ed) (1905) *The royal forests of England.* London: Methuen
70a Coysh, A. W. and Henrywood, R. K. (1982) *Dictionary of blue and white printed pottery 1780–1880.* UK: Antique Collectors Club
71. Crane, E. (1961) The Worshipful Company of Wax Chandlers. *Bee Wld* 42(3): 63–71
72. ——— (1963) The Prince, the Inquisition and the bee book that was never published. *Bee Wld* 44(1): 43–4

Bibliography

73. ——— (1971) Frameless movable-comb hives in beekeeping development programmes. *Bee Wld* 52(1): 33–7
74. ——— (1975) *Honey: a comprehensive survey.* London: William Heinemann Ltd and International Bee Research Association
75. ——— (1977a) The shape, construction and identification of traditional hives. *Bee Wld* 58(3): 119–27
76. ——— (1977b) Beehives, bees and beekeepers. *Proc. XXVI int. Beekeep. Congr.*: 183–9
77. ——— (1978) *Bibliography of tropical apiculture.* London: International Bee Research Association
77a ——— (1979) Directory of the world's beekeeping museums. *Bee Wld* 60(1): 9–23.
78. ——— (1980a) *A book of honey.* Oxford: Oxford University Press
79. ——— (1980b) Apiculture. Chapter 10 (pp. 261–94) from *Perspectives in world agriculture.* Farnham Royal, UK: Commonwealth Agricultural Bureaux
80. ——— (1981) Bee houses. *Bee Wld* 62(2): 43–5
81. ——— (1982a) Britannia's skep. *Bee Wld* 63(1): 47–9
82. ——— (1982b) The evolution of protective clothing for beekeepers. *Bee Wld* 63. In press
83. Crane, E. and Graham, A. J. (1982) *Hives of the Ancient World.* To be published
84. Cunnington, Mr and Mrs B. H. (1913) Casterley Camp. *Wilts. archaeol. Mag.* 38: 53–105
85. Dams, L. R. (1978) Bees and honey-hunting scenes in the Mesolithic rock art of eastern Spain. *Bee Wld* 59(2): 45–53
86. Darchen, B. and Darchen, R. (1978) Le comportement constructeur des abeilles sociales. *Courr. C.N.R.S.* No. 30: 38–45
87. Dathe, G. (1870) *Lehrbuch der Bienenzucht.* Bensheim: J. Ehrhard & Co.
87a David, A. R. et al. (1984) *Kahun.* Warminster, Wilts.: Aris & Phillips
88. Davies, N. de (1944) *The tomb of Rekhmire at Thebes.* New York
89. Della Rocca, l'Abbé (1790) *Traité complet sur les abeilles . . .* Paris: Bleuet Père
90. Desborough, V. F. (1955) Bee boles and beehouses. *Archaeol. cantiana* 69: 90–5
91. ——— (1956) Further bee boles in Kent. *Archaeol. cantiana* 70: 237–40
92. ——— (1958) Further note on Kentish bee holes. *Archaeol. cantiana* 72: 189–91
93. ——— (1960) More Kentish bee boles. *Archaeol. cantiana* 74: 91–4
94. Dickson, J. H. (1978) Bronze age mead. *Antiquity* 52: 108–13
95. Diekmann, H. (1963) Frühgeschichtliche Klotzbeuten im Germanenhof (1. Jahrhundert) zu Oerlinghausen im Teutoburger Walde. *Westf. Bienenztg* 76(1): 9–10
96. Donaldson, T. L. (1827) *Pompeii.* London
97. Dudley, P. (1721) An account of a method lately found out in New-England, for discovering where the bees hive in the woods, in order to get their honey. *Phil. Trans. R. Soc.* 31(367): 148–50
98. Dummelow, J. (1973) *The Wax Chandlers of London. A short history of the*

Worshipful Company of Wax Chandlers, London. London: Phillimore & Co.

99. [Dunbar, W.] (1840) *The natural history of bees.* Edinburgh: W. H. Lizars (See p. 180)
100. Duruz, R. M. (1953) The honey-pot game. *Bee Wld* 34(5): 90–3
101. Duruz, R. M. and Crane, E. E. (1953) English bee boles. *Bee Wld* 34(11): 209–24
102. Dzierzon, J. (1848) *Theorie und Praxis des neuen Bienenfreundes* ... published by the author
103. Eberle, M. W. (1979/80) Lucas Cranach's Cupid as honey thief paintings: allegories of syphilis? *Comitatus* 10: 21–30
104. Edgell, G. H. (1949) *The bee hunter.* Cambridge, MA, USA: Harvard University Press
105. Edwardes, E. Tickner (1907) *The bee-master of Warrilow.* London: Pall Mall
106. ——— (1908) *The lore of the honeybee.* London: Methuen
107. Erup, O. (1956) L'apiculture en Espagne. *Gaz. apic.* 57(592): 83–5
108. ——— (1957) Alte Mauer-Bienenstände in Spanien. *Arch. Bienenk.* 34: 1, 9, 10
109. Erup, O. and Armbruster, L. (1958) Nochmals spanische Bienenmauern. *Arch. Bienenk.* 35: 32–3
110. Evans, Sir Arthur (1935) *The palace of Minos.* Vol. IV. London: Macmillan *Ill. Lond. News* (29 June): 1162–5
110a Evans, E. (1966) *Prehistoric and early Christian Ireland: a guide.* London: B. T. Batsford Ltd

Filleul, P. V. M. (1851) See Country Curate (67)

111. Forster, K. A. (1975) *Sammlung Karl August Forster: Die Biene.* Küsnacht-Zürich: published by the author
112. Fraser, H. M. (1931) *Beekeeping in antiquity.* London: University Press; 2nd ed. 1951
113. ——— (1950) The story of the progress of beekeeping before 1800. *Bee Wld* 31(5): 33–8
114. ——— (1951) *Anton Janscha on the swarming of bees.* Foxton: Apis Club
115. ——— (1955) Beekeeping in the British Isles before 1500. *Bee Wld* 36: 177–86, 223–6, 248
116. ——— (1958) *History of beekeeping in Britain.* London: Bee Research Association
117. French, K. (1960) Wide variety of bee boles can still be found in Devon. *Beekeeping, Devon* 26(3): 37–8
118. Fürtwängler (1896) Referred to by L. Armbruster (1952)[14]
119. Galton, D. (1971) *Survey of a thousand years of beekeeping in Russia.* London: Bee Research Association
120. ——— (1980) Personal communication
121. Gardner, E. (1930) A triple-banked enclosure on Chobham Common. *J. Surrey archaeol. Soc.* 35: 105–13
122. Gaunt, A. (1977) Where the bee sucked ... *Yorks. Life* (Oct.): 50–2
123. Gautier, A. (1977) De oudste Europese bijenkorf. [The oldest European bee hive] *Maandbl. vlaam. Imkersb.* 63(9): 326–8
124. Gayre, G. R. (1948) *Wassail! in mazers of mead.* London: Phillimore & Co.

Bibliography

125. Geffcken, H. (1979) Strohkörbe und Geräte der Lüneburger Heideimker im Imkereimuseum des Celler Bieneninstituts. Pp. 130–44 from *Bienenmuseum und Geschichte der Bienenzucht*. Bucharest: Apimondia
126. Gerstung, F. (1919, 1926) *Der Bien und seine Zucht*. Berlin: Pfenningstorff 5th, 7th ed.
127. Gimbutas, M. (1974) *The gods and goddesses of Old Europe*. London: Thames & Hudson
128. Goetze, G. (1953) Personal communication
129. Golding, R. (1847) *The shilling bee book* . . . London: Longman & Co.
130. Goodall, E. (1959) Rock paintings in Mashonaland. In *Prehistoric rock art of the Federation of Rhodesia and Nyasaland*. Ed. R. Summers, Salisbury, Rhodesia: National Publications Trust Rhodesia and Nyasaland
130a Gordon, D. H. (1960) *The prehistoric background of Indian culture*. 2nd ed. Bombay: N. M. Tripathi
131. Graham, A. J. (1975) Beehives from Ancient Greece. *Bee Wld* 56(2): 64–75
132. ——— (1978) The Vari House—an addendum. *A. Br. Sch. Archaeol. Athens* 73: 99–101
133. Gunda, B. (1968) Bee-hunting in the Carpathian area. *Acta ethnogr. hung.* 17: 1–62
134. Gutmann, B. (1922) Die Imkerei bei den Dschagga. *Arch. Anthrop., Braunschw.* 19(1): 8–35
135. Guy, R. D. (1972) The honey hunters of Southern Africa. *Bee Wld* 53(4): 159–66
136. Haccour, P. (1961) Le plus grand rucher collectif du monde. *Abeilles et Fleurs* (92): 4–6
137. Hadfield, M. (1958) Bee houses. *Gdnrs' Chron.* (21 June): 1 p.
138. Hämäläinen, A. (1909) Tšeremissien mehiläisviljelyksestä. [Cheremessian honey hunting.] *J. Soc. finno-ougrienne, Helsinki* 26   *In Finnish*
139. ——— (1934) Lisätietoja mordvalaisten ja tšeremissien vanhasta mehiläisviljelyksestä. *Suom. MuinaismYhd. Aikak.* 40   *In Finnish*
140. Handasyde, Pseud. Emily Handasyde Buchanan (1907) The haunted garden. In *The four gardens*. London: T. N. Fowles
141. Hansmann, C. and Hansmann, L. (1959) *Viel köstlich Wachsgebild*. München: F. Bruckmann
142. Harrer, H. (1953) *Seven years in Tibet*. London: Hart-Davis
143. Hartlib, S. (1655) *The reformed commonwealth of bees*. London: Giles Calvert
144. Hernández-Pacheco, E. (1924) Las pinturas prehistóricas de las cuevas de la araña (Valencia). *Mem. Com. Invest. paleont., Madr.* No. 34
145. Herrod-Hempsall, W. (1930) *Bee-keeping new and old described with pen and camera*. Vol. 1. London: British Bee Journal
146. ——— (1937) *Bee-keeping new and old described with pen and camera*. Vol. 2. London: British Bee Journal
147. Hikscher, E. (1887/1888) Zur Geschichte der Bienenzucht. *Leipzig. Bienenztg* 2: 146–8, 162–4, 178–80; 3: 3–5, 19–21, 36–8, 50–1, 65–6, 81–3, 98–9, 113–14, 129–30, 145–7, 161–3, 177–80
148. Historia (1966) Bee towers in Ireland. *Bee Craft* 48(6): 74–5
149. Höhnel, W. (1959/60) Geschichte der Honigkuchenbäckerei. *Bienenzucht*

12: 334–6, 363–4; 13: 4–7
150. Hood, S. (1976) The Mallia gold pendant: wasps or bees? Pp. 59–72 from *Tribute to an antiquary: essays presented to Marc Fitch by some of his friends*. Ed. Emmison & Stephens, London
150a Hoy, R. (1788) *Proper directions how to manage bees in Hoy's octagon box bee hives*. London: Honey Warehouse
151. Huber, F. (1792) *Nouvelles observations sur les abeilles*. Geneva: Barde, Manget & Co.
152. Huish, R. (1815) *A treatise on the nature, economy, and practical management, of bees; . . .* London: Baldwin, Cradock, & Joy
152a Ichikawa, M. (1980) [Beekeeping of the Suiei Dorobo in East Africa.] *Kikan-Jinruigaku* 11(2): 117–52 In Japanese
152b —— (1981) Ecological and sociological importance of honey to the Mbuti net hunters, Eastern Zaire. *Afr. Stud. Monogr.* 1: 55–68
153. Ifantidis, M. D. (1983) The movable-nest hive—forerunner of the movable comb hive. *Bee Wld* 64: 79–87
153a Inglesent, H. (1974) Personal communication
154. Inoue, T. (1981) Japanese native honeybees in Tsushima island. *Honeybee Sci* 2(1): 19–22
155. International Bee Research Association (1979) *British bee books: a bibliography 1500–1976*. London: International Bee Research Association
156. —— (1980) Mediaeval manuscripts. 45 colour slides from the Bodleian Museum, Oxford. *IBRA Slide No. 1*
Janscha, A. (1775) See Münzberg, J. (214)
157. Jenn, R. A. Y. (1969) Bee-boles . . . a feature of British beekeeping of the past. *Am. Bee J.* 109(11): 427
158. Jenyns, F. G. (1886) *A book about bees . . .* London: Wells Gardner, Darton, & Co.
159. Jewsewjew, T. J. and Erdödi, J. (1974) Bienenzucht bei den Tscheremissen. *J. Soc. finno-ougrienne* 73
160. Johansson, T. S. K. and Johansson, M. P. (1967) Lorenzo L. Langstroth and the bee space. *Bee Wld* 48(4): 133–43
161. —— (1972) Bee-library of the late Rev. L. L. Langstroth. *Bee Wld* 53(1): 22–7
162. Jones, J. (1843) *The eclectic hive . . .* Hereford: Times Office
163. Jones, J. E. (1976) Hives and honey of Hymettus. Beekeeping in Ancient Greece. *Archaeology* 29(2): 80–91
164. Jones, J. E., Graham, A. J. and Sackett, L. H. (1973) An Attic country house below the cave of Pan at Vari. *A. Br. Sch. Archaeol. Athens* 68: 355–452
165. Jung-Hoffmann, I. and Jelinski, M. (1979) 3000 Jahre alter Klotzstülper oder Opferbrunnen? *Allg. dt. Imkerztg* 13(10): 306–9
166. Kardara, C. (1961) Dyeing and weaving works at Isthmia. *Am. J. Archaeol.* 65: 261–6 + pl. 81
167. Keith, G. (1811) *Agricultural survey for Aberdeenshire*.
168. Kelly, F. and Charles-Edwards, T. (1983) *Bechbretha (Bee-laws)*. Dublin: Dublin Institute for Advanced Studies. In press
169. Kelly, M. T. (1897) Saint Gobnata, and her hive of bees. *J. Cork hist.*

Bibliography

archaeol. Soc. 3: 100–6
170. Kenward, H. (1981) Personal communication
171. Keys, J. (1796) *The antient bee-master's farewell* . . . London: Robinson
172. Kiangsi Province Beekeeping Research Institute (1975) [Manual of beekeeping.] Peking: Agricultural Publishing House *In Chinese*
173. Kigatiira, K. I. (1974) Hive designs for beekeeping in Kenya. *Proc. ent. Soc. Ont.* 105: 118–28
174. Kigatiira, K. I. and Morse, R. A. (1979) The construction, dimensions and siting of log hives near Nairobi, Kenya. Pp. 53–8 from *Beekeeping in rural development. Unexploited beekeeping potential in the tropics: with particular reference to the Commonwealth*. London: Commonwealth Secretariat and International Bee Research Association
175. Kitchell, K. F. (1981) The Mallia 'wasp' pendant reconsidered. *Antiquity* 55: 9–15 + 1 pl.
175a Kither, G. Y. (1982) Personal communication
176. Knox, R. (1681) *An historical relation of Ceylon* . . . Glasgow: MacLehose
177. Kostecki, R. and Jelinski, M. (1979) Entwicklungsformen der polnischen Bienenzucht im Lichte des Freiland-Museums in Swarzedz. Pp. 94–9 from *Bienenmuseum und Geschichte der Bienenzucht*. Bucharest: Apimondia
178. Krünitz, J. G. (1774) *Das Wesentlichste der Bienen-Geschichte und Bienen-Zucht*. Berlin: Joachim Pauli
179. Kuény, G. (1950) Scènes apicoles dans l'ancienne Égypte. *J. Near E. Stud.* 9: 84–93
180. LaFleur, R. A., Matthews, R. W. and McCorkle, D. B. (1979) A re-examination of the Mallia insect pendant. *Am. J. Archaeol.* 83: 208–12
181. Landa, Diego de (1978) *Yucatán before and after the conquest*. New York: Dover Publications. Translation of 16th century Spanish text
182. Langstroth, L. L. (1853) *Langstroth on the hive and the honey-bee, a bee keeper's manual*. Northampton, MA, USA: Hopkins, Bridgman & Co.
182a ——— (1857) *A practical treatise on the hive and honey-bee*. 2nd ed. New York: C. M. Saxton & Co.
183. Larwood, J. and Hotten, J. C. (1866) *The history of signboards*. London: Chatto & Windus. 11th ed. 1900
183a Lavelle, D. (1976) *Skellig: an island outpost of Europe*. 2nd ed. Dublin: O'Brien Press
184. Lawson, W. (1618) *A new orchard and garden . . . with The country housewife's garden for hearbes of common use . . . as also the husbandry of bees, with their several uses and annoyances*. London: printed by B. Alsop for R. Jackson
185. Leek, F. F. (1975) Some evidence of bees and honey in Ancient Egypt. *Bee Wld* 56(4): 141–8
186. Legros, E. (1969) *Sur les types de ruches en Gaule romane et leurs noms*. Liège, Belgium: Musée de la Vie Wallonne
187. Lehmann, H. (1965/66) Ein dreitausendjähriger 'Klotzstülper' aus Berlin-Lichterfelde. Ein Beitrag zur Geschichte der Bienenhaltung. *Berl. Blätter Vor- und Frühgeschichte* 11: 45–98
188. ——— (1966) Ein Bienenstock aus Berlin-Lichterfelde mit Opfergaben der

Lausitzer Kultur. *Acts VII int. Congr. prehist. protohist. Sci., Prague* 1: 720–2
189. Lerner, F. (1963) *Aber die Biene nur findet die Süssigkeit*. Düsseldorf: Econ Verlag
190. Lesowski, L. (1977) Beeswax—out of the sand. *Daily Colonist* (23 Jan.): 8
191. Liger, L. and Torre, D. F. de la (1720) *Economia general de la casa del campo.* Madrid: Juan de Ariztia
192. Linnus, F. (1939) Eesti vanem mesindus. I. Metsamesindus. *Eesti Rahva Muuseumi Aastaraamat, Tartu* 12–13: 495 pp. In Estonian
193. Liversidge, J. (1968) *Britain in the Roman Empire.* London: Routledge & Kegan Paul
194. Loudon, J. C. (1822) *Encyclopedia of gardening.* London: Longman Hurst
194a ——— (1840) *The cottager's manual.* London: Baldwin and Cradock
195. Love, J. R. B. (1936) *Stone-age bushmen of today.* London and Glasgow: Blackie & Sons Ltd
196. Lyle, Sir Oliver (1960) *The Plaistow story.* London: Tate & Lyle Ltd
197. Maandschrift voor Bijenteelt (1958) Uit de Jaagkieps. *Maandschr. Bijent.* 60(12): 198–9
198. Marchenay, P. (1979) *L'Homme et l'abeille*. Paris: Berger-Levrault
199. Marshall, J. (1966) Pot hives. *Brit. Bee J.* 94(4128): 311–12
200. Mazak, K. S. (1975) Bać odrzańska ma około 2055 lat. [Hive from about 2055 years ago.] *Pszczelarstwo* (11): 18–21
201. Mazel, A. D. (1981) Personal communication
202. Mazurkiewicz, J. (1958) Zabytek prawa bartnego w Wierzchowiskach z ostatnich lat Rzeczypospolitej szlacheckiej. [A charter of the beekeeping law in Wierzchowiska from the last year of the elective monarchy.] *Czas. prawn.* 10(2): 291–302
202a Mellaart, J. (1963) Excavations at Çatal Hüyük, 1962. *Anatolian Stud.* 13: 43–103 + 27 pl.
203. Mellor, J. E. M. (1928) Beekeeping in Egypt: Part I. An account of the Beladi craft . . . *Bull. Soc. ent. Égypte* 12: 17–33 (printed 1929)
204. Meyers, A. F. (1977) Beeswax and politics in Morocco, 1697–1701. *Bee Wld* 58(4): 153–60
205. Michener, C. D. (1974) *The social behavior of the bees: a comparative study.* Cambridge, MA, USA: Belknap Press
206. Michener, C. D. and Michener, M. H. (1951) *American social insects: a book about bees, ants, wasps and termites.* New York, Toronto and London: D. Van Nostrand Co., Inc.
207. Mills, J. W. and Gillespie, M. (1969) *Studio bronze casting.* London: Maclaren
209. Molnàr, B. (1957) Méhkeresés, es mézzsakmányolás Domaházán. [Search for wild bees and honey hunting in Domaháza.] *Ethnographia* 68(3): 483–91
210. Moody, T. (1979) Ghana and Nepal: beeswax used in lost-wax brass-casting. Pp. 185–9 from *Beekeeping in rural development. Unexploited beekeeping potential in the tropics: with particular reference to the Commonwealth.* London: Commonwealth Secretariat and International Bee Research Association

Bibliography

211. Moore, Sir Jonas (1707) *England's interest or, The gentleman and farmer's friend, shewing . . . of the husbandry of bees, and the great benefit thereby.* 4th ed. London: J. How
212. Müller, A. von (1964) Die jungbronzezeitliche Siedlung von Berlin-Lichtefelde. *Berl. Beitr. Vor- und Frühgeschichte* 9
213. Münz Zentrum (1980) *Auktion XL. Westfälische Privatsammlung im Spiegel der Kulturgeschichte.* Cologne: Münz Zentrum
214. Münzberg, J. (1775) *Hinterlassene vollständige Lehre von der Bienenzucht.* Vienna: Joseph Münzberg
215. Muzeji Radovljiške Občine (1973) *Čebelarski muzej v Radovljici.* [Apicultural Museum in Radovljica.] Radovljici: Čebelarski Muzej
216. Nachtsheim, H. (1921) Naturröhren als Naturbienenwohnungen. *Arch. Bienenk.* 3(1/2): 81+1 pl.
217. Naile, F. (1976) *America's master of bee cultures: The life of L. L. Langstroth.* Ithaca, NY, USA: Cornell University Press. Reprint of 1942 ed., with new Foreword by R. A. Morse
218. Neighbour, A. (1865) *The apiary; or, Bees, bee-hives and bee-culture . . .* London: Kent & Neighbour; 3rd ed. 1878
219. Nicolaidis, N. J. (1959) [Beekeeping: modern intensive methods.] 3rd ed. Athens: published by the author   In Greek
220. Nivaille, J. (1978/79) Le type d'abeille dans le monnayage grec. *Bull. trimestr. Cercle d'Études numismatiques* 15(4): 61–9; 16(1): 1–8, (2): 21–8. All reprinted in *Gaz. apic.* (859): 205–15 (1979)
221. Nobbs, E. (1969) Make your own skep and revive a lost art. *Leafl. V.B.B.A.* No. 8
222. Nutt, T. (1832) *Humanity to honey bees.* Wisbech, Cambs.: Leach
223. O'Kelly, M. J. et al. (1950 or later, undated) *St. Gobnait of Ballyvourney.* Publisher unknown
224. Opie, I. and Opie, P. (1969) *Children's games in street and playground.* Oxford: Clarendon Press
225. Pager, H. (1971) *Ndedema.* Graz, Austria: Akademische Druck- u. Verlagsanstalt
226. ——— (1973) Rock paintings in Southern Africa showing bees and honey hunting. *Bee Wld* 54(2): 61–8
227. ——— (1974) The magico-religious importance of bees and honey for the rock painters and Bushmen of Southern Africa. *S. Afr. Bee J.* 46(6): 6–9
228. ——— (1975a) *Stone Age myth and magic as documented in the rock paintings of South Africa.* Graz, Austria: Akademische Druck- u. Verlagsanstalt
229. ——— (1975b) Honeycombs, ladders and bees in the rock art of Africa. Unpublished report to International Bee Research Association
230. ——— (1976) Cave paintings suggest honey hunting activities in Ice Age times. *Bee Wld* 57(1): 9–14
231. ——— (1981) Personal communication
232. Payne, J. H. (1838) *The apiarian's guide.* 2nd ed. London: Simpkin & Marshall
233. Petie, B. (1974) Bees or birds? *Rhod. Prehist.* 6(12): 2–3
234. Petrov, E. M. (1980) [The Bashkir tree-hole bee.] Ufa: Bashkirskoe

Knizhnoe Izdatel'stvo *In Russian*
235. Pettigrew, A. (1870) *The handy book of bees*. Edinburgh and London: Blackwood
236. Pinto, E. H. (1969) *Treen and other wooden bygones*. London: G. Bell & Sons
238. Potter, B. (1908) *The tale of Jemima Puddle-duck*. London: F. Warne & Co. Ltd
239. Pourasghar, D. (1982) Personal communication
240. Pruess, K. P. (1973) Postage stamps showing bees and hives. *Bee Wld* 54(2): 53–6
241. Purchas, S. (the younger) (1657) *A theatre of politicall flying-insects*. London: R.I. for Thomas Parkhurst
242. Pyle, R., Bentzien, M. and Opler, P. (1981) Insect conservation. *A. Rev. Ent.* 26: 233–58
243. Ransome, H. M. (1937) *The sacred bee in ancient times and folklore*. London: George Allen & Unwin Ltd
244. Rashad, S. E. and El-Sarrag, M. S. A. (1978) Beekeeping in Sudan. *Bee Wld* 59(3): 105–11
245. Rauhala, A. (1953) *Suomen mehiläishoidon varhaisvaiheet*. Helsinki
     In Finnish
246. Reilly, D. R. (1953) *Portrait waxes*. London: Batsford
247. Richards, O. W. (1953) *The social insects*. London: Macdonald
248. ——— (1974) The Cretan 'hornet' pendant. *Antiquity* 48: 222
249. ——— (1975) The Cretan 'hornet'. *Antiquity* 49: 212–13
250. Richardson, E. W. (1916) *A veteran naturalist, being the life and work of W. B. Tegetmeier*. London: Witherby & Co.
251. Ritterman, V. (1953) Stone hives in a Dalmatian island. *Bee Wld* 34(9): 177–9
252. Rivals, C. (1980) *L'Art et l'abeille: ruches décorées en Slovenie*. Paris: Les Provinciades
253. Robertson, R. and Gilbert, G. (1979) Bee boles. Pp. 5–7, 23, 37–8 from *Some aspects of the domestic archaeology of Cornwall*. Cornwall: Institute of Cornish Studies and Cornwall Committee for Rescue Archaeology
254. Robinson, F. E. (1977) Personal communication
255. Rodenberg (1926) Wozu gebrauchten die alten hannoverschen Heidimker das Imkerbeil? *Bienenw. Zbl.* 62(11): 308–9
256. Roys, R. L. (1972) *The Indian background of colonial Yucatán*. Norman, OK, USA: University of Oklahoma Press
257. Rupp, K. (1959) Albania—méhészete. [Beekeeping in Albania.] *Méhészet* 7(12): 229
258. Ruttner, F. (1977) Ein Bienenkorb von der Nordseeküste aus prähistorischer Zeit. *Allg. dt. Imkerztg* 11(9): 257–63
259. ——— (1979a) *Historische Entwicklung des Bienenstockes*. Bucharest: Apimondia
260. ——— (1979b) Minoische und Altgriechische Imkertechnik auf Kreta. Pp. 209–29 from *Bienenmuseum und Geschichte der Bienenzucht*. Bucharest: Apimondia
261. ——— (1980) The golden pendant of Malia, the Congress badge—prehistoric vestige and early evidence of beekeeping. *Proc. XXVII int.*

Bibliography

        Beekeep. Congr.: 191–6
262. Savage, G. (1968) *A concise history of bronzes*. London: Thames & Hudson
262a. Säve–Söderbergh, T. (1957) *Private tombs at Thebes*. London: Oxford University Press
263. Scheybalová, J. (1974) Perníkářství. [Gingerbread making.] *Muzejní a Vlastivědná Práce* 12(3): 146–72
264. Schier, B. (1939) *Der Bienenstand in Mitteleuropa*. Leipzig: S. Hirzel; 2nd ed. 1972
265. Schmidt, R. (1926) Imkerbeile. *Bienenw. Zbl.* 62(10): 268–77
266. Schwarz, H. F. (1948) Stingless bees (Meliponidae) of the Western Hemisphere. *Bull. Am. Mus. nat. Hist.* No. 90
268. Schwärzel, E. (1972) Das letzte Zeidelgericht zu Feucht. *Imkerfreund* 27(8): 255–60
269. ——— (1981) Das Zeidelwesen in Deutschland. *Imkerfreund* 36: 13–15, 41–3, 73–4, 125–8, 205–7, 275–7, 311–12, 347–50, 393–6, 413–16, 459–71
270. Seeley, T. (1977) Measurement of nest cavity volume by the honey bee (*Apis mellifera*). *Behavl Ecol. Sociobiol.* 2(2): 201–27
271. Seeley, T. D. and Morse, R. A. (1977) Dispersal behavior of honey bee swarms. *Psyche* 84(3/4): 199–209
272. Segschneider, E. H. (1977) *Imkerei des Osnabrücker Landes*. Osnabrück: German Federal Republic: Kulturgeschichtliches Museum
273. Seyffert, C. (1930) *Biene und Honig im Volksleben der Afrikaner*. Leipzig: Voigtländer
274. Silberrad, R. E. M. (1976) *Bee-keeping in Zambia*. Bucharest: Apimondia
275. Simpson, J. (1969) The amounts of hive-space needed by colonies of European *Apis mellifera*. *J. apic. Res.* 8(1): 3–8
276. Slater, L. G. (1969) *Hunting the wild honey bee*. Olympia, WA, USA: Terry Publishing Co.
277. Smith, D. A. (ed.) (1966) *John Evelyn's manuscript on bees from Elysium Britannicum*. London: Bee Research Association
278. Smith, F. G. (1960) *Beekeeping in the tropics*. London: Longmans
279. Sokolov, V. (1974) [Heraldry and beekeeping.] *Pchelovodstvo* 94(9): 44–5 *In Russian*
280. Sooder, M. (1952) *Bienen und Bienenhalten in der Schweiz*. Basel: G. Krebs
281. Southerne, E. (1593) *A treatise concerning the right use and ordering of bees*. London: Thomas Orwin for Thomas Woodcocke
282. Spiller, J. (1949) A straw skep carved on a church corbel. *Bee Wld* 30(2): 10
283. Staniforth, A. (1981) *Straw and straw craftsmen*. Princes Risborough, Bucks.: Shire Publications Ltd
283a. Stevens, K. (1969) Beekeeping in the Maltese islands. *Glean. Bee Cult.* 97(2): 102–8, 119
284. Stevenson, J. (1978) *The catacombs*. London: Thames & Hudson
284a. Strickland, S. S. (1982) Honey hunting by the Gurungs of Nepal. *Bee Wld* 63(4): 153–61
285. Struthers, J. (1951) The Stewarton hive: a monument of craftsmanship. *Scott. Beekpr* 27(12): 239–41
286. Sumner, H. (1910) *The book of Gorley*. Southampton: H. M. Gilbert

287. ———— (1917) *The ancient earthworks of the New Forest*. London: Chiswick Press
288. ———— (1931) The 'bee-garden', Holt Heath. Pp. 32–5 from *Local papers, archaeological and topographical. Hampshire, Dorset and Wiltshire.* London: Chiswick Press
289. Tanner, L. E. and Nevinson, J. L. (1936) On some later funeral effigies in Westminster Abbey. *Archaeologia* 85: 169–202
290. Taylor, C. N. (1936) *Odyssey of the islands*. New York: C. Scribner's Sons
291. Taylor, H. (1838) *The bee-keeper's manual* . . . London: Groombridge
292. Tegetmeier, W. B. [c. 1860] *Bees, hives and honey*. London: W. Allan
293. Thomaides, X. (1979) Beekeeping in Ancient Greece. *Apiacta* 14(3): 97–108
294. Toth, G. (1981) The hivestones of Hungary. *Am. Bee J.* 121(3): 202–3
295. Townley, M. B. (1964) Saint Ambrose. *Bee Wld* 45(4): 162–3
296. Tozzer, A. M. (1941) Landa's Relación de las cosas de Yucatán. *Pap. Peabody Mus.* 18: references to bees on pp. 156, 193, 194, 197
297. Tozzer, A. M. and Allen, G. M. (1910) Animal figures in the Maya Codices. *Pap. Peabody Mus.* 4(3): references to bees on pp. 298–301 and pl. 2
298. Trevise, P. (1973) *Last days of a wilderness*. London: Collins
299. Turnbull, C. M. (1961) *The forest people*. London: Jonathan Cape
300. Turnbull, W. H. (1958) *One hundred years of beekeeping in British Columbia 1858–1958*. Vernon, Canada: BC Honey Producers' Association
301. Tusser, T. (1557) *A hundreth good pointes of husbandrie*. London: R. Tottel
302. Tyson, B. (1982) Architecture of Lakeland beekeeping. *Country Life, Lond.*: 171(4409): 408–9
303. Vellard, J. (1939) *Une civilisation du miel: les Indiens guayakis du Paraguay*. Paris: Librairie Gallimard
304. Verhagen, R. (1974/75) Un aspect de l'apiculture du passé. *Gaz. apic.* 75(5): 106–10; 76(3): 56–8
305. Vernon, F. G. (1974) Collection of historical and contemporary beekeeping material. Chapter 11 (pp. 103–11) from *Bee Research Association, 1949–1974: a history of the first 25 years*. London: Bee Research Association
306. ———— (1979) Beekeeping in 1269–1270 at Beaulieu Abbey in England. *Bee Wld* 60(4): 170–5
307. ———— (1981) *Hogs at the honey pot*. Steventon, Basingstoke: Bee Books New & Old
308. Vrána, J. (1976) Historie včelařství a Bělověžská Pušča. [The history of apiculture and Bialowiesa Forest [Poland].] *Ziva* 24(3): 102–3
309. Watson, J. K. (1981) *Bee-keeping in Ireland: a history*. Dublin: Glendale Press
310. Weaver, E. C. and Weaver, N. (1981) Beekeeping with the stingless bee (*Melipona beecheii*) by the Yucatecan Maya. *Bee Wld* 62(1): 7–19
311. Weir, S. (1973) *The Gonds of central India*. London: Trustees of the British Museum
312. Weiske, W. (1971) Etwas über Zeidelweiden, über Waldbienenzucht, umstrittene Gebiete und Gerichtsbarkeit der Zeidlergessellschaften. *Imkerfreund* 26(9): 275–7
313. Wellmann, K. F. (1979) *A survey of North American Indian rock art*. Graz, Austria: Akademische Druck- u. Verlagsanstalt

Bibliography

314. Wheler, Sir George (1682) *A journey into Greece*. London: T. Cademan
315. Whiston, J. W. (1963) Bee boles at West Bromwich manor-house. *Trans. Lichfield archaeol. hist. Soc.* 4: 47–51
316. ———— (1969) Bee-boles at Olveston Court, Glos. *Trans. Bristol & Glos. archaeol. Soc.* 87: 144–8 + 1 pl.
317. ———— (1971/72) Bee-boles at Pipe Ridware Hall Farm, Staffs. *Trans. S. Staffs. archaeol. hist. Soc.* 13: 43–5 + 1 pl.
318. Wildman, T. (1768) *A treatise on the management of bees*. London: Cadell for the author
319. Williams, H. (1950) *Ceylon, pearl of the east*. London: Robert Hale
320. Wordsworth, D. (1971) *Journals*. Ed. M. Moorman, London: Oxford University Press
321. Zander, E. (1923) *Handbuch der Bienenkunde. V. Die Zucht der Biene*. 2nd ed. Stuttgart: Eugen Ulmer
322. Zander, E. (1941) *Beiträge zur Herkunftsbestimmung bei Honig. III. Pollengestaltung und Herkunftsbestimmung bei Blütenhonig* (Ergänzungsband). Leipzig: Liedloff, Loth & Michaelis
323. Zoller, D. (1972) Die Ergebnisse der Graburg Gristede. *Neue Ausgrabungen und Forschungen in Niedersachsen* 7
324. Zymbragoudakis, C. (1979) The bee and beekeeping in Crete. *Apiacta* 14(3): 134–8

Index

Pages indicated in bold type include an illustration. Names of historical personages are indexed, but contemporary authors are listed only in the Bibliography. Most place names are indexed by the name of the country. The many place names in England are, however, indexed by county, and there are separate entries for Wales, Scotland and Ireland.

Afghanistan, 61, **194–5**
Africa, 8, 12, 16, 18, 23–7, 53–69, **59**, 72, 246
—, North, 53, 65, 76, 194
  see also individual countries
Albania, 192
alcoves, *see* bee alcoves
Alexander the Great, 240
Algeria, 25, 65, 68
Alps, 41, **65**, 91, 155, 216
Ambrose, St, 214–18, 220
America, 8, 18, 33–4, 56, 63, 246, 336–7
—, Central, 33, 57, 60–4, 72, 246
—, North, 33, 94, 115, 172
  see also individual countries
animal hides used for hive, **70**
apiaries:
  enclosures, 73, **74–5**, 88–90
  origin, 18, 35, 79, 85–6, 88, 91
  walled, in Spain, 72–3, **74–6**
Apis, see under bee/bees (honeybees)
Arabia, 48, **60**
Argentina, 336
Aristaeus, **214**, 224
Aristomachus of Soli, 51
Aristotle, 200
arms, *see* heraldry
Ashanti gold weights, 246
Asia, 8, 11, 16, 18, 27–30, 56, 67, 72, 91, 115, 244, 337
  see also individual countries
aspect (direction faced):
  bee boles, 134–7, 150, 156
  hives, 75, 113, 134–7
  nests of wild bees, 136–7
Australia, 16, 18, 32, 226, 242, **243**, 244
Austria, 167, 187, 189, **222**, 242, 329
Avon, 128, 150, 234, 252
axe, beekeeper's, 28, **86**
Azores, **112**

back-opening (upright) hives, 91, 97, **98**, 189–90, 208, **209**
Bagster, Samuel, 119
Bali, 57, 69, **70**
Balkans, 91, 192
bamboo leaves, hives of, 58
banana leaves, hives using, **57**, 69
Bannkorb, 97, **110**
Barberini, Maffeo, 218–19, 226
bark hives, **64**, 66, 69, 95–7
  see also cork hives
Bashkir, 79
basket hives, 11, 18, **49**, 54, 66, 91, 99
  see also skep/skeps
batik, 244
bears, **78**, 214, **217**
Bedfordshire, 173, 236, 314, 318
bee/bees (honeybees):
  Apidae/Apinae, 18n
  Apini, 18n
  Apis species, 18
  — —, distribution, 13, **14–15**
  Apis cerana, 28, 50, 69, **70**, **94**, 113, 194–5
  Apis dorsata, 28, **29–31**
  Apis florea, **157**, 158
  Apis mellifera, 13, 39, 69
  Bombinae, 18n
  distribution, 13, **14–15**
  'killer', 16
  killing to take honey, 18, 91, 105–6
  life history, 10, 13–18
  nests, *see* nests, honeybee
  queen, *see* queen (honeybee)
  representations of, 213–38
  as symbol/emblem, 9, 18, **43**, 218, 220, **226**, 227, 230, 232–8, **233**, **237**
bee/bees (others):
  Apidae/Apinae, 18n
  distribution, 13, **14–15**

Heriades, 137
stingless, *see* Meliponini
bee alcoves, 152, 160, 163–7, **164—5**, 310–13
bee beds, 88–9
bee boles:
  general, **117–59**, 160–2, 164n, 167, 171–2, 178, **248–9**, 250–309
  accessible to public, 121–3, 151
  aspect (direction faced), 134–7, 150, 156
  condition/preservation, 129–31
  cross-bars, 119, 125, 144, 148, **149**
  date of construction, 132–3
  destruction, 153
  distance apart, 147
  distribution, 152–5, **248–9**
  doors to, 149–50
  in garden walls, *see* garden, site for bees
  height above ground, 144–7
  location on property, 119–20, 124–7
  locking devices, 119, 125, 144, 148, **149**
  names for, 132
  number per property, 127–8, 156
  in orchard walls, *see* orchard, site for bees
  origin, 160–1
  packing used in, 143–4, 149, 176
  rebated, **149**
  records/register, 118, 121, 162, 250–309
  shape, 139–44, 157
  size, 139–44, 149, 156
  tiered, **130–1**, 145, **146**, 156
  types of walling with, *see* walls containing bee boles
  winter, 176

353

Index

bee butts, 180, 182
bee cellars, 163, 172
bee ceremony, **62**, 63
bee gardens, 88–90
　see also garden, site for bees
bee garth/gardd, 132, 165
bee goddess, **229**, 232
bee gods, **63**
bee houses, 53, 75–6, 97–8, 118, 127, 152, 163, **166–72**, 182, **248–9**, 314–17
　see also winter: bee storage
— —, German type, **187**, 188, **189–91**
— —, —, distribution, **190**
bee hunting, *see* honey hunting
bee markets, **104**
bee shelters, **103**, 117, 127, 144, 152, **163**, 167, 176, **177–81**, 182, **185–6**, 187, **192–3**, **206**, 217, **224**, 225–6, **227**, **242**, **248–9**, 321–7
bee space, *see* comb spacing
bee storage, winter, 127, 171–6, **173–5**, 318–20
bee towers, 184–5
bee walks, 82–3
bee walls, 72–3, **74–6**, 132, 160
　see also entries under walls
Beehive Dreaming, 32
Beehive Inns, 232–5
beekeepers' protective clothing, 10, 28–9, 33, **111**, **213**
beekeeping, origin, 35–6, 51, 56–7, 79, 86–7, 99, 115–16
　see also more specific beekeeping entries
beeswax, 45, 62–3, 81, 85, 93, 124, 240–6
—, harvest from hive, 19, 78–9, 82
—, uses, 18, 42, 53, 63, 82, 106, 212, 240–2, **243–5**, 246, 335
Belgium, **66**, 67, **68**, **100**, 101, 239, 330
bell glasses/jars, **108**, 109, 202, **207**
Berkshire, **117**, 121, 128, 170, **171**, 220, 234, 252, 314, 335
Berlepsch, Baron August von, 190, 209
Bevan, Dr Edward, 142, 168, 177, **211**
Bible, 21, 53, 224–5
boat, hives moved by, 42, 220
bombard (hive), 57
Bombinae, 18n
Bonner, James, 128, 143
bortnik, 78–83
bottle-shaped hives, 72
Brać, **112**, 113, 192, 197
Brachystegia, 64
Brazil, 16, **69**
Breughel, Pieter, 224
brick hives, 52, 54, 71, 238

brick walls, *see under* walls containing bee boles
Brigitta, St, 224
Britain, 8, 17, 101, 107
—, bee boles, 117–55, 160–6, **248–9**, 252–306
—, bee houses, etc., 166–85, **248–9**, 311–28
—, museums, 335–6
　see also England, Scotland, Wales
Bronze Age finds, 93, 213, 238, 241
Buckinghamshire, **124**, 142, 215, 236, 252, 321, 335
bumble bees (*Bombus*), 18n
burial in honey, 240
Burma, 113
Butler, Rev. Charles, 126–7, 142, 149, 220

cabinet-maker's hives, **202–9**
Cambridgeshire, 218, 234–5
Cameroon, **66**, 67
Canada, 33, 336
candles, 53, 63, 241–2
cane hives, 54, **55–6**, **67**, 76
cannon hive, **48**, 57, 72
cap for skep, **108**, 109, 116, **125**, 202
Carniola, **52–3**, 65, **186**, 187, 189, 336
Carpathians, 23, 79
catacombs, 224–5
Caucasus, **58**, 59, 91
Cellini, Benvenuto, 245
ceramics, **225**
Cesi, Prince, 219
Chad, 66
Chaga tribe, 64
chalk walls, **151**
charter boles, **159**
Cheremessia, **78**, 79–80, **81**, **83**
Cheshire, **164**, 165, 234, 236, 253, 310
chestnut, sweet (*Castanea*), 95
Childeric, King, 230
China, 54, 57, 59, 67, 245
church properties, bee boles etc., in, **124–5**, 133, **135**, 184, 190–1
clay hive extensions, **46**, **49**, 72
— hives, 17, 35, 45–51, **46–8**, 54, **57**, 58, 67n, **71–3**, 75 **112–13**, 114, 182, 194, 197, **198**, 200–1, 229
—, used in hive construction, 65, 76, 101, 197
climate, effects of, 40, 115–16, 153–6
clooming hives, 52, **66**, 67, 76, **100–1**, 117, 149, 197
clunch walls, 166
cob walls, **133**, 137, 138, **140**, **145**, 166

coconut shells for hives, 69, **70**
coiled-work used for hives, 58, 66, **68**, 99, 210, **211**
　see also skep/skeps: straw, coiled
coins, 231–2, **233**
collateral hives, 168, **169**, 182, **183**, **205–7**, 210
collection of honey, *see* honey, removal from hive; honey hunting
Colombia, 63, 246
colony, honeybee:
　dividing, **157**, 197, 200
　natural history, 16
Columella, 21, 52, 54, 56, 71, 76, 135, 148
comb spacing, 13, 41, 48, 197, 205, 207–10, **211–12**
'combing' on inner hive surface, 47–8
concrete hives, 113
conical hives, **76**, **100**, 101, 214, **215**
cork hives, 52–3, 64–5, 95, **96**, 107, 113
cork oak (*Quercus suber*), 53, 65
Cornwall, 119–54, 173–6, **175**, 234, 253–5, 318
Corsica, 54, **65**, 227
Cosimo, Piero di, 221
Cotton, W. C., 136
Cowan, T. W., 136, 149
Cranach, Lucas, 221, **222**
Crete, 17, **48**, 72, 115, 197, **198**, 210
—, Ancient, 8, 45, 48, **57**, 200–2, **228**, 229, 232
cross-bars to bee boles, 119, 125, 144, 148, **149**
crownpiece, **100**
Cumbria, 119–62, **144**, 176–82, **178**, **180**, 234, 255–62, 310, 318, 321–3
cylindrical hives, 17, 39, **55**, 57–63, **64**, 66–8, **72**, 73, **74**, 75, **112**, 113, 210
Cyprus, **72**, 191
Czechoslovakia, 79, 84, 189, 240, 330

dances, honey-hunting, 26
David, St, 220
Denmark, **104**, 189, 221, **222**, 241, 330
Derbyshire, 121, 132, 150, 234, 262–3
Devon, 119–55, **133**, **140**, **145**, 173, 177, 181–2, 263–8, 310, 318, 323
Domesday Book, 87–8
donkeys, 42, 51, 107, 221
doors in upright hives, 97–8, **208**
— to bee boles, 149–50

Index

— —, bee nests in trees, **77–80**, 81–4, 87, **89**, **97**, 116
dormouse fattener, **114**, 115
Dorset, 90, **108**, 133, **134**, 220, 234, 269
dovecotes, *see* pigeons, nests for
driving bees, **106**, 107, **111**
dung used for hives, 51, 52, 54, 58, 67, 76, 101, 117, 194
Dürer, Albrecht, 221
Durham, 165, 234, 310
Dzierzon, Dr J., 97, 189, 209

earth (general), hives of, 71, 111
earthenware hives, *see* hives, by material: clay
eclectic hives, 168, **205**
Egypt, 17, 36–43, **40**, 60, 63, 71–2, 239, 242
— —, Ancient, 8, 10, **34–8**, 39–42 **43**, 51, 116, 200, 210, 218, 238–9, 242, 244–5
eke, 107, **108**
Elba, 227, 236
emblem, *see* bee/bees: as symbol/emblem; symbolism
encaustic paintings, 244
England:
  alcoves, 163, **164**, 165–7, **248–9**, 310–13
  bee boles, 118–62, **117–51**, **248–9**, 252–89
  bee houses, **166**, 167–8, **169–72**, **248–9**, 314–17
  bee shelters, 176, **177–80**, 181–2, **248–9**, 321–5
  forest beekeeping, 87–8, **89**, 90
  hives, 91, **203**, **205–7**, 209–10, **211**
  museums, 335
  skeps, **101–10**, 111, 117, **122**, 141–4, 147, 150, 155, 164–8, 173–4, **175–80**, 181–3
  winter storage, 171–6, **173**, **175**, **248–9**, 318–19
  see also individual counties
Essex, **131**, 151, 155, 158, 181–2, 234, 269, 323
Estonia, 79
Ethiopia, 51, 52, **56**, 66
Europe, 8, 16, 19, 21, **92**, 110, **190**, 210, 217, 228, 242
— —, northern, 10–11, 23, 56, 94 115, 117, 189
— —, western, 56, 117–18, 214
  see also individual countries
Evelyn, John, 136, 202
extensions to horizontal hives, **46**, 48, **49**, 72
Exultet Rolls, **44**, 53, **54**, 65, 115n, 213

fennel/ferula used for hives, 17, 52, 54, **55**, 187
Finan Cam, St, 220
Finland, 11, 79, 330
Finno-Ugrian people, 79–82
fixed-comb hives, management of, 41, 196, 200
Flanders, 215, 224
Floris, Frans, 224
flower parts used for hives, 69, 111
forest beekeeping, 11, **77–89**, 90, **92**, 94, 189, 208
'formlings', 24–5
frame hives, 207, **208–9**, 210
  see also movable-frame hives
France:
  bee boles and other recesses, 120, 123, **154**, 155, **156–7**, 180, 190–1, 308–9
  beeswax, 244
  hives/skeps, **95**, 101, 110, **214–15**, 217
  monks/nuns, 180, 220
  museums, 330–1
  representations of bees and hives, **214–15**, 227, 230, 232, 236, **237**
  sculptures and models, **214–15**, 242

Gambia, **68**
garden, site for bees, **118–19**, 120, **124**, 125–7, **129**, 133, **146**, 147, **151**, 156, 159, **166**, 167–8, **177–8**, 182, 184, 194
gardens, bee, 88–90
geese, recesses for, 158, 160
gems, 231–2
German language (and bee houses), 187–90
Germany:
  bee houses/shelters, **103**, 171, **184–7**, 188, **189–91**
  beeswax, 243–4
  forest beekeeping, 11, 79–80, 84, **85–6**, 92
  gingerbread moulds, 239–40
  hives, 91–2, **93**, 97–9, 209
  museums, 331–2
  representations of bees and hives, **213**, 215–16, **217**, 221, **222–7**, 232, **233**, 236
  skeps, **99**, **100**, **103**, 110, 183, **184–5**, 215–16, **217**, 224, **227**, **233**
Ghana/Gold Coast, 246
gingerbread, 238–40
gipsies' skeps, 99, **108**, **110**
'girth' for skep-making, 102–3
glass, stained, 219, **220**, 221
Gloucestershire, 119–32, 178–80, **179**, 182, 202, 234, 236, 269–70, 323, 328

Gobnet, St, 216, 220
gods, bee, **63**, **229**, 232
golden ornaments, 18, 63, **228**, 229, **230**, 246
Golding, Robert, 142, 210–11
Gonds, 29
gourd hives, **69**
grass hives, 58
Greece, 45, **46**, 120, **157**, 193, **196**, 197, **198–9**, 200, **201**, 209–10, 309
— —, Ancient, 8, 41, 45, **46**, 47–51, 197, 200–2, 216, 229, **230**, 231–2, **233**, 240
  see also Crete
Grünewald, Mathias, 221, 226
Guinea, 66
Guinea-Bissau, 58
Gurungs, **30–1**

hackles, **101**, **104**, 105, 117, **199**, 200
Hampshire, 87–8, **89**, 90, **106**, **110**, 114, **122**, 123–60, **146**, 220, 231, 240, 270
hawks, recesses for, 158
hazel (*Corylus*), 79
heat, effect on bees, 16, 40, 116
heated hives, 185, 194–5
heated walls, 159, 195
heather (*Calluna vulgaris*), 88, 103, 153, 183, 238
hens, recesses for, 158
heraldry/arms, 217, 235–6, **237**, 238
Hereford & Worcester, 87, **101**, **106**, 128, **129**, 155, 167–8, **169**, 171, **178**, **205**, 216, 234, 271, 314–15, 324
Heriades truncorum, 137
Hertfordshire, 234, 271
hieroglyphs, Egyptian, **43**, 218
Himalayas, 61, 91, 194
hive parts/accessories:
  extension (clay), **46**, 49, 72
  extension rings, **46**, 48, **49**, 107
  honey chamber, *see* honey chamber of hive
  see also under skep/skeps
hives, by beekeeping feature or operation:
  back-opening (upright), 91, 97, **98**, 189–90, 208, **209**
  collateral, 168, **169**. 182, 183, **205–7**, 210
  'combing' on, 47–8
  doors in (trees), **77–80**, 81–4, 87, **89**, **97**, 116
  — — (upright hives), 97–8, 208
  eclectic, 168, **205**
  fixed-comb, management of, 41, 196, 200

355

Index

frame, 207, **208–9**, 210
heated, 185, 194–5
insulation of, 52, 64–5, 95, **198**
leaf, 207, **208**, 210
movable-comb, 48, 92,
   196–9, 200–2, 210, **211–12**
movable-frame, 9, 17, 91–2,
   111, 127, 167, 189, 196,
   202, 210, **211–12**
movable-nest, 202
moving (migrating), 42, 51,
   55, 89, 127, 183, 220
observation/transparent, 52,
   54, **208**, 210
ownership marks, 78–9, 81,
   83
protection of, 57, 67, **105**,
   111, 113, 117, 119, 125–6,
   135–6, 144, 148, **149**, **151**,
   153–5
tiered boxes, 188, 202–5,
   203, 210
top-bar, **157**, **196**, 197,
   198–9, 200–1, 209–10,
   211–12
ventilation of, 16, **71**, 188
hives, by country, *included
   under* country entries
hives, by material:
   animal skins (hides), **70**
   bamboo leaves, 58
   bark, **64**, 66, 69, 95–7
   brick, 52, 54, 71, 238
   cane, woven, 54, **55–6**, **67**, 76
   clay/earthenware/pottery/
      terracotta, 17, 35, 45–51,
      46–8, 54, **57**, 58, 67n,
      71–3, **75**, **112–13**, 114, 182,
      194, 197, **198**, 200–1, 229
   coconut shells, 69, **70**
   concrete, 113
   cork, 52–3, 64–5, 95, **96**, 107,
      113
   earth general, 71, 111
   fennel/ferula, 17, 52, 54, **55**,
      187
   flower parts, 69, 111
   gourd, **69**
   grass, 58
   leaves, **57**, 58, **69**, 111
   log, 17, 52–67, **57–62**, **78**, 79,
      85, 92, **93–5**, 96, **97–8**, 208
   mud, 17, **35–8**, 39, **40**, 41–3,
      49, **51**, 54, 71–2, 111, 194,
      210
   nutshells, 69, **70**
   plant stems, **55**, 66–8, 111
   reeds, 67, 110–11, **112**, 113
   —, woven, 67, 194
   stone, 71, **112**, 113, **192**, 197
   straw, **68**, 210, **211**
      see also skep/skeps: straw
   tuff, **192**
   water and other pots, 58, **71**,
      72, **73**, 91, 194

wicker (woven), **27**, 52, 54,
   55, **66**, 67, **157**, 194
   196, **198–9**, 200, 210
   see also skep/skeps:
      wicker
wooden boards, **44**, **52–4**, **65**,
   94, 186
hives, by method of
   construction:
   basket, 11, 18, **49**, 54, 66, 91,
      99
      see also skep/skeps
   cabinet-maker's, 202–9
   clay used, 65, 76, 101, 197
      see also hives, by material:
         clay
   clay used for making
      extension, 46, 49, 72
   clooming, 52, **66**, 67, 76,
      100–1, 117, 149, 197
   coiled-work, 58, 66, **68**, 99,
      210, **211**
   dung used, **51**, 52, 54, 58, 67,
      76, 101, 117, 194
   fronts painted, **52–3**, 65
   mud used, 52, 66–7, **195**
      see also hives, by material:
         mud
   ornamented, **52–3**, 60, 65, **71**,
      73, 97, **98**, **101**, 110
   plaited-work, 66, **68**
   woven, **27**, **49**, 52, 54, **55–6**,
      66–7, 76, **99–101**, 117,
      157, 194, 196, **198–9**, 200,
      210
hives by shape:
   general, 39–40, 47–8, **56**,
      60, 67, 101, **211**, 219
   bottle-shaped, **72**
   conical, **76**, **100**, 101, 214,
      215
   cylindrical, 17, 39, **55**,
      57–63, **64**, **66–8**, **72**, 73, **74**,
      75, **112**, 113, 210
hives, measurements of, 17, 54,
   60, 64, 67, 75–6, 93, 142, 192,
   198, **211**
   see also hives, size (volume)
hives, named:
   bombard, 57
   cannon, **48**, 57, 72
   Huber, 208, 210–11
   Langstroth, 210, **211**, 212
   leaf, 207, **208**, 210
   Mew's, **203**
   Munn, 211
   Nutt, 182, **183**, **206–7**, 328
   pipe, 41, 51
   Stewarton, **204**, 205
   tunnel, 65
   Vicat, 207
hives, origin of, 35–6, 51, 56–7,
   99, 115–16
   see also beekeeping, origin

hives, siting of:
   general, 62–3, 69, **74–6**, 113
      see also entries under bee
         boles
   aspect (direction faced), 75,
      113, 134–7
   density (number per unit
      area), 155
   disposition:
      horizontal, 37–43, **44–76**,
         111, 115–16, 193, 210
      horno, 73–6
      jacente, 73–6, **75**
      peon, 73, **75**
      upright (vertical), 53, 56,
         66, 68, 73, **75**, 79, 91–2,
         93–113, 114–16, 208
   distance apart, 147–8
   in gardens, *see* garden, site
      for bees
   height above ground, 144–7
   in house walls, **50**, 61, 76,
      191, **194–5**
   hung from or fixed in tree,
      60–1, **64**, **68–70**, **78**, 83,
      85–6
   in loess cliff, 193
   in orchard, *see* orchard, site for
      bees
   tiers, **40**, 54, **72**, 74, 75–6, 115n
hives, size (volume) of, 16–18,
   39–40, 55, 58, 142–3
   see also hives, measurements
hives, stingless bees in, 35, **62**,
   63–4, **69**, **73**
hives, wrong identification of,
   114, 115, **201**
Honduras, 62
honey, ancient, 238–40
—, burial in, 240
—, collection from wild nests,
   see honey hunting
—, pressed, 106
—, removal from hive, **35–8**,
   41, **44**, 51, 60, 62, 67n, 68,
   69, 75–6, 91, 97, 105–7, 109,
   116, 186, 195, 197, 200, 205,
   207
—, squeezed, 106
—, storage in hive/supers, 13,
   16, 19, 115–16, 202–8
—, symbol of sweetness, 216,
   225
—, treatment after removal
   from hive, 36–7, 106, 109
—, from wasps, 33, 63
—, world production of, 212
honey badger, 61
honey chamber of hive, 48, 58,
   107, **108**, 109, **125**, 202,
   204–5, **209**, 210
honey containers, 21–8, **22**, **27**,
   29–31, **35–8**, 41, 45, 66, 84,
   238, **239**
honey feast cakes, 42, 239

356

Index

honey hunting:
 general, **frontispiece**, 8,
  19–34, **20**, **22–3**, 25, 27,
  29–31, 61–3, 72, 77, 87
 climbing rope, 26, **27**, 28,
  29–31, **78**, 79–80, **81**–3
 containers, 21–8, **22**, 27,
  29–31
 dances, 26
 forbidden, 30
 ladders, *see* ladders for
  honey collecting
 songs, 26
 trap boxes, 34
honey pots, *see* honey
 containers
honey yields, 81–2, 87
honeybee, *see* bee/bees
 (honeybee)
— colony, *see* colony,
 honeybee
'honey-tongued', 216
Hong Kong, **67**
Hooke, Robert, 202
horizontal hives, 37–43, **44–76**,
 111, 115–16, 193, 210
hornets/wasps, 33, 109, 229
horno (hive), 73–6
house walls, hives in, **50**, 61,
 76, 191, **194–5**
houses for hives, *see* bee houses
Huber, François, 187, 208,
 210–11
'humanity to honeybees', 91,
 207
Humberside, 167, 315
Hungary, 21, 79, 189, 192, 207,
 229, 240, 332

IBRA, 12, 118, 162, 215, 238,
 335
Ice Age, 21
illuminated manuscripts, 10,
 44, 53, **54**, **100**, 115n, 213,
 221
India, 27, 29, **31**, 35, 113, 337
—, Kashmir, **37**, 45, **47**, **49–50**,
 54, **194–5**
Indonesia, **29**
—, Bali, 57, 69, **70**
—, Sumatra, 29
Indus valley, 245
industry/thrift, exemplified by
 bees/skep, 18, 102, 218,
 227–8, 232, 235
Ingres, J. A. D., 227
inn signs, 232–5
insulating properties of hives,
 52, 64–5, 95, **198**
International Bee Research
 Association, *see* IBRA
Iran, 51
Iraq, 76
Ireland, 8, 87n, 118–53, **129**,
 139, **161**, 162, 172–4, 184–5,
 216, 220, 234–5, 239,
 248–9, 306–8, 320
Iron Age finds, 114, 202
Isle of Man, 131, 152, 158,
 171, **173**, 271–2, 319
Israel, **71**, 72, 337
Italy, **44**, 53, **54–5**, **115**, **192**,
 214, 216, 218, **219**, 220–1,
 224, 226, 241–2, **245**, 333
 see also Rome, Ancient
Ituri Forest, 26

jacente (hive), 73–6, **75**
Janscha, Anton, 187–8
Japan, 94, 337
jewellery, **228**, 229, **230**, 231–2,
 246
Jordan, **71**
Julbernardia, **64**

Karelia, 11, 79
Kashmir, **37**, 45, **47**, **49–50**, 54,
 194–5
Kent, 119–55, **135**, **141**, 165,
 210, 234, 239, 272–4, 310–11,
 315
Kenya, 12, 17, **27**, **60–1**, 212
Ker (Kerr), Robin, 205
Keys, John, 126, 136, 142, 177
killing bees to take honey, 18,
 91, 105–6
kingship, bee as symbol, **43**,
 218, 227
Korea, 11, 91, **94**

ladders for honey collecting,
 20, 21–9, **22–3**, **30–1**, 84, 145,
 182
Lancashire, 131, 152, 154–5,
 178, 181, 234, 236, 274–5,
 311, 324
Landa, Bishop Diego de, 62
Langstroth hive, 210, **211**, 212
Langstroth, Rev. L. L., 210–12
latitude, effects of, 115
Lawson, William, 107, 119,
 126, 172
leaf hive, 207, **208**, 210
leaves used for hives, **57**, 58,
 69, 111
Lebanon, **55**
Leicestershire, 122, 234, 243,
 275, 311
Libya, 25
Liebig, Curt, 225
limes (*Tilia*), 79–80, 83, 109,
 238
Lincolnshire, 122, 125, 142, 234,
 275
lion and bees, 21, 225, 232
Lithuania, 79
loess cliff, hives in, 193
log hives, 17, 52–67, **57–62**, **78**,
 79, 85, 92, **93–5**, 96, **97–8**, 208

London/Greater London,
 122–59, 168, **169**, 217, 220–1,
 230–5, **231**, 241–4, **242**, 246,
 276, 315
Loudon, J. C., 119, 148, 165
Louis XII, King, 227
Luxembourg, **100**, 236, **237**

Maillet, de, 42
Malawi, 25, 69
Mali, **27**, **66**
Malta, 39, 48, **71–2**, 115, 194
Manchester, Greater, 234, 236,
 276
manuscripts, *see* illuminated
 manuscripts
Marot, Jean, 227
Mauritius/Rodrigues/Agalega,
 57
Maya peoples, 62–3
mazers, 239
mead, 81, 105–6, 238–9
measurements of hives, *see*
 hives, measurements of
medals, 231–2, **233**
Medoc, St, 220
Meliponini (stingless bees),
 14–15, 18, 32–3, 35, 57, 61,
 62, 63–4, **69**, 72, **73**, 242, 246
Mellivora capensis, 61
Menorca, **67**
Merseyside, 234, 236
Mew, Rev. William, 202
Mew's hive, **203**
Mexico (Yucatán), **62–3**, 64,
 73
Michelangelo, 245
migration by colony, 116
— of hives, 51, 89, 127, 183
Mimusops Schimperi, 42
monkeys, 19
Moore, Sir Jonas, 145, 148
Mormon skep symbol, 218, 227
Morocco, 25, 48, **49**, **55**, 65,
 193, **195**, 244
movable-comb hives, 48, 92,
 196–9, 200–2, 210, **211**–12
movable-frame hives, 9, 17,
 91–2, 111, 127, 167, 189, 196,
 202, 210, **211**–12
movable-nest hives, 202
moving hives, 42, 51, **55**, 89,
 127, 183, 220
Mozambique, **70**
mud hives, 17, **35–8**, 39, 40,
 41–3, **49**, **51**, 54, **71–2**, 111,
 194, 210
mud used in hive construction,
 52, 66–7, 195
Munn, W. A., 209–11
muros de abejas, 72, 73, **74–6**
museums, general, 43, 45, 80,
 86, 97, 113–15, 131, 214, 218,
 229–30, 232, 238–46

357

Index

—, beekeeping, 12, 26, 80, 329–37

Namibia, 24–6, 34
Napoleon, 220, **226**, 227, 230
National Trust, 121, 130, 184
needlework, 218, **226**, 227–8
Nefertiti, 244
Neighbour, George, 108–9
neolithic period, 35, 134, 158, 160, 229
Nepal, **30–1**
nests, honeybee, **3**, 13, 17, 21–34, **23**, **25**, **29–31**, 53, 77–8, 225
—, —, finding, 21, 26, 32, 78
Netherlands, **101**, **104**, **107**, 111, **163**, 183, 215, 217–18, 220, 225, 237, 242, 333
Neuserre, 36
New Forest, 87–90, **89**, **110**
New Zealand, 16
niches, *see* bee boles
Nigeria, 66, 69
Norfolk, 122, **146**, 154–5, 159, 177, **180**, 181, 219, **220**, 234, 276, 324
Norse influences, 160–2
Northamptonshire, 128, 133, 145, **166**, 168, 234, 276–7, 315
Northumberland, 122–55, 164, **226**, 234, 277–8, 311
Norway, **66**, 333
Nottinghamshire, 315
numismatics, 231–2, **233**
Nutt hive, 182, **183**, **206–7**, 328
Nutt, Thomas, 182, 207
nutshells, hives using, 69, **70**

oak, cork (*Q. suber*), 53, 65
oaks (*Quercus*), 79, 87, 93, **198**
observation hives, 52, 54, **208**, 210
olive (*Olea europaea*), 240–1
Oman, 60, **157**, 158
orchard, site for bees, 119, 124–6, 171, 194
origin of apiaries, 18, 35, 79, 85–6, 88, 91
— — bee boles, 160–1
— — beekeeping and hives, 35–6, 51, 56–7, 79, 86–7, 99, 115–16
ornamental hives, 52–3, **60**, 65, **71**, **73**, 97, **98**, **101**, **110**
osiers, *see* willows
Ostade, Adriaen van, 225
ownership marks, 78–9, 81, **83**
Oxfordshire, 150, 155, 173, 182, **183**, 221, 234, 279, 316, 319, 328

Pacific area, 16

packing skeps for winter, 143–4, 149, 176, **181**
pain d'épice, 239
painted hive fronts, **52–3**, 65
paintings, rock, **frontispiece**, 19–33, **20**, **22–3**, **25**, **31**, 160
—, wall (in tombs), **35–8**
— with bees as theme, **213**, 221–6, **222–4**, **227**
Pakistan (Indus valley), 245
palaeolithic period, 19, 23, 28, 243
Palladius, 52
palm:
  cibe (*Borassus flabellifer*), 58
  coconut (*Cocos nucifera*), 69, **70**
  date (*Phoenix dactylifera*), 60
  doum (*Hyphaene thebaica*), **69**
  oil (*Elaeis guineensis*), 58
  tara (*Raphia*), 58
peacocks, recesses for, 158
peon (hive), 73, **75**
Pepys, Samuel, 202
petroglyphs, 24, 33
Pettigrew, Alfred, 142–3, 147
Philippines, 28, 244
Philiscus of Thasos, 51
pictographs, 33
pigeons, nests for, 158, **165**, 166
pines (*Pinus*), 79, 83
pipe hives, 41, 51
plaited-work hives, 66, **68**
plant stems, hives from, 52, 54, **55**, 66–8, **110**, 111, **112**, 113, 187
Pliny, 51–2, 55
Poland, **77**, 79, **80**, **83**, 84, **97–8**, 189, 239, 333–4
pollen grains, 42, 238
Pope Urban VIII, 218, 224
poplars (*Populus*), 79, 83
Portugal, 53, 65, 95–7, **96**, 113, 334
pottery hives, *see* hives by material: clay
Poussin, Nicolas, 221
prehistoric recesses, 160, **161**
pressed honey, 106
primates (monkeys), 19
primitive man and predecessors, **frontispiece**, 9, 19, **20–3**
Prokopovich, Peter, 98, 208–9
propolis, 80
protection of hives/skeps, 57, 67, **105**, 111, 113, 117, 119, 125–6, 135–6, 144, 148, **149**, **151**, 153–5
protective clothing, 10, 28–9, 33, **111**, **213**
public houses, 232–5
Purchas, Samuel, 142–3
pygmies, 26
Pyrenees, 91, 95

queen (honeybee), 16–17, **107**, 109, 200, 205, 229
queen cage, **107**
queen excluder, 212
Quintilian, pseudo-, 52

raffia (*Raphia*), 58
Rameses III, 42
recesses, not for bees, 158, **159**, 160–2, **161**
reed hives, 67, 110–11, **112**, 113, 194
Réunion, **57**
Rhodes, 229
rock paintings, **frontispiece**, 19–33, **20**, **22–3**, **25**, **31**, 160
Romania, 23, 79, 237, 334
Rome, Ancient, 8, 17, 21, 41, 45–56, 65, 67, **113–115**, 148, 160, 194, 200, 232, 238, **239**, 243–4
rope for reaching nests/hives, 26, **27**, 28, **29–31**, **78**, 79–80, **81–3**
Rubens, Peter Paul, 224
Russia, *see under* U.S.S.R.
Rwanda, 67

Sachsenspiegel, 115n
Sahara, 25, 65, **195**
—, Western, 25
saints, beekeeping, 214–18, 220, 224
see also individual names
Scirpus reed, 43, 218
Scotland:
  alcoves, 164–5, **248–9**, 312–13
  bee boles, 118–54, **123**, **125**, **130**, **147**, **149**, **151**, **159**, 162, **248–9**, 292–306
  bee houses, 167, **248–9**, 317
  bee shelters, 177–80, **181**, **248–9**, 326–8
  Bronze Age burial, 238
  hives/skeps, **125**, 128, 141–4, **181**, 182, **204**, 205
  inn signs, 234–5
  museum, 131, 335
  prehistoric Orkney sites, 160, **161**
  winter storage of bees, 173–4, **248–9**, 319–320
sculptures, 213, **214–15**, 216, **217**, 218, **219**
sedges, 43, **110**, 218
shape of hives, *see* hives, by shape
sheltering hives, *see* protection of hives/skeps
shelters, *see* bee shelters
Shropshire, 122, 171, **172**, 182, 235, 279, 316, 324, 328
Sicily, 54, **55**, 241
siting of hives, *see* hives, siting of

358

Index

size of hives, *see* hives, size (volume) of
skep/skeps:
  general, 17, 91, **92**, **99–111**, 116, 142, 164, 181, **213**, 214, **215**, **220**, **223–4**, 225–6, **227**
  cap, **108**, 109, 116, **125**, 202
  crownpiece, **100**
  eke, 107, **108**
  gipsy, 99, **108**, **110**
  girth for making, 102–3
  hackle, **101**, **104**, 105, 117, **199**, 200
  operating, 105–7, 111, **122**, 149, 166–7, 182–3, 214
  packing for winter, 143–4, 149, 176, **181**
  protection, *see* protection of hives/skeps
  reed, 110–11
  sedge, **110**
  size, *see* hives, size (volume) of
  stand (base), **104**, **108**, 117, 132, 148, 150, 155, 165 179, 182, **183**
  straw, coiled, 99, **102–11**, 117, **122**, **125**, 175
  as symbol, 102, 216–24, 225, 227, 232, 236, **237**, 238
  wicker, 66–7, **99–101**, 117, 214, **215**
Slav peoples, 79
smoke/smoking bees, **frontispiece**, 23, 28–9, 47–8, 84, 91
snecks, 105
Somerset, **118–19**, 120, 132, 150, 177, 181, 214, 235, 279–80, 311, 316, 324
songs, honey-hunting, 26
South Africa, 23–5, **23**, **25**
Southerne, Edmund, 107, 128, 142, 145, 148
Soviet Union, *see* U.S.S.R.
Spain, 19, **20**, 21, **22**, 53, **63**, 65, 72, 73, **74–6**, 95, **96**, 97, 107, **112**
Spanish bee walls, 72, 73, **74–6**
squeezed honey, 106
Sri Lanka, 28, 72
Staffordshire, 119, 151, 167, 235, 280, 316
stained glass, 219–21, **220**
Stewarton hive, **204**, 205
stingless bees, *see* Meliponini
 —, hives for, 35, **62**, 63–4, **69**, 73
stings/stinging by bees, 28–9, 62, 182, 207, 221, **222**
Stone Age, 19, 23, 28, 35, 134, 158, 160, 229, 243
stone hives, 71, **112**, 113, **192**, 197

stone walls, *see* walls containing bee boles
straw hives, **68**, 210, **211**
  see also skep/skeps: straw, coiled
Stülper, 91
Sudan, 66, **69**
Suffolk, 159, 235, 280, 311
sulphuring bees, 106
Sumatra, 29
Sumerians, 245
sunshine important for bees, 135–6
Surrey, 90, 181, 235, 280, 324
Sussex, 87, 155, 159, 167, **170**, 221, 235, 281, 316–17
swarm control, 17, 127, 197
swarms, 16–17, 35, 55, 84, 126, 153, 165, 216, 225, 228, 234
 —, hiving, 41, **44**, 54, 224, **242**
 —, tanging, 224, **242**
Sweden, **111**, 172, 224, 334
sweetness, honey symbol of 216, 225
Switzerland, 110, 179, 189, 208, 210, 220, 225, 237, 240, 334–5
symbolism of bees/hives/skeps, 9, 18, **43**, 102, 216–32, **226**, **233**, 234–6, **237**, 238
syphilis, 221

tanging swarms, 224, **242**
Tanzania, 25, **64**
tectiforms, 24
Tegetmeier, W. B., 168
termites, 61
terracotta hives, *see* hives by material: clay
Tharaka tribe, **27**, **60**
theft, prevention against, 126, 148, 188, 194
  see also bee boles: locking devices
thrift, *see* industry
Tibet, 30
tiered boxes as hives, 188, 202–5, **203**, 210
tiers, hives in, **40**, 54, **72**, 74, 75–6, 115n
Togo, 69
tombs, Egyptian, **35–8**, 42, 238
top-bar hives, **157**, **196**, 197, **198–9**, 200–1, 209–10, **211–12**
top-bars, 116, 197, 200–1, 205, 209–12
trade tokens, 231–2, **233**
transparent hives, 52, 54
trap boxes, 34
tree beekeeping, *see* forest beekeeping
trees, hives sited in, **60–1**, 64, **68–70**, 78, 83, 85–6
Trigona, **14–15**

 — wax, 243
tuff hives, **192**
Tunisia, 25, 241
tunnel hives, 65
Turkey, 35, 51, 67, **82**, 194
Tutankhamun, 42
Tyne & Wear, **150**, 235, 281

Ukraine, 79, 84, **229**
United Kingdom, *see* Britain, England, Ireland, Scotland, Wales
upright hives, 53, 56, **66**, 68, 73, **75**, 79, 91–2, **93–113**, 114–16, 208
Urals, 11, 79
Urban VIII, Pope, 218, 224
U.S.A., **94**, 111, 210, **211**, 212, 218, 221, 227, 230, 232, 244, 336–7
U.S.S.R.:
  general, 237, 335–6
  Bashkir, 79
  Caucasus, **58**, 59, 91
  Cheremessia, **78**, 79–80, **81**, **83**
  Estonia, 79
  Karelia, 11, 79
  Lithuania, 79
  R.S.F.S.R. (Russia), 81–4, 97–8, 172, **209**
  Ukraine, 79, 84, **229**
  Urals, 11, 79
  Volga region, 11, 79–80

Varro, 1, 52, 54–5, 67
Vatican, **44**, 218–20, 226
ventilation of hives, 16, **71**, 188
vertical hives, 53, 56, **66**, 68, 73, **75**, 79, 91–2, **93–113**, 114–16, 208
Vicat, Mme, 207
Vikings, 160–2
Virgil, 52, 214, 226
Vitex gaumeri, 62
Volga region, 11, 79–80
volume of hives, *see* hives, size (volume) of

Wales:
  alcoves, **164**, 165, **248–9**, 312
  bee boles, 118–54, **146**, **248–9**, 289–92
  bee houses, 167, **168**, 182, **183**, **248–9**, 317, 328
  bee shelters, 177, 325–6
  bees sent to Ireland, 220
  gingerbread moulds, 239–40
  inn signs, 234–5
  museum, 335
  Nutt hive, shelter, 182, **183**, 328
  winter bee house, **174**, **248–9**, 319

359

Index

walls containing bee boles:
  general, 117–20, 125, 129–30, 132–48, 150–1, 153
  see also bee walls
  brick, **117**, **122**, **124**, **129**, 131, 133, **135**, **137**, 138–40, 141, **146**–7, **150**, 160, 166
  chalk, **151**
  clunch, 166
  cob, **133**, 137, 138, **140**, **145**, 166
  stone, **118**–**19**, **123**, **125**, **129**–**59**, 161, 165–6
walls, heated, 159, 195
walls, hives in house, **50**, 61, 76, 191, **194**–**5**
Warwickshire, 122, 128, **129**, 147, 149, 221, 236, 281, 325
wasps/hornets, 33, 109, 229
wasps, honey from, 33, 63
watch cocks, **231**
water and other pots used as hives, 58, **71**, 72, **73**, 91, 194
wax, *see* beeswax
Wax Chandlers, 241–2

West Midlands, 119–58, 221, 235, 239–40, 281
Western Sahara, 25
Wheler, Sir George, 196–200, 210
White, Stephen, 205
wicker hives/skeps, **27**, 52, 54, **55**, **66**, 67, **99**–**101**, 117, **157**, 194, 196, **198**–**9**, 200, 210, 214, **215**
Wildman, Thomas, 147–8
willows (*Salix*), 79, 83, 196, 200
Wiltshire, **113**, 114, 150, **151**, 160, 235, 282, 312
winter:
  bee storage, 127, 171–6, **173**–**5**, 188, 318–20
  cellars, 163, 172
  packing, 143–4, 149, 176, **181**
wintering bees, 105, 135, 143–4, 185, 188
wooden (board) hives, **44**, **52**–**4**, **65**, 94, 186
Wordsworth, Dorothy and William, 178

woven hives/skeps, **27**, **49**, 52, 54, **55**–**6**, **66**–**7**, **76**, **99**–**101**, 117, **157**, 194, 196, **198**–**9**, 200, 210, 214, **215**
Wren, Sir Christopher, 202

yacente (hive), 73–6, **75**
Yorkshire, 102, 106–7, 119, 143, 155, 162
—, North, **102**, **104**–**5**, 106, **108**, 122–45, **130**, **138**–**40**, 152, 156, 179, 181–3, 217, 235, 282–7, 312, 319, 325
—, South, 128, 150, 235, 287
—, West, 132, 158, 235, 288–9, 325
Yugoslavia (Brać), **112**, 113, 192, 197
Yugoslavia (Carniola), **52**–**3**, 65, 158, **186**, 187, 189, 336

Zaire (Ituri Forest), 26
Zambia, 64
Zeidler, 78, 84–5, **86**
Zimbabwe, **frontispiece**, 23–5